The Physics of Composite and Porous Media

The Physics of Composite and Porous Media

T.J.T. Spanos
Norman Udey

CRC Press
Taylor & Francis Group
Boca Raton London New York

CRC Press is an imprint of the
Taylor & Francis Group, an **informa** business

CRC Press
Taylor & Francis Group
6000 Broken Sound Parkway NW, Suite 300
Boca Raton, FL 33487-2742

First issued in paperback 2019

ISBN-13: 978-1-4987-4670-0 (hbk)
ISBN-13: 978-0-367-87530-5 (pbk)

Library of Congress Cataloging-in-Publication Data

Names: Spanos, T. J. T., author. | Udey, Norman, author.
Title: The physics of composite and porous media / Tim Spanos and Norman Udey.
Description: Boca Raton, FL : CRC Press, Taylor & Francis Group, [2017] |
Includes bibliographical references and index.
Identifiers: LCCN 2017022980| ISBN 9781498746700 (hardback ; alk. paper) |
ISBN 1498746705 (hardback ; alk. paper) | ISBN 9781498746724 (e-Book) |
ISBN 1498746721 (e-Book)
Subjects: LCSH: Porous materials. | Composite materials. | Thermodynamics.
Classification: LCC QC173.4.P67 S58 2017 | DDC 620.1/16--dc23
LC record available at https://lccn.loc.gov/2017022980

Visit the Taylor & Francis Web site at
http://www.taylorandfrancis.com

and the CRC Press Web site at
http://www.crcpress.com

Contents

Preface

This book starts with the description of a number of two-component problems in composite media. Seismic wave propagation in two-component composite elastic media, seismic wave propagation in fluid-filled porous media and two-phase flow in a rigid porous media may all be solved by the same volume averaging techniques and physical arguments. Here, *physical argument* refers to the constraints that basic laws of physics (conservation of mass, Newton's laws, conservation of energy, the principle of equivalence, etc.) must hold at all scales. While these constraints may seem trivially obvious, it is observed that they are too often violated in models of porous media used by engineers and groundwater hydrologists.

An interesting observation was made once the equations for these two-component systems were constructed. The equations of motion that describe seismic wave propagation in a porous medium predict an additional wave process, that is, a wave process composed of incompressible elastic deformations of the matrix coupled to the incompressible motions of the fluid moving through the matrix, much like blood moving through the veins of a body. Furthermore, if the matrix is highly stressed, this highly non-linear wave may extract mechanical energy from the stress field of the matrix. Also, the equations that describe multiphase flow through porous media predict two instability processes when a low viscosity fluid displaces a high viscosity fluid. The first instability process is viscous fingering, in which the front becomes very sharp and then the low viscosity fluid channels through the high viscosity fluid along channels called *viscous fingers*; the size of these channels may be specified from a stability analysis. The second instability process is one where a connate saturation of the displacing fluid is present in the porous medium; the front may initially broaden in an unstable fashion and then each saturation contour propagates at a constant speed described by a wave equation very similar to the wave equations that described the propagation of porosity–pressure waves. A fascinating experimental observation is then made: a low viscosity fluid is injected into a porous medium containing a high viscosity fluid (e.g. heavy oil) with a connate saturation of the displacing fluid (e.g. water). If that displacement is done under steady-state conditions, the injected fluid will channel through the displaced fluid, forming viscous fingers. However, if the displacement is done using porosity–pressure wave propagation, then dispersion (separation of the front and the saturation contours propagating as waves) will always dominate.

It is this coupling that leads to the next theoretical development. To physically describe this behaviour requires the solution of a three-body problem. Of course, the most simplistic approach would be to construct the equations of motion for the components and look at their interactions. This approach is severely flawed. The origin of this flaw may be seen by considering another two-component problem, miscible flow in a porous medium where the components mix at two scales: the molecular scale (diffusion) and the pore scale (mechanical mixing). When this problem is constructed, both the concept of the component fluids and their volume fractions lose physical meaning. This leads to the understanding that both the component velocities and volume fractions only appeared to be thermodynamic variables in the previous two-component problems because they were proportional to the actual thermodynamic variables. This same observation is then made in the case of the three-body problem by a process of elimination. What is discovered is that if there are three components there are four equations of motion. These four equations describe one in-phase motion and three out-of-phase motions. Here, the components may never be considered independently or in pairs. Without this conclusion the three-body problem cannot be solved and Onsager's relation cannot be constructed. As soon as the three-body problem is solved, the n-body problem is obvious.

This is in fact similar to an n-body solution I obtained in 1973 in the context of charged spinning masses in a stationary electromagnetic vacuum field in general relativity.

One of the most important points in this book is to illustrate specific calculations and show how simplifying assumptions may be applied to numerical calculations. Sadly, Norman passed away on 11 November 2015, almost a year and a half before the completion of this book. His continued work in this area would have greatly improved the examples in this book. However, hopefully it is a good starting point for students who wish to study this subject area.

This book was written in its entirety at Wavefront Technology Solutions, a publicly traded company with head offices in Edmonton, Alberta, Canada. Most of this book is based on research into patented technologies funded by Wavefront Technology Solutions, Inc. (Wavefront). The foundations upon which this book is written are based on the book *The Thermophysics of Porous Media*, published by Chapman & Hall/CRC Press Monographs and Surveys in Pure and Applied Mathematics. The origins of this research began with funding at the Physics Department at the University of Alberta from the Alberta Oil Sands Technology and Research Authority (AOSTRA) in the late 1970s and early 1980s.

Tim Spanos
Norman Udey

Acknowledgement

The majority of this book describes research funded by Wavefront Technology Solutions.

This book contains a great deal of copyrighted material previously published by the authors. This work is being republished in this book with the permission of Elsevier, Wiley, Taylor & Francis, Proceedings of the Royal Society of London, Springer, Canadian Journal of Physics and Wavefront Technology Solutions. A summary of the parts of the book this work appears in and the articles they came from is given below. The section on seismic wave propagation in a composite elastic medium of Chapter 2 contains material from *Transp. Porous Med.* (2009), 79, 135–148. The plots on the reflection transmission problem in Chapter 3 are taken from *Wave Motion* (1992), 16, 1–16. Much of Chapter 4 is based on *Proc. R. Soc. Lond. A* (1993), 443, 247–255. Chapter 5 contains plots from Chapter 1, 'Relative Permeability', in M.A.D. Viera, P.N. Sahay, M. Coronado, and A.O. Tapia (eds.), *Mathematical and Numerical Modeling in Porous Media: Applications in Geosciences*, CRC Press, July 24, 2012. Chapter 5 also contains plots from *Can. J. Chem. Eng.* (1985), 63, 735–746. Chapter 6 contains plots from Chapter 16, 'Coupled Porosity and Saturation Waves In Porous Media', MAD Viera, PN Sahay, M Coronado, AO Tapia (eds.), *Mathematical and Numerical Modeling in Porous Media: Applications in Geosciences* CRC Press, July 24, 2012 as well material and figures are taken from Wavefront Technology Solutions documents and marketing material. Chapter 7 contains material and plots from *Transp. Porous Med.* (1993), 10, 1–41. Chapter 11 contains material and plots from *Proc. R. Soc. Lond. A* (1999), 455, 3565–3587; plots from *Can. J. Phys.* (1999), 77, 473–470; and plots from *Transp. Porous Med.* (1999), 35, 37–47.

Authors

Tim Spanos received his MSc in black hole dynamics in general relativity and his PhD in geodynamics at the University of Alberta. He then held an AOSTRA postdoctoral fellowship, followed by an AOSTRA chair in physics and then a faculty position in physics, all at the University of Alberta. In 1999, with Brett Davidson (the president and CEO of Wavefront) and Maurice Dusseault (a faculty member at the University of Waterloo), he co-founded a company, PE-TECH, which later became the publicly traded company Wavefront. He is currently a research physicist at Wavefront, working mainly on the material covered in this book.

Norman Udey received his PhD in the thermodynamics of mixtures in general relativity and a special certificate in computing science (specialization in software design), both at the University of Alberta. Following this degree, Norman received an AOSTRA post-doctoral fellowship working with Tim Spanos. After this fellowship ended, he worked as a research associate in seismology for four years. Norman then spent the next six and a half years in industry as a systems analyst and software developer. In 2001 he came to work for Wavefront on the physical processes and applications related to the patents held by Wavefront until the time of his passing in 2015.

CHAPTER 1

Introduction

BACKGROUND

Composite media are composed of structure and materials mixed at various scales. Some examples of composite media are most structures in the Earth's interior, granular media at the Earth's surface, plant materials such as wood, animal material such as bone, building material such as concrete and modern materials made with carbon fibre.

The theory of porous and composite media is an area of research that has been studied by engineers, geologists, geophysicists and applied mathematicians for many decades. Many unphysical models have been used for a very long time. Some of these models have been useful in the study of quasi-static processes. In the study of multiphase flow, Muskat proposed a model using equations analogous to Darcy's equation (Muskat 1937, 1949). These equations are still in use by some petroleum engineers today but have no relationship to physical reality. These equations model the highly non-linear processes of multiphase flow in porous media by separate linear equations, thus completely ignoring the dynamic interactions of the fluids. Furthermore, the dynamic pressure difference between the fluids is modelled by a zeroth order constraint, leading to a theory that is physically self-contradictory (see Chapter 5). In the study of miscible flow associated with groundwater contaminants, geologists and engineers have used the convection diffusion equation (which they sometimes call the *advection diffusion equation*) (Perkins and Johnston 1963). This is an equation that was constructed to describe molecular diffusion with phases mixing at only one scale, the molecular scale. Many authors have recognized this failure and have proposed alternatives (e.g. Koch and Brady 1987; Quintard and Whitaker 1994). The most common explanation for deviations from convection diffusion models is heterogeneities and fractures. As is shown in this book, if miscible phases are mixed at various scales, as in the case of a porous medium, then the theory describing the fluid motions is given by coupled Fokker–Planck equations. Furthermore, diffusion processes can be an order of magnitude or more larger than that predicted by the convection diffusion equation, which is consistent with observations.

In the case of deformations of porous media, the theory most commonly used by engineers is the Biot–Gassmann theory (Gassmann 1951; Biot 1956; Biot and Willis 1957).

In spite of all of its flaws, this theory has been quite useful to civil engineers as well as some other areas of engineering. The main flaw with this theory is that as it is constructed it is unphysical. The main advantage of this theory is that in the static limit the equations are of the correct form and can be made into a physical theory (in the case of statics) by reinterpreting the variables (see de la Cruz et al. 1993). Thus, experimentalists have been able to construct many relationships between theory and experiment in static experiments. The original construction of this theory gives the elastic matrix a preferential roll by describing it in the context of equilibrium thermodynamics. The fluid is then introduced as an external constraint through Darcy's equation, which is self-contradictory since fluid flow involves non-equilibrium thermodynamics and the components are coupled. Another serious error is that the role of porosity is not accounted for. This is not a serious error for statics; as is shown by de la Cruz et al. (1993), the Biot strain may be reinterpreted to include the role of porosity and it then becomes consistent with equilibrium thermodynamics. However, it cannot be generalized to describe dynamic processes because it is not consistent with non-equilibrium thermodynamics. The procedure to make the Biot-Gassmann theory into a physical theory for two-component dynamic processes would be to follow through the construction given in this book and introduce porosity changes at the outset. In its present form, this theory describes motions that do not conserve mass and violate Newton's third law. As a result, it is also not possible to construct physical boundary conditions for this theory (see Chapter 3, 'Boundary Conditions' section).

Other descriptions of porous media have been constructed in the context of thermodynamic interactions in the Earth's lithosphere (Steefel et al. 2005). In this theory porosity changes may occur due to deformations, chemical reactions or thermal processes (Connolly 1997). As is the case for the theory presented in this book, these descriptions predict the propagation of porosity waves in the earth (Malvoisin et al. 2015).

Some other dynamic descriptions of composite materials are made in describing a broad range of processes (Brennen 2005), including landslides (Straub 1997), volcanic activity (McTigue 1986; Natale 1998), fluidized beds (Harris and Crighton 1994), shock waves in porous media (Spiegelman 1993a, 1993b), earthquake interactions (Lemoin et al. 2001), earthquake fluid-level interactions in general (Montgomery and Manga 2003), allowing for viscoplastic behaviour of rocks (Connolly and Podladchikov 2015), etc.

THEORY

This book studies the form that physical equations take when materials are mixed at various scales. As a starting point, the following definitions of scale will be used:

Microscale (molecular scale)

Physical interactions described in terms of molecular dynamics.

Mixing of the phases is described by molecular diffusion.

Macroscale (pore scale)

Physical processes described in terms of continuum equations.

Fluid motions: Navier–Stokes equation.

Solid motions: Elasticity equation, plus frictional sliding of grains, etc.

Miscible fluid interactions: Convection diffusion equation, chemical potential, mass fractions of the component fluid phases.

The component phases for simple motions are described by linear equations, which interact according to prescribed boundary conditions or chemical mixing.

Megascale (laboratory scale)

The component fluid and solid phases cannot be described independently. At this scale, the averaged behaviour of the component phases and their interactions is being described. The phases at this scale are described by a non-linear classical field theory, which accounts for both molecular scale interactions of the phases and the macroscale interactions.

Mesoscale (intermediate scale)

At this scale, the description of the motions is too close to the scale of the structure for a continuum theory to be constructed. For example, acoustic wave propagation would need to be described in terms of scattering theory.

Gigascale (crustal scale)

Any scale larger than the megascale where additional structure has been introduced.

When macroscopic mixtures are observed at the megascale, one type of process that occurs is channelling, where the majority of a flow occurs along pathways, such as in landslides, when the majority of the flow occurs along narrow pathways, or in porous media, where the displacing fluid channels through the displaced fluid. Another observation, which is a process that will be discussed in this book, is the case where the flow occurs uniformly (i.e. the components interact uniformly at the macroscale), which in the case of the displacement of one phase by another is referred to as *dispersion*. So one of the points to be discussed is the conditions under which these different flow processes occur.

In order to gain additional understanding of the problem that must be addressed in going up in scale, consider the following thought experiment. Consider cubic membranes the size of the small squares shown in Figure 1.1 containing a fluid.

Assume the membranes in the upper half with the light grey colour contain water and the membranes in the lower half with the darker grey colour contain glycerine. In the first experiment, allow all of the membranes to dissolve and wait until the fluids are completely mixed at the molecular scale, as shown in Figure 1.2.

Here, it is observed that the fluids mix with each other at the molecular scale.

The thermodynamic description is given by the chemical potential and the mass fractions of the fluids. Here the mixing occurs at the molecular scale and the energy involved is the chemical energy. This process involves only molecular-scale thermodynamics.

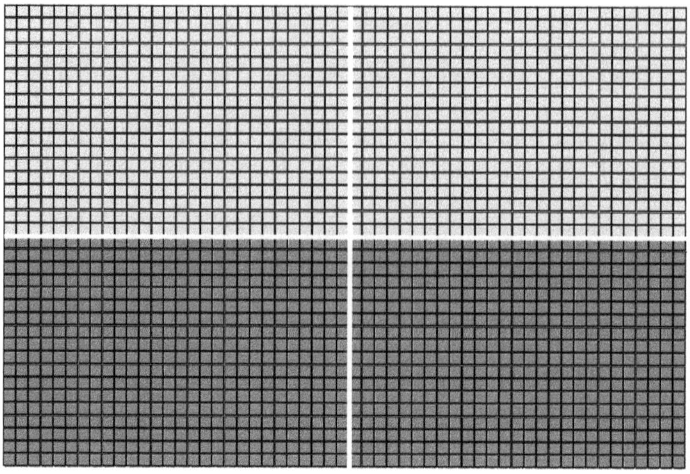

FIGURE 1.1 Fluid is contained in membranes the size of the small squares shown. The upper membranes contain water and the lower ones contain glycerine. The smallest scale at which the motions may be described is the scale of the four large squares.

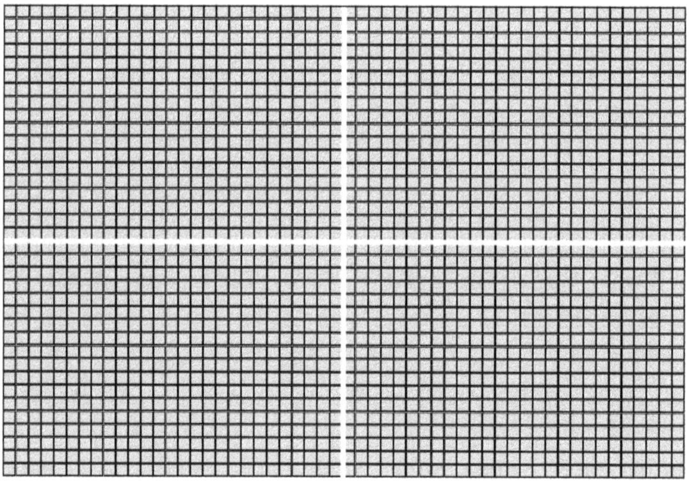

FIGURE 1.2 When the membranes dissolve, the fluids mix at the molecular scale. The mixing is described by the chemical potential and the mass fractions of the fluids.

Now assume that the membranes all remain intact and the fluids are mixed by moving the membranes around in a semi-random manner, as shown in Figure 1.3.

Now, before analysing this experiment, perform another one in which the membranes are moved around in orderly fashion, as shown in Figure 1.4.

In the last two experiments, mixing occurs at macroscale, which could be referred to as the *pore scale*, and the results are observed at a much larger scale.

The mixing that occurs in the last two experiments has absolutely nothing to do with molecular-scale thermodynamics. The mixing occurs though mechanical energy

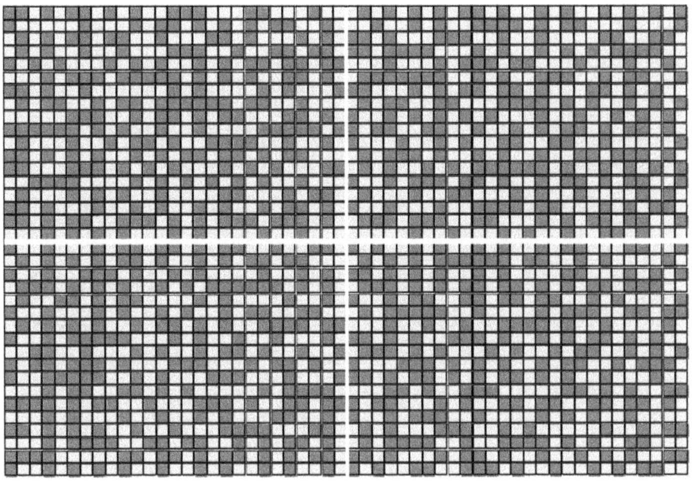

FIGURE 1.3 The membranes are moved around randomly so that approximately the same number of membranes containing each liquid are in each of the large squares.

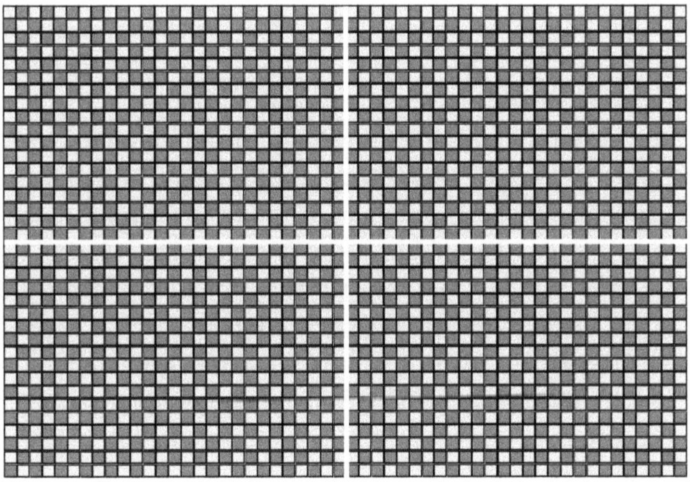

FIGURE 1.4 The membranes are moved around systematically so that they alternate along each line and column in each of the large squares.

at the scale of the membranes, and the fluids can be described independently by their volume fractions. Furthermore, their volume fractions are mixed differently in Figures 1.3 and 1.4. If the thermodynamics is now considered at the scale of the larger cubes shown by the white lines, then the mixing is described by the volume fractions and an order potential (see Chapter 4). If these processes are now generalized to consider the dynamic mixing of the phases, then saturation (the volume fractions of the fluids) becomes a dynamic variable.

In general the phases involved may be gasses, liquids or solids and not only can they flow but wave propagation such as seismic waves may be observed. So the next matter to be discussed is how these processes may be described. This work starts with the construction of the equations describing the behaviour of a number of physical processes in multicomponent systems. For example, the steady-state flow of a fluid through a homogenous porous matrix may be described by Darcy's equation. This result may be established through fairly straightforward experimental procedures (see Chapter 5). This result may also be established starting with the well-known equations describing the fluid flow and boundary conditions at the pore scale and then going up in scale using volume averaging (Anderson and Jackson 1967; Slattery 1967; Whitaker 1967; Newman 1969). This example is very important in that it establishes a relationship between theory and experiment. In fact, physical theory is a description of experimental observations or a prediction of experimental results subject to the constraints of the assumptions made in the construction of the theory. Each equation in a physical theory is a description of an experiment or a set of experiments. For example, consider the following equation when describing a compression associated with a porous medium.

$$\eta - \eta_o = \delta_s \nabla \cdot \vec{u}_s - \delta_f \nabla \cdot \vec{u}_f \tag{1.1}$$

Here, the left-hand side of this equation describes the change in the volume fraction of space occupied by the fluid (the porosity). The right-hand side describes a dimensionless physical parameter times the dilatational motion of the solid minus another dimensionless physical parameter times the dilatational motion of the fluid. This equation describes how porosity changes when a porous medium is compressed under static conditions. The associated experiments may be done in an infinite number of different ways depending on how the fluid pressure is constrained during the compression (see Hickey 1994). Any two of these experiments uniquely specify the values of the parameters δ_s and δ_f. Those values are the same no matter which experiments are chosen (Hickey 1994; Hickey et al. 1995). That is in fact a requirement for a quantity to be a physical parameter, for example, viscosity, bulk modulus and shear modulus are physical parameters because they may be related to experimental observations in the same fashion and are uniquely determined.

The special case where both components are elastic solids is considered in Chapter 2.

Here, Equation 1.1 is a static equation and the experiments discussed above are static experiments. However, this book is about dynamic processes, so the concept of statics will be discarded at this point except as a point of comparison. An example of a dynamic dilation is a seismic P wave. So in order to describe a dilatational seismic wave, the dynamic equation associated with changes in porosity is

$$\frac{\partial \eta}{\partial t} = \delta_s \nabla \cdot \vec{v}_s - \delta_f \nabla \cdot \vec{v}_f \tag{1.2}$$

So the obvious question to be asked is, 'What is the difference between δ_s and δ_f in Equations 1.1 and 1.2?' The answer is that different processes are being described in the two cases. There is nothing new about this observation. For example, if a bulk modulus is measured, it is necessary to specify the physical conditions under which it was measured in order to get a unique result. For example, isothermal or adiabatic measurements will yield different results. In the case of Equation 1.1, δ_s and δ_f may be specified by drained and undrained compression experiments (Zimmerman 1991) or from any of a continuous family of compression experiments in between these two extremes (Hickey et al. 1995). This result is of zeroth order in bulk modulus experiments (see Chapter 3).

$$\delta_s = (1-\eta_o)\frac{(K_s - K_{ud})(K_s - K_d)}{(K_s - K_f)[K_s - (1-\eta_o)K_d]} \tag{1.3}$$

$$\delta_f = \eta_o \frac{K_s(K_{ud} - K_f) - (1-\eta_o)K_d(K_s - K_f)}{(K_s - K_f)[K_s - (1-\eta_o)K_d]} \tag{1.4}$$

Here, K_s is the solid bulk modulus, K_f is the fluid bulk modulus, K_{ud} is the undrained bulk modulus (no fluid motion is allowed in or out of the matrix during the compression) and K_d is the drained bulk modulus (the fluid is allowed to escape and the fluid pressure remains constant during the compression). Equation 1.2, however, also includes local fluid flow associated with the dilatational motion, which is affected by the solid shear modulus, the shear viscosity of the fluid and the bulk viscosity of the fluid. This causes a correction to Equations 1.3 and 1.4 in the form of a dynamic term of the order of 1/K (i.e. δ_s and δ_f have values close to those in Equation 1.1 but deviate slightly due to dynamics of fluid flow associated with seismic compressions) (see Chapter 3).

$$\delta_s = (1-\eta_o)\frac{(K_s - K_{ud})(K_s - K_d)}{(K_s - K_f)[K_s - (1-\eta_o)K_d]} + \frac{\Xi(\mu_s, \mu_f, \xi_f)}{[K_s - (1-\eta_o)K_d]} \tag{1.5}$$

$$\delta_f = \eta_o \frac{K_s(K_{ud} - K_f) - (1-\eta_o)K_d(K_s - K_f)}{(K_s - K_f)[K_s - (1-\eta_o)K_d]} + \frac{\Theta(\mu_s, \mu_f, \xi_f)}{[K_s - (1-\eta_o)K_d]} \tag{1.6}$$

where μ_s, μ_f, ξ_f are the solid shear modulus, the fluid shear viscosity and the fluid bulk viscosity, respectively. However, now things begin to change: upon constructing a complete set of physical equations (Equations 3.67 to 3.73) for the dynamics of a fluid-filled porous medium, the equations now predict another wave process. As it turns out, this predicted wave process is fairly easy to observe in a laboratory. What is predicted is that there is a limit at which the fluid behaves in an almost incompressible fashion. In this limit, fluid flow is coupled to elastic deformations of the matrix, and the fluid moves through the porous medium like blood through the veins of a body. This differs completely from the flow associated with seismic compressions, which predicts local and compressional flow, in which the porous medium returns to its original configuration after the wave passes. It also differs from the steady-state flow described by Darcy's equation, where the matrix simply acts as an external constraint on the motion of the fluid. Since this is a different physical process and involves almost no compression of the fluid, δ_s and

δ_f must have values that are obtained from an experiment that describes the associated motions (see Chapter 3).

$$\delta_s = (1-\eta_o)\frac{(K_s - K_d)}{[K_s - (1-\eta_o)K_d]} + \frac{A(\mu_s, \mu_f)}{K_d} \tag{1.7}$$

$$\delta_f = \eta_o \frac{(1-\eta_o)K_d}{[K_s - (1-\eta_o)K_d]} + \frac{B(\mu_s, \mu_f)}{K_d} \tag{1.8}$$

Here, a drained dilatational motion of the solid is described, which causes deformations of the matrix without solid compressions and is coupled to fluid flow. In this limit the undrained bulk modulus does not contribute. Here, it is important that the dilatational motions occur at the rate specified by the wave process. In the context of the thermodynamics, it is observed that the values of δ_s and δ_f are constrained by Onsager's relations (see Chapter 4). At this point it is useful to observe a result obtained prior to the analysis of the thermodynamics (Hickey 1990). He observed that the equations of motion gave unphysical results unless δ_s and δ_f were constrained to specific values (Figure 1.5).

Here, it may now be observed that the upper right-hand corner of the shaded area yields values for δ_s and δ_f that describe static compressions and slight deviations from those values describe seismic waves. The lower left-hand corner yields the values for porosity waves. The equation referred to as Equation 50 in this plot is

$$\frac{\delta_s}{\delta_f} = \frac{K_s}{K_f} \tag{1.9}$$

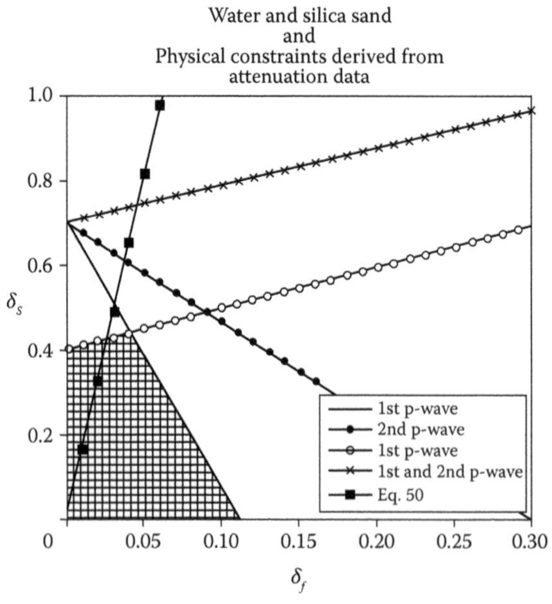

FIGURE 1.5 The values of the δ's for which the equations of motion give physical results for the first P wave are given by the shaded region of this plot.

Another difference between the porosity-pressure waves and the seismic waves is that the seismic waves are non-linear and the porosity-pressure waves are highly non-linear. This becomes an important point when the matrix is highly stressed as is the case for the Earth's tectonic stress field. In this case, if irreversible motions of the matrix are triggered as the porosity-pressure waves pass, then these waves may extract mechanical energy from the stress field of the matrix. If this energy is less than the energy lost to viscous dissipation, then the wave propagates further than it would have without this extracted mechanical energy. If this extracted mechanical energy is equivalent to the energy lost to viscous dissipation, then the porosity-pressure waves propagate as solitons. If this extracted mechanical energy is greater than the energy lost to viscous dissipation at some position, then the porosity-pressure waves may induce rock bursts or seismic events. In all cases a redistribution of the fluid and fluid pressure occurs, which causes porosity pressure diffusion to occur after the wave passes. This may be observed experimentally in Figure 3.11, where these processes have been slowed by using glycerine as the pore fluid.

In Chapter 4, the equilibrium thermodynamics of porous media is described. Thus, constraints obtained in this chapter are only strictly true for static experiments or dynamic motions of perfectly elastic materials. So a slight correction should be expected for near-equilibrium dynamic processes. This chapter illustrates that in the case of static deformations porosity change may be incorporated into the strains. This is the transformation that when incorporated into the Biot strains allowed that theory to be consistent with physical theory in the static limit (see de la Cruz et. al. 1993). As well, reciprocity is introduced and constraints are placed on δ_s and δ_f by the Onsager relations subject to equilibrium thermodynamics.

In Chapter 5, the case of two immiscible fluid phases in a rigid porous matrix is considered. It is observed that for the interaction of the two fluid phases, saturation (the volume fraction of a fluid phase) plays a role similar to the role played by porosity in the interaction between a single fluid and the solid.

$$\frac{\partial \eta_1}{\partial t} = \delta_2 \nabla \cdot \mathbf{v}_2 - \delta_1 \nabla \cdot \mathbf{v}_1 \tag{1.10}$$

A new wave process is predicted and observed, the propagation of saturation waves. These waves are straightforward to observe experimentally and they are observed to be strongly coupled to porosity waves. Porosity waves are observed to suppress viscous fingering and enhance sweep efficiency. This chapter explores the conditions under which viscous fingering or dispersion becomes the dominant flow process when an adverse mobility ratio is present. In addition, the stability of steam–water front and the effect of phase transitions on the stability of the front are considered. An analysis of the coupling between porosity waves and saturation waves (dispersion) using simplifying assumptions is discussed in Chapter 6. However, in later chapters, it is observed that the construction up to that point in the book is limited to two-component models mixed only at the macroscale. For more complex systems the concept of volume averaging as a physical tool must be discarded. Upon taking the equations of motion for the component phases in a three-component system (a solid matrix and two fluids) and the volume fractions expressed by the porosity and saturation equations, it is observed that solutions of the porosity and saturation wave equations are only possible in the limit where the waves are decoupled. This is interesting because simple experiments illustrate that they are in general strongly coupled. Another clue comes from the thermodynamics in that it is not possible to construct Onsager's relations from the equations of motion for the three-component phases and the equations for porosity and saturation. Yet another clue comes from the equations for miscible flow when the phases are mixed at various scales; here the concepts of component velocities and volume fractions lose physical meaning.

In Chapter 7 it is shown that the equations of immiscible flow may be rewritten in the form of miscible flow with negligible molecular diffusion when surface tension is taken to be zero. In this limiting case, it is shown that such flow may be described by a Fokker–Planck equation. In Chapter 8 the equations of miscible flow in porous media may be formally constructed in terms of momentum fluxes and mass fractions. The resulting description is expressed in terms of two coupled Fokker–Planck equations, and it is shown that, in the limiting case where molecular diffusion becomes negligible, one of these equations becomes identical to the Fokker–Planck equation obtained in Chapter 7 and the other Fokker-Planck equation vanishes.

In Chapter 9, the non-equilibrium thermodynamics of porous media (or in general multicomponent systems mixed at various scales) is constructed. This description is required to be expressed in terms of mass fractions and momentum fluxes.

In Chapter 10, the equations of motion for a three-phase system are constructed. It is observed that there is an additional degree of freedom that was not present in the case of the construction in terms of the component phases using volume averaging and the volume fractions (porosity and saturation) obtained from dilatational experiments. This occurs because the interactions of the three phases yield four momentum fluxes. Thus, an additional degree of freedom is obtained, which allows for the construction of Onsager's relations. In the special case where the phases are only mixed at the macroscale, these momentums and mass fractions may be rewritten in terms of the component phases and volume fractions (the porosity and saturation). Thus, the system of equations may also be cast in this form with one more equation than was obtained from volume averaging and dilatational experiments. Here, the three phases cannot be observed independently at the megascale, and their interactions yield one in-phase momentum flux and three out-of-phase momentum fluxes. Thus, four momentum equations are required to specify Newton's second law at the megascale. Once this three-body problem is solved, then the generalization to n-bodies is trivial.

Chapter 11 presents a Lorentz invariant thermal lattice gas computation. Here, particles are placed on a lattice and the motion of one lattice unit per time step is defined as the speed of light. The lattice points are used as a bookkeeper, much like the centre of a volume averaging sphere, the difference being that unlike volume averaging, where a volume average is ascribed to every point in space ,the lattice points define a discrete space–time. Thus, here the particles are allowed to populate a discretized space–time that is equivalent to all particles and their properties in a volume of space being ascribed to a point in that volume, the lattice point. Unlike early models where the particles were constrained to move on the lattice (e.g. Frisch et al. 1986), the particle motions are independent of the lattice in this computation. As a result, there is no exclusion principle (i.e. any number of particles may occupy the volume ascribed to a given lattice site). It is shown that the energy momentum tensor and relativistic Boltzmann equation can be derived from the collision and propagation rules. Thus, as in all other constructions in this book the objective is to go from the fundamental to the applied. The physics should arise naturally from the particle behaviour. The next step is to compare this computation to a more simplistic one that models non-relativistic elastic collisions. When this is done, it is observed that, if the average particle velocity is less the 0.1 times the speed of light, the two computational models are indistinguishable. As well, the possibility of a particle propagating at greater than the speed of light in the non-relativistic model is so rare (maybe once in an hour or so of computation) that such events may be simply discarded without affecting the results.

In thermal lattice gas computations, flow conditions and the boundary conditions are extremely important because of the thermal nature of the computation. In order to mimic a pressure gradient when modelling fluid flow, momentum is added to each particle at each iteration. This introduces a problem because energy is being added to the system and thus the temperature is increased. However, temperature may be kept constant by introducing a thermodynamic boundary condition. This allows models of Poiseuille flow to be constructed. Turbulence, relativistic shock waves, etc., are straightforward to construct but are not presented here. Next, the thermodynamic boundary condition is introduced randomly throughout the medium and Darcy flow is observed. Future constructions that readers are encouraged to consider are non-ideal gas behaviour and phase transitions.

EXPERIMENTAL RESULTS (LABORATORY AND FIELD)

Porosity and saturation waves have been observed in laboratory experiments (e.g. Wang et al. 1998; Davidson et al. 1999; Zschuppe 2002). In the early experiments it was observed that the tighter the sand pack and the more highly stressed it was, the better the flow enhancement. These media were encased in materials like concrete, steel and Plexiglas. Any contact with materials like rubber or neoprene was shown to damp out the porosity waves. Using pressure measurements it was observed that wave motions occurred in the porous medium moving about 40 times slower than a P wave. These waves attenuated due to viscous dissipation, and the higher the viscosity of the fluid, the more quickly the waves attenuated. Also it was observed that the higher the fluid viscosity, the slower the waves propagated. These waves could be generated by either a sudden increase or a sudden decrease in pressure. For a one-darcy permeability and a 30% porosity, with silica sand and water, the effectiveness of the pressure pulses could be tuned such that the most effective pulses have a rise time of about 1/10th of a second followed by a relaxation time back to the original pressure of about 1 second. The most effective laboratory pulser was constructed using two solenoids, one producing the sudden pressure increase and the other venting back to the original pressure (see Chapter 3). Other laboratory experiments have demonstrated that porosity waves may be used to suppress viscous fingering and enhance the distribution of injected fluids through dispersion. This may be done by pulsing in a lower viscosity fluid such as a surfactant and causing it to spread throughout the medium or by injecting a high viscosity fluid such as a polymer and then pulsing a low viscosity fluid through it, causing the high viscosity fluid to spread through the medium (see Chapter 5). These experiments clearly illustrate that porosity waves enhance flow in porous media and strongly couple with saturation waves to make dispersion the dominant displacement process when a lower viscosity fluid is used to displace a higher viscosity fluid.

Many field applications have been developed to remediate environmental contaminants, to enhance oil recovery and to improve production in older wells (e.g. Dusseault et al. 1999a, 1999b, 2000, 2002; Samaroo 1999; Spanos et al. 2003). These processes may also be used to disperse chemicals into an inhomogeneous porous medium. This has proven to be a useful method of environmental remediation. Some of the environmental clean-ups this process has been applied to are gas stations where tanks have leaked, trichloroethylene remediation processes involving the dispersion of nano-iron particles, creosote removal using surfactants, light non-aqueous phase liquid recovery levels have been observed to be increased by more than 250% over standard injection processes (Environmental Technology Verification program, Government of Canada licence #ETV 00012, March 28, 2001) (Grey et al. 2001), and many other environmental projects involving the injection of chemicals to neutralize groundwater contaminants. Oil field applications involve enhanced oil recovery and workovers to re-establish flow to older wells. In the case of workovers, water, chemicals or acids are injected uniformly into the formation surrounding the well to re-establish the original permeability reduced by plugging of flow paths during production. Without pulsing, the injected fluids tend to flow into the formation along high permeability pathways, bypassing the blockages. This was illustrated in one case by a successful acid workover in a well in which two previous acid workovers without pulsing had failed. Of course, this is only one example, but this technology has now been in commercial use around the world for 19 years with no technical failures. The original application of this technology and its primary use was to enhance oil recovery. Many years ago, oil was produced using the reservoir's existing pressure to cause flow from a production well. This was called *primary production*. Once the reservoir pressure dropped, water was injected into the formation in injection wells in an attempt to re-establish reservoir pressure. This was called *secondary production*. Of course, the injected water channelled through the formation to the production wells, with usually 70%–80% of the original oil in place still in the reservoir. Therefore, other injection technologies were then employed involving various chemicals (surfactants, polymers, supercritical CO_2, etc.); this was called *tertiary recovery*. Of course it was eventually recognized that much of the production done during primary and secondary recovery was damaging the reservoir and making tertiary processes less effective. This led to the terminology of *enhanced oil recovery*, which may be implemented at any stage of the production in an attempt to maximize the recovery of oil. This has made reservoir modelling an important tool in planning the production technologies to be used in reservoirs.

OTHER OBSERVATIONS

The following observations are consistent with the predictions of the theory presented here, but only anecdotal correlations have been made at this stage. This appears to be a fruitful area for future research.

The very first constructions of the porosity wave were made in 1993. It was noticed at that time two earthquake pairs had been observed at CICESE in Mexico. In one case, the second earthquake occurred at exactly the time that the porosity wave would have arrived. In the second case, the second earthquake occurred about 30 seconds after it would have passed. Since that time many anecdotal observations have been made of earthquake interactions. However, the following discussion relates only to published observations of earthquakes.

Earthquakes appear to affect fluid motions. In the United States, water level fluctuations were recorded in 716 wells after the Alaska earthquake of magnitude of 8.5 in 1964. (Note that the Alaska earthquake was the largest earthquake in the Northern Hemisphere in the previous century.) It is well documented that the Alaska earthquake also had effects on water level recorders in many other countries around the world. This is of course consistent with the soliton predictions of this theory, where fluid is moved great distances by extracting mechanical energy from the Earth's tectonic stress field. Similar observations were made following the Japan earthquake in 2011. There have also been many observations of both increases and decreases in pressure and flow rate in the earth at distances several hundred kilometres from earthquake epicentres (Roeloffs 1988). Changes in oil production, induced seismicity, changes in the height of water wells, etc., have all been observed. It has been observed that the Landers earthquake triggered earthquake swarms more than 1280 km away in Yellowstone National Park (USA), as well as other jolts near Mammoth Lake, California, and Yucca Mountain, Nevada (Hauksson et al. 1993). In the Yellowstone National Park area, geyser activity and micro-earthquake swarms increased after the Denali Fault earthquake (Alaska) on November 2, 2002 (Husen et al. 2004). The park is about 3100 km distant from the Denali fault epicentre.

Dogo Hot Spring, situated in Matsuyama City, Ehime Prefecture, Japan, has had groundwater level or discharge at the spring decrease four times during the past eight or nine Nankai earthquakes. These are large interplate earthquakes that occur in the western part of the Nankai Trough at intervals of 100–200 years (Satoshi and Koizumi 2007).

Associated with the 2004 earthquake off the west coast of northern Sumatra, changes in groundwater levels were observed at many observation stations in Japan that were more than 5000 km from the hypocentre (Kitagawa et al. 2006). At 38 of the 45 observation stations, there were changes in groundwater levels. At the 10 observation stations in which the Ishii-type borehole strain instruments were established, changes in crustal strains were also observed. A major part of the changes in crustal strains and groundwater levels or pressures were dynamic oscillations due to a seismic wave.

Distant as well as local earthquakes have induced groundwater-level changes persisting for days to weeks at Long Valley Caldera, California (Roeloffs et al. 2003).

Water-level/pressure data recorded at a dense network of 16 wells of depths ranging from 23 to 201 m within 400 m of the Tono Mine in Gifu Prefecture, central Japan, have been studied in search of possible earthquake-related changes (King et al. 1999).

The post-seismic change in the groundwater level following the 1999 (M_w = 7.5) Chi-Chi earthquake in central Taiwan was recorded by a network of 70 evenly distributed hydrological stations over a large alluvial fan near the epicentre. Four types of post-seismic responses were distinguished (Wang et al. 2004).

The 2008 Wenchuan earthquake occurred at the border of the Indo-Australian Plate and the Eurasian Plate. Following the earthquake, persistent changes in groundwater level were observed in more than 60 monitoring wells in Taiwan located at the boundary between the Eurasian Plate and the Philippine Sea Plate (Lee et al. 2012).

This only represents a very small sample of the observations of fluid level changes that have been observed following earthquakes. Fluid level changes have also been correlated as a precursor to earthquake activity (Roeloffs 1988) indicating the possibility of earthquake interactions (i.e. one earthquake being a trigger for a subsequent earthquake) (Xiaolong 2011).

Based on physical theory (Spanos 2002; also see Chapter 4), a tsunami-like process (a soliton) is predicted to propagate through the Earth's crust with a range of speeds from 100 to about 300 m/sec depending on the crustal properties and the wavelength of the solitary wave. In a study looking to see if earthquake interactions could be occurring at this speed, Rashid (2006) found 232 sources for possible single or multiple earthquake interactions that fit with this prediction. These observations are for the year 2003 and for earthquake sources over a magnitude of 4.3. Note that these are only earthquakes that would be triggered immediately after the solitary wave passes; as noted above, fluid level changes before many earthquakes are observed long before the main earthquake is triggered. Another interesting result that came from this study was that apparent earthquake interactions in the Himalayas did not fit with this theory. Upon closer inspection they appeared to be interacting vertically at the speed of a Stoneley wave. In Chapters 2 and 3 it is shown that when a medium is heterogeneous (its properties changing with depth) these surface waves (in homogeneous media) may propagate as highly non-linear body waves. This is of course only one possible explanation of these interactions and much more research is required before a definitive explanation may be given. However, this illustrates that often the most interesting experimental results are those that do not fit with expectations.

Following the loading of reservoirs, micro-seismic activity has been observed to diffuse outward from the reservoir with a diffusion constant of about 10^5 cm^2/sec (Talwani and Acree 1985). Geilikman et al. (1993) have shown that this is the same diffusion constant predicted by the porosity-pressure diffusion process described in this book.

A similar pressure diffusion process may be observed in the lab by suddenly opening a flow valve (i.e. suddenly changing the pressure) at one end of a porous medium and then observing the pressure changes across the porous medium as it evolves to Darcy flow (see Chapter 3). This process may be made much easier to observe by using a high viscosity fluid such glycerine, thus slowing the process.

SUMMARY

Dynamic fluid motions have been observed after earthquakes over thousands of kilometres. Pressure diffusion has been observed to cause micro-seismic activity after the loading of man-made reservoirs. Oil production and chemical injection has been observed to be able to be controlled by the pulsing of fluids. These observations have been made in commercial enhanced oil recovery processes and environmental remediation. Laboratory experiments have led to insight into the dynamic processes at work. In the following chapters the underlying physical description of the processes responsible for these observations at the various scales are investigated. The correlations of the theory discussed in this book to enhanced oil recovery and environmental remediation have been firmly established. However, applications to larger-scale geophysical observations are anecdotal at present.

REFERENCES

Anderson, T.B., and Jackson, R., 1967, Fluid mechanical description of fluidized beds equations of motion, *Ind. Eng. Chem. Fundam.*, 6, 527–539.

Biot, M.A., 1956, Theory of propagation of elastic waves in a fluid saturated porous solid, I, Low-frequency range, *J. Acoust. Soc. Am.*, 28, 168–178.

Biot, M.A., and Willis, D.G., 1957, The elastic coefficients of the theory of consolidation, *J. Appl. Mech.*, 24, 594–601.

Brennen, C.E., 2005, *Fundamentals of Multiphase Flows*, Cambridge University Press, Cambridge.

Connolly, J.A.D., 1997, Devolatilization-generated fluid pressure and deformation-propagated fluid flow during prograde regional metamorphism, *J. Geophys. Res.*, 102, 18149–18173.

Connolly, J.A.D., and Podladchikov, Y.Y., 2015, An analytical solution for solitary porosity waves: Dynamic permeability and fluidization of nonlinear viscous and viscoplastic rock, *Geofluids*, 15, 269–292.

Davidson, B., Spanos, T.J.T., and Dusseault, M.B., 1999, Laboratory experiments on pressure pulse flow enhancement in porous media, Proceedings of the CIM Regina Technical Meeting, October 1999.

de la Cruz, V., Sahay, P.N., and Spanos, T.J.T., 1993, Thermodynamics of porous media, *Proc. Ror. Soc. Lond. A*, 443, 247–255.

Dusseault, M.B., Davidson, B., and Spanos, T.J.T., 1999a, A new workover tool for CHOP wells, Proceedings of the CSPG and Petroleum Society, Calgary, June 14–18, paper 99–77, pp. 1–9, 1999a.

Dusseault, M.B., Davidson, B., and Spanos, T.J.T., 2000, Removing mechanical skin in heavy oil wells, SPE International Symposium on Formation Damage, Lafayette, Louisiana, 23–24 February 2000.

Dusseault, M.B., Shand, D., Meling, T., Spanos, T.J.T., and Davidson, B.C., 2002, Field applications of pressure pulsing in porous media, Proceedings of the 2nd Biot Conference on Poromechanics, Grenoble France (August), Balkema, Rotterdam, pp. 639–645.

Dusseault, M.B., Spanos, T.J.T., and Davidson, B., 1999, A dynamic pulsing workover technique forwells, *Proceedings of the CIM Regina Technical Meeting.*

Frisch, U., Hasslacher, B., and Pomeau, Y., 1986, Lattice-gas automata for the Navier-Stokes equation, *Phys. Rev. Lett.*, 56(14), 1505–1508.

Gassmann, F., 1951, Elastic waves through a packing of spheres, *Geophysics*, 16(4), 673–685.

Geilikman, M.A., Spanos, T.J.T., and Nyland, E., 1993, Porosity diffusion in fluid-saturated media, *Tectonophysics*, 217, 111–115.

Grey, P., Davidson, B.C., and Macdonald, A., 2001, Dramatic LNAPL recovery at an Ontario manufacturing facility, *Environ. Sci. Eng.*, 18(1), 22–24.

Harris, S.E., and Crighton D.G., 1994, Solitons, solitary waves, and voidage disturbances in gas-fluidized beds, *J. Fluid Mech.*, 266, 243–276.

Hauksson, E., Jones, L.M., Hutton, K., and Eberhart-Phillips, D., 1993, The 1992 landers earthquake sequence: Seismological observations, *J. Geophys. Res.*, 98(B11), 19835–19858.

Hickey, C.J., 1990, Seismic Wave Propagation in Porous Media, M.Sc. Thesis in Physics, University of Alberta.

Hickey, C.J., 1994, Mechanics of porous media, PhD dissertation in Physics, University of Alberta.

Hickey, C.J., Spanos, T.J.T., and de la Cruz, V., 1995, Deformation parameters of permeable media, *Geophys. J. Int.*, 121, 359–370.

Husen, S., Wiemer, S., and Smith, R.B., 2004, Remotely triggered seismicity in the Yellowstone National Park region by the 2002 *Mw* 7.9 Denali Fault earthquake, Alaska, *Bull. Seism. Soc. Am.*, 94(6B), S317–S331.

King, C.-Y., Azuma, S., Igarashi, G., Ohno, M., Saito, H., and Wakita, H., 1999, Earthquake-related water-level changes at 16 closely clustered wells in Tono, Central Japan, *J. Geophys. Res.*, 104(B6), 13073–13082.

Kitagawa, Y., Koizumi, N., Takahashi, M., Matsumoto, N., and Sato, T., 2006, Changes in groundwater levels or pressures associated with the 2004 earthquake off the west coast of northern Sumatra (M9.0), *Earth, Planets Space*, 58(2), 173–179.

Koch, D.L., and Brady, J.F., 1987, A non-local description of advection-diffusion with application to dispersion in porous media, *J. Fluid Mech.*, 180, 387–403.

Lee, T.P., Chia, Y., Yang, H.Y., Liu, C.Y., and Chiu, Y.C., 2012, Groundwater level changes in Taiwan caused by the Wenchuan earthquake on 12 May 2008, *Pure Appl. Geophys.*, 169(11), 1947–1962.

Lemoin, A, Madariaga, R., and Campos, J., 2001, Evidence for earthquake interaction in central Chile' the July 1997–September 1998 sequence, *Geophys. Res. Lett.*, 28(14), 2743–2746.

Malvoisin, B., Podladchikov, Y.Y., and Vrijmoed, J.C., 2015, Coupling changes in densities and porosity to fluid pressure variations in reactive porous fluid flow: Local thermodynamic equilibrium, *Geochem. Geophys. Geosyst.*, 16, 4362–4387.

McTigue, D.F., Thermoelastic response of fluid-saturated rock, *J. Geophys. Res.*, 91/89, 9533–9542, 1986.

Montgomery, D.R., and Manga, M., 2003, Streamflow and water well responses to earthquakes, *Science*, 300(27), 2047–2049.

Muskat, M., 1937, *The Flow of Homogeneous Fluids Through Porous Media*, McGraw-Hill, New York.

Muskat, M., 1949, *Physical Principles of Oil Production,* McGraw-Hill, New York.

Natale, G., 1998, The effect of fluid thermal expansivity on thermo-mechanical solitary shock waves in the underground of volcanic domain, *Pure Appl. Geophys.*, 152, 193–211.

Newman, S.P., 1977, Theoretical derivation of Darcy's Law, *Acta. Mech.*, 25, 153–170.

Perkins, T.K., and Johnston, O.C., 1963, A review of diffusion and dispersion in porous media, *Soc. Petroleum Eng. J.*, March, 70–84.

Quintard, M., and Whitaker, S., 1994, Convection, dispersion and interfacial transport of contaminants: Homogeneous porous media, *Adv. Water Resour.*, 17(4), 221–239.

Rashid, S., 2006, Seismic coupling and hydrological responses, MSc thesis, University of Waterloo.

Roeloffs, E.A., 1988, Hydrologic precursors to earthquakes: A review, *PAGEOPH,* 126, 175–209.

Roeloffs, E.A, Sneed, M., Galloway, D.L., and Hughs, J., 2003, Water-level changes induced by local and distant earthquakes at Long Valley Caldera, California, *J. Volcanol. Geotherm. Res.*, 127(3), 269–303.

Samaroo, M., 1999, Pressure pulse enhancement: Report on the first Reservoir Scale Experiment conducted by PE-TECH Inc. in Section 36 of Wascana Energy Inc.'s Morgan Field Lease., MSc thesis, University of Waterloo.

Satoshi, I., and Koizumi, N., 2007, Earthquake-related changes in groundwater levels at the Dogo Hot Spring, Japan, *Pure Appl. Geophys.*, 164, 2397–2410.

Slattery, J.C., 1967, Flow of viscoelastic fluids through porous media, *AIChE J.*, 13, 1066–1071.

Spanos T.J.T., 2002, *The Thermophysics of Porous Media,* Monographs and Surveys in Pure and Applied Mathematics, Chapman & Hall/CRC Press, Boca Raton, FL.

Spanos, T.J.T., Davidson, B.C., Dusseault, M.B., Shand, D., and Samaroo, M., 2003, Pressure pulsing at the Reservoir Scale: A new IOR approach, *J. Can. Petroleum Technol.*, 42, 1–13.

Spiegelman, M., 1993a, Flow in deformable porous media. Part 1. Simple analysis. *J. Fluid Mech.*, 247, 17–38.

Spiegelman, M., 1993b, Flow in deformable porous media. Part 2, Numerical analysis—The relationship between shock waves and solitary waves, *J. Fluid Mech.*, 247, 39–63.

Steefel, C.I., DePaolo, D.J., and Lichtner, P.C., 2005, Reactive transport modeling: An essential tool and a new research approach for the Earth sciences, *Earth Planet. Sci. Lett.*, 240, 539–558.

Straub, S., 1997, Predictability of long run out landslide motion: Implications from granular flow mechanics, *Geologische Rundschau,* 86(2), 415–425.

Talwani, P., and Acree, S., 1985, Pore pressure diffusion and the mechanism of reservoir-induced seismicity, *PAGEOPH,* 122, 947–965.

Wang, C.-Y., Wang, C.-H., and Kuo, C.-H., 2004, Temporal change in groundwater level following the 1999 ($M_w = 7.5$) Chi-Chi earthquake, Taiwan, *Geofluids*, 4(3), 210–220.

Wang, J., Dusseault, M.B., Davidson, B., and Spanos, T.J.T., 1998, Fluid enhancement under liquid pressure pulsing at low frequency, Proceedings UNITAR Conference on Heavy Oil and Tar Sands, Beijing, PRC, 1–7.

Whitaker, S., 1967, Diffusion and dispersion in porous media, *AIChE. J.*, 13, 420–427.

Xiaolong, S., Yaowei, L., and Hongwei, R., 2011, Fluid level changes and earthquake interactions, *Geod. Geodynam.*, 2(4), 33–39.

Zimmerman, R.W., 1991, *Compressibility of Sandstones, Developments in Petroleum Science,* 29th edn., Chilingarian, Elsevier, Amsterdam.

Zschuppe, R., 2001, Pulse flow enhancement in two-phase media, MSc thesis, University of Waterloo.

CHAPTER 2

Elastic Wave Propagation in a Composite Medium

OBJECTIVES OF THIS CHAPTER

Here, an introduction to the non-linear field theory presented in this book is given in the context of equilibrium thermodynamics. In the case of perfectly elastic deformations, mechanical energy is conserved. However, if more than one elastic material interacts at a scale orders of magnitude less than the scale at which deformations are being described, then their independent motions and interactions are coupled in the description of the megascopic motions. Thus a very simple introduction to non-linear field theory is obtained.

The first problem considered is the Rayleigh wave—a porous half space composed of a single elastic material with the pores containing a vacuum and the boundary in contact with a vacuum. It is then observed that if the porosity varies with depth in the medium then the Rayleigh wave may also propagate as a body wave. This occurs because as the depth changes going into the interior of the matrix, the porosity gradient appears as boundary conditions throughout the interior of the medium.

When two or more elastic solids interact at the pore scale, their coupled interactions at the pore scale must be accounted for when describing deformations at much larger scales (megascale). What is observed is that not only are two coupled equations of motion obtained but a new dynamic variable and associated equation is obtained that describes how their volume fractions change in time. This differs substantially from standard seismic analyses which present slight deviations from elasticity (Yilmaz 2000).

THE RAYLEIGH WAVE FOR A POROUS MEDIUM

Here, it is assumed that the porous medium is a uniform and homogeneous half space composed of an elastic matrix. The pores contain a vacuum and the porous half space in contact with a vacuum at the boundary. The equations of motion for the porous medium are given by

$$\rho_s \frac{\partial}{\partial t}\vec{v}_s = K_s \vec{\nabla}(\vec{\nabla}\times\vec{u}_s) - \frac{K_s}{(1-\eta_o)}\vec{\nabla}\eta + \frac{\mu_M}{(1-\eta_o)}\left[\nabla^2 u_s + \tfrac{1}{3}\vec{\nabla}(\vec{\nabla}\cdot\vec{u}_s)\right] \tag{2.1}$$

and

$$\eta - \eta_o = \delta_s \vec{\nabla} \cdot \vec{u}_s \tag{2.2}$$

Using the identity $\nabla^2 \vec{u}_s = \vec{\nabla}(\vec{\nabla} \cdot \vec{u}_s) - \vec{\nabla} \times \vec{\nabla} \times \vec{u}_s$ a single equation of motion may be written as

$$\rho_s \frac{\partial^2}{\partial t^2} \vec{u}_s = \left(K_s + \frac{4}{3} \frac{\mu_M}{(1-\eta_o)} - \frac{K_s}{(1-\eta_o)} \delta_s \right) \vec{\nabla}(\vec{\nabla} \cdot \vec{u}_s) - \frac{\mu_M}{(1-\eta_o)} \vec{\nabla} \times \vec{\nabla} \times \vec{u}_s \tag{2.3}$$

Taking the divergence of this equation yields the equation of motion for a P wave

$$\rho_s \frac{\partial^2}{\partial t^2} (\vec{\nabla} \cdot \vec{u}_s) = \left(K_s + \frac{4}{3} \frac{\mu_M}{(1-\eta_o)} - \frac{K_s}{(1-\eta_o)} \delta_s \right) \nabla^2 (\vec{\nabla} \cdot \vec{u}_s) \tag{2.4}$$

which may be written as

$$\nabla^2 (\vec{\nabla} \cdot \vec{u}_s) - \frac{1}{\alpha^2} \frac{\partial^2}{\partial t^2} (\vec{\nabla} \cdot \vec{u}_s) = 0 \tag{2.5}$$

where the P wave velocity is given by

$$\alpha^2 = \frac{\left((1-\eta_o) K_s + \frac{4}{3} \mu_M - K_s \delta_s \right)}{(1-\eta_o) \rho_s} \tag{2.6}$$

Taking the curl of Equation 2.3 yields the equation of motion for an S wave:

$$\rho_s \frac{\partial^2}{\partial t^2} (\vec{\nabla} \times \vec{u}_s) = \frac{\mu_M}{(1-\eta_o)} \vec{\nabla} \times \vec{\nabla} \times \vec{\nabla} \times \vec{u}_s \tag{2.7}$$

Using the identity $\vec{\nabla} \times \vec{\nabla} \times \vec{u}_s = \vec{\nabla}(\vec{\nabla} \cdot \vec{u}_s) - \nabla^2 \vec{u}_s$, this equation may be written as

$$\nabla^2 (\vec{\nabla} \times \vec{u}_s) - \frac{1}{\beta^2} \frac{\partial^2}{\partial t^2} (\vec{\nabla} \times \vec{u}_s) = 0 \tag{2.8}$$

where the S wave velocity β is given by

$$\beta^2 = \frac{\mu_M}{(1-\eta_o) \rho_s} \tag{2.9}$$

Now consider displacements of the form

$$u_x^s = A \exp(-bz) \exp[ik(x-ct)] \tag{2.10}$$

$$u_y^s = 0 \tag{2.11}$$

$$u_z^s = B\exp(-bz)\exp[ik(x-ct)] \tag{2.12}$$

where $\vec{u}^s = \vec{U}^s \exp(-i\omega t)$

$$\vec{U}^s = \vec{\nabla}\phi^s + \vec{\nabla}\times\vec{\psi}^s \tag{2.13}$$

$$\vec{\nabla}\cdot\vec{\psi}^s = 0 \tag{2.14}$$

Thus

$$\phi^s = \frac{1}{ik}A\exp(-bz)\exp(ikx) \tag{2.15}$$

$$\psi_y^s = -\frac{1}{b}B\exp(-bz)\exp(ikx) \tag{2.16}$$

Substituting Equations 2.15 and 2.16 into the equations of motion (Equations 2.4 and 2.8)

$$\left(\nabla^2 + \frac{\omega^2}{\alpha^2}\right)\phi^s = 0 \tag{2.17}$$

and

$$\left(\nabla^2 + \frac{\omega^2}{\beta^2}\right)\vec{\psi}^s = 0 \tag{2.18}$$

yields

$$b_\alpha = k\left(1 - \frac{c^2}{\alpha^2}\right)^{1/2} \tag{2.19}$$

and

$$b_\beta = k\left(1 - \frac{c^2}{\beta^2}\right)^{1/2} \tag{2.20}$$

where $c = \dfrac{\omega}{k}$. Thus

$$\vec{u}^s = \vec{\nabla}[\phi^s\exp(i\omega t)] + \vec{\nabla}\times[\vec{\psi}^s\exp(i\omega t)] \tag{2.21}$$

where

$$\phi^s = \frac{1}{ik} A\exp(-b_\alpha z)\exp(ikx) \tag{2.22}$$

$$\psi_y^s = -\frac{1}{b_\beta} B\exp(-b_\beta z)\exp(ikx) \tag{2.23}$$

The boundary conditions for a free boundary yield

$$\tau_{iz}\,|_{z=0} = 0 \tag{2.24}$$

which may be rewritten as (evaluated at $z = 0$)

$$\left[(1-\eta_o)K_s\left(1-\frac{\delta_s}{(1-\eta_o)}\right)-\frac{2}{3}\mu_m\right]\vec{\nabla}\cdot\vec{u}^s + 2\mu_m\frac{\partial u_z^s}{\partial z} = 0 \tag{2.25}$$

$$\frac{\partial u_x^s}{\partial z} + \frac{\partial u_z^s}{\partial x} = 0 \tag{2.26}$$

Substituting Equation 2.21 into Equations 2.25 and 2.26 yields

$$\left[(1-\eta_o)K_s\left(1-\frac{\delta_s}{(1-\eta_o)}\right)-\frac{2}{3}\mu_m\right]\nabla^2\phi^s \tag{2.27}$$

$$+\,2\mu_m\frac{\partial^2}{\partial z^2}\phi^s + 2\mu_m\frac{\partial^2\psi_y^s}{\partial x\,\partial z} = 0$$

$$2\frac{\partial^2\phi^s}{\partial x\,\partial z} - \frac{\partial^2\psi_y^s}{\partial z^2} + \frac{\partial^2\psi_y^s}{\partial x^2} = 0 \tag{2.28}$$

Substituting for ϕ^s and ψ_y^s in Equations 2.27 and 2.28 yields

$$\left[(1-\eta_o)K_s\left(1-\frac{\delta_s}{(1-\eta_o)}\right)-\frac{2}{3}\mu_m\right]\left(k^2-b_\alpha^2\right)A\exp(-b_\alpha z) \tag{2.29}$$

$$-\,2\mu_m b_\alpha^2 A\exp(-b_\alpha z) + 2\mu_m k^2 B\exp(-b_\beta z) = 0$$

$$-\,2b_\alpha A\exp(-b_\alpha z) + b_\beta B\exp(-b_\beta z) - \frac{k^2}{b_\beta}B\exp(-b_\beta z) = 0 \tag{2.30}$$

Substituting Equation 2.19 into 2.29 yields

$$\left[(1-\eta_o)K_s\left(1-\frac{\delta_s}{(1-\eta_o)}\right)-\frac{2}{3}\mu_m\right]\frac{c^2}{\alpha^2}A\exp(-b_\alpha z) \tag{2.31}$$

$$-2\mu_m k^2\left(1-\frac{c^2}{\alpha^2}\right)A\exp(-b_\alpha z)+2\mu_m k^2 B\exp(-b_\beta z)=0 \tag{2.32}$$

and substituting Equations 2.19 and 2.20 into Equation 2.30 yields

$$-2k^2\left(1-\frac{c^2}{\alpha^2}\right)^{1/2}\left(1-\frac{c^2}{\beta^2}\right)^{1/2}A\exp(-b_\alpha z)-\frac{\omega^2}{\beta^2}B\exp(-b_\beta z)=0 \tag{2.33}$$

Thus, Equations 2.17, 2.18, 2.31 and 2.32 along with the relation $c=\frac{\omega}{k}$ yield four equations for the four unknowns ω, *c*, *A* and *B*. Note in seismology it is common to write these relations in terms of the lame constants λ and μ, where $K_s=\lambda+(2/3)\mu$ for a single elastic solid. Then assuming λ = μ yields a Poisson ratio of 1/4. This allows for a specific solution where the wave speed is $c_R=0.9194\sqrt{\frac{\mu}{\rho}}$.

THE EFFECT OF A POROSITY VARYING WITH DEPTH

Consider a porous half space with a free surface and the pores occupied by a vacuum. Now assume that the porosity varies with depth such that it decreases exponentially with depth. The basic equations describing this medium are given by

$$\eta_o(z)=\eta_o e^{-\gamma z} \tag{2.34}$$

Continuity equation:

$$(1-\eta_o(z))\frac{\rho-\rho_o}{\rho_o}+\partial_j((1-\eta_o(z))u_j)-(\eta(z)-\eta_o(z))=0 \tag{2.35}$$

Equations of motion:

$$(1-\eta_o(z))\rho_o\frac{\partial v_i}{\partial t}=\partial_k\tau_{ik} \tag{2.36}$$

$$\tau_{ik}=(1-\eta_o(z))\sigma_{ik}=K[(1-\eta_o(z))\partial_j u_j-(\eta(z)-\eta_o(z)-u_z\beta\eta_o(z))]\delta_{ik} \tag{2.37}$$

$$\eta-\eta_o+u_z^s\frac{\partial\eta}{\partial z}=\Delta_{ik}^s u_{ik}^s \tag{2.38}$$

$$u_{ik}^s=\frac{1}{2}\left(\frac{\partial u_i^s}{\partial x_k}+\frac{\partial u_k^s}{\partial x_i}\right) \tag{2.39}$$

$$\Delta_{ik}^s=\begin{pmatrix}\Delta_{xx}^s & 0 & \Delta_{xz}^s\\ 0 & \Delta_{xx}^s & \Delta_{xz}^s\\ \Delta_{xz}^s & \Delta_{xz}^s & \Delta_{zz}^s\end{pmatrix} \tag{2.40}$$

Now note that all deformations are elastic and equilibrium thermodynamics applies. The values of Δ^s_{ik} may therefore be determined from straightforward static experiments. Note that all boundaries within the porous medium and at the upper surface are free surfaces and therefore are traction free. Consider the motion

$$u_x = \mathrm{Re}\{A\exp(-bz)\exp[ik(x - c(z)t)]\} \tag{2.41}$$

$$u_y = 0 \tag{2.42}$$

$$u_z = \mathrm{Re}\{B\exp(-bz)\exp[ik(x - c(z)t)]\} \tag{2.43}$$

Rewriting the equations of motion in terms of compression and shear waves yields

$$\left(\nabla^2 + \frac{\omega^2}{\alpha^2}\right)\phi^s = 0 \tag{2.44}$$

$$\alpha = \frac{K_s\left(1 - \dfrac{\delta_s}{(1-\eta_o)}\right) + \dfrac{4}{3}\dfrac{\mu_M}{(1-\eta_o)}}{(1-\eta_o)\,\rho_s(z)} \tag{2.45}$$

$$\left(\nabla^2 + \frac{\omega^2}{\beta^2}\right)\vec{\psi}^s = 0 \tag{2.46}$$

where the compressional wave velocity is

$$\alpha(z) = \sqrt{\frac{K(z) + \frac{4}{3}\mu(z)}{\rho(z)}} \tag{2.47}$$

and the shear wave velocity is

$$\beta(z) = \sqrt{\frac{\mu_M(z)}{(1-\eta_o)\rho_s(z)}} \tag{2.48}$$

$$\rho(z) = \rho_s(1 - \eta(z)) \tag{2.49}$$

Here, $K(z)$ and $\mu(z)$ are the bulk modulus and shear modulus of the medium, respectively. Since the pore spaces are occupied by a vacuum, both of these parameters are primarily dependent on the shear modulus of the solid. Note that compressions of the matrix will primarily result in a change in porosity.

These two waves interact at the boundary, yielding a Rayleigh wave, and due to the changing porosity they interact throughout the interior of the medium as well, yielding an additional body wave.

At the boundary of the porous medium

$$\tau_{zz}\,|_{z=0} = 0 \tag{2.50}$$

$$\tau_{xz}\,|_{z=0} = 0 \tag{2.51}$$

Upon substituting into the equations of motion, a body wave is observed that couples compressional and rotational motions. In the limit as $\gamma \to 0$, this body wave vanishes and the only solution that remains is that of a Raleigh wave.

SEISMIC WAVE PROPAGATION IN A COMPOSITE ELASTIC MEDIUM

The case of elasticity involves reversible deformations and may be described by equilibrium thermodynamics. So, as an introduction to a megascopic description of composite media such as the Earth's interior, the case of elastic solids mixed at the scale of sand grains is considered. Here, elasticity theory describes the physical behaviour of each of the elastic materials at the macroscale (pore scale). The elastic solids interact with each other across the boundaries between them according to prescribed boundary conditions given by Newton's third law and conservation of mass. However, this behaviour must be described at a scale orders of magnitude larger in scale. What is observed is that the two elastic components would propagate sound waves at different speeds but now must come to a compromise. They do so by propagating two compressional waves and two shear waves associated with in-phase and out-of-phase motions of the component elastic materials. This chapter serves as a simplistic introduction to composite materials, and the associated equilibrium thermodynamics is described in Chapter 4; in all of the subsequent chapters, equilibrium thermodynamics must be discarded.

Pore-Scale Equations

Inside the elastic solids, the pore-scale equations of motion are given by

$$\frac{\partial^2}{\partial t^2}\left[\rho_A u_i^A\right] = \frac{\partial}{\partial x_k}\sigma_{ik}^A + B_i^A \tag{2.52}$$

where

$$\sigma_{ik}^A = -K_A\alpha_A(T_A - T_0)\delta_{ik} + K_A u_{jj}^A\delta_{ik} + 2\mu_A\left(u_{ik}^A - \frac{2}{3}\delta_{ik}\,\partial_j u_j^A\right) \tag{2.53}$$

are the elastic stress tensors for the solid. Here A = 1, 2 distinguishes between the solid components and capital Latin letters are not summed over. T_0 is the ambient temperature, and $T_A(\vec{x},t)$ is the actual temperature in solid A. The temperature difference $T_A - T_0$ is treated as a first order quantity; ρ_A, $\vec{u}^A$, K_A, μ_A and α_A are the mass density, displacement, bulk modulus, shear modulus and thermal expansion coefficient, respectively, for material A.

$$u_{ik}^A = \frac{1}{2}\left(u_{i,k}^A + u_{k,i}^A\right) + \text{second order in } \vec{u}^A \tag{2.54}$$

$\vec{B}^A$ represents the body forces acting on the solid A by external forces such as gravity and will be assumed to be 0 in the following discussion.

The linearized equation of heat transfer in each solid medium is as follows (Landau and Lifshitz 1975):

$$\rho_A c_v^A \frac{\partial T_A}{\partial t} + \alpha_A K_A T_A \frac{\partial}{\partial t} \nabla \cdot \vec{u}^A - \kappa_A \nabla^2 T_A \tag{2.55}$$

where c_v^A is the heat capacity of solid A at constant volume and κ_A is the thermal conductivity of solid A.

The equations of continuity are given by

$$\frac{\partial \rho_A}{\partial t} + \nabla \cdot (\rho_A \vec{v}^A) = 0 \tag{2.56}$$

The mechanical boundary conditions between the two elastic solids are

$$\vec{u}^1 = \vec{u}^2 \tag{2.57}$$

$$\sigma_{ik}^1 n_k = \sigma_{ik}^2 n_k \tag{2.58}$$

and the boundary condition on temperature is

$$\kappa_1 \nabla T_1 = \kappa_2 \nabla T_2 \tag{2.59}$$

Construction of the Megascopic Equations

The megascopic continuum equations constructed here describe the deformations of a porous elastic matrix whose pores are completely filled with another elastic material. Furthermore, these deformations are assumed to occur at a scale orders of magnitude larger than the scale at which elastic materials are mixed. The medium is also assumed to be megascopically homogeneous and isotropic. *Thermomechanical coupling* refers to the first-order heating of the phases from compression and the expansion/contraction of the phases due to heating and cooling.

The fractional volume change in the interior of each elastic solid during deformation, $\nabla \cdot \vec{u}^A$, may be written as

$$\nabla \cdot \vec{u}^A = -\frac{(\rho_A - \rho_A^o)}{\rho_A^o} \tag{2.60}$$

Taking the volume average of each side

$$\eta_A^o \frac{(\bar{\rho}_A - \rho_A^o)}{\rho_A^o} = -\nabla \cdot \frac{1}{V} \int \vec{u}^A dV - \frac{1}{V} \int_{A_{AB}} \vec{u}^A \cdot d\vec{s} \tag{2.61}$$

where B = 1, 2 and B ≠ A.

Here, $\vec{u}^A \cdot d\vec{s}$ is the volume swept out by the displacement $\vec{u}^A$ of the boundary surface element.

$$\frac{1}{V}\int_{A_{AB}} \vec{u}^A \cdot d\vec{s} = -(\eta_A^o - \eta_A) \tag{2.62}$$

Thus

$$\eta_A^o \frac{(\bar{\rho}_A - \rho_A^o)}{\rho_A^o} = -\eta_A^o \nabla \cdot \bar{\vec{u}}^A + (\eta_A^o - \eta_A) \tag{2.63}$$

Taking the volume average of the following pore-scale equation for each solid (in the absence of thermomechanical coupling)

$$\frac{(\rho_A - \rho_A^o)}{\rho_A^o} = \frac{(p_A - p_o)}{K_A} \tag{2.64}$$

this yields

$$\frac{(\bar{\rho}_A - \rho_A^o)}{\rho_A^o} = \frac{(\bar{p}_A - p_o)}{K_A} \tag{2.65}$$

Combining the megascopic continuity equation for solid A with this equation yields

$$\frac{1}{K_A}(\bar{p}_A - p_o) = -\nabla \cdot \bar{\vec{u}}^A + \frac{(\eta_A^o - \eta_A)}{\eta_A^o} \tag{2.66}$$

If thermomechanical coupling is included, then Equation 2.65 becomes

$$\frac{(\rho_A - \rho_A^o)}{\rho_A^o} = \frac{(p_A - p_o)}{K_A} + \alpha_A (T_A - T_o) \tag{2.67}$$

Taking the average of this equation yields

$$\frac{1}{K_A}(\bar{p}_A - p_o) = -\nabla \cdot \bar{\vec{u}}^A + \frac{(\eta_A^o - \eta_A)}{\eta_A^o} + \alpha_A (T_A - T_o) \tag{2.68}$$

The volume average of the equations of motion yields

$$\begin{aligned}\eta_A^o \rho_A \frac{\partial^2 \bar{\vec{u}}_i^A}{\partial t^2} &= \eta_A^o K_A \, \partial_i (\nabla \cdot \bar{\vec{u}}^A) + K_A \nabla \eta_A \\ &\quad + \eta_A^o \mu_A [\nabla^2 \bar{\vec{u}}_i^A + \tfrac{1}{3} \partial_i (\nabla \cdot \bar{\vec{u}}^A)] \\ &\quad - \eta_A^o K_A \alpha_A \, \partial_i \bar{T}_A - I_i^{(A)} + \mu_A \, \partial_k I_{ik}^{(A)}\end{aligned} \tag{2.69}$$

The integrals (denoted by I) over the solid–solid interface in the above equations represent the coupling between the constituents, and representative expressions in terms of megascopic observables may be uniquely obtained through physical arguments. These expressions introduce the majority of the megascopic parameters in the theory, except for porosity. These area integrals are not all independent but are related due to the pore-scale boundary conditions. The area integrals given by

$$I_i^{(A)} = -\frac{1}{V}\int_{A_{AB}} [\sigma_{ik}^A + p_o\delta_{ik}]n_k\, dA \tag{2.70}$$

are related, due to continuity of stress at the pore-scale interface (Newton's third law), by

$$I_i^{(1)} = -I_i^{(2)} \tag{2.71}$$

The area integrals given by

$$I_{ik}^{(A)} = \frac{1}{V}\int_{A_{AB}} (u_k^A n_i + u_i^A n_k - \frac{2}{3}u_j^A n_j \delta_{ik})dA \tag{2.72}$$

are related by

$$I_{ik}^{(1)} = -I_{ik}^{(2)} \tag{2.73}$$

Taking the volume average of the heat equation yields

$$\eta_A^o \rho_A c_v^A \frac{\partial \bar{T}_A}{\partial t} + T_o K_A \alpha_A [\frac{\partial \eta_A}{\partial t} + \eta_A^o \frac{\partial}{\partial t}\nabla \cdot \bar{\vec{u}}^A]$$

$$-\eta_A^o \kappa_A \nabla^2 \bar{T}_A - \kappa_A \nabla \cdot \vec{J}^{(A)} - J^{(A)} = 0 \tag{2.74}$$

where

$$\vec{J}^{(A)} = \frac{1}{V}\int_{A_{AB}} (T_A - T_o)d\vec{A} \tag{2.75}$$

and

$$\vec{J}^{(1)} = -\vec{J}^{(2)} \tag{2.76}$$

due to continuity of temperatures of the two components on the interface. The two area integrals

$$J^{(A)} = \frac{1}{V}\int_{A_{AB}} \kappa_A \nabla T_A \cdot d\vec{A} \tag{2.77}$$

are related by

$$J^{(1)} = -J^{(2)} \tag{2.78}$$

due to continuity of heat flux.

The integral $I_{ik}^{(A)}$ is the force (per unit volume) exerted on one elastic solid by the other elastic solid across the interface due to compressional or shear motion. From the point of view of the megascopic continuum equations, it is a body force. An additional term, proportional to the relative acceleration, may also be presented (Landau and Lifshitz 1975; de la Cruz and Spanos 1989): $\rho_{12}\frac{\partial^2}{\partial t^2}(\vec{\bar{u}}^1 - \vec{\bar{u}}^2)$. Now, note that the equations in their current form do not satisfy the principle of equivalence. In the presence of gravity there will be an induced buoyancy force acting on one solid by the other, say $-\rho_b g_i$. However, a uniformly accelerating frame can simulate gravity. Since relative acceleration is an invariant, another linear combination of accelerations is needed. Including gravity, this additional term is of the form $\rho_b\left(\frac{\partial^2 u_i^m}{\partial t^2} - g_i\right)$, where $\frac{\partial^2 u_i^m}{\partial t^2}$ is the acceleration of the megascopic medium

$$\frac{\partial^2 u_i^m}{\partial t^2} = \frac{\eta_1^o \rho_o^1}{\rho_o^m}\frac{\partial^2 \bar{u}_i^1}{\partial t^2} + \frac{\eta_1^o \rho_o^2}{\rho_o^m}\frac{\partial^2 \bar{u}_i^2}{\partial t^2} \tag{2.79}$$

and $\rho_o^m = \eta_1^o \rho_o^1 + \eta_1^o \rho_o^2$ is the mass density of the megascopic medium. When gravity is switched off, this term has the form $\rho_b \frac{\partial^2 u_i^m}{\partial t^2}$.

The area integral $I_{ik}^{(A)}$ may be expressed in megascopic form in the present case as follows: According to Equation 2.69, $\mu_A I_{ik}^{(A)}$ is the piece needed to fully determine the megascopic solid stress tensor (which will be denoted by $\bar{\sigma}_{ij}^A$). It can be shown quite generally (de la Cruz et al. 1993; Spanos 2002) that the dependence of $\bar{\sigma}_{ij}^A$ on deformation $\bar{u}_{ij}^A$ and $\eta_A - \eta_A^o$ occurs only through the combination

$$\bar{u}_{ij}^{A'} - \bar{u}_{ij}^A + \frac{1}{3}\delta_{ij}(\eta_A - \eta_A^o)/\eta_A^o \tag{2.80}$$

where

$$\bar{u}_{ij}^A = \frac{1}{2}(\partial_i \bar{u}_j^A + \partial_j \bar{u}_i^A) \tag{2.81}$$

Here, the symmetric tensor $I_{ik}^{(A)}$ has the general form

$$I_{ik}^{(A)} = \varsigma_A \eta_A^o[\partial_k \bar{u}_i^A + \partial_i \bar{u}_k^A - \frac{2}{3}\delta_{ik}\partial_j \bar{u}_j^A] + \varsigma_A' \delta_{ik}\bar{u}_{jj}^A + \varsigma_A'' \delta_{ik}(\eta_A - \eta_A^o) \tag{2.82}$$

where ς_A, ς_A' and ς_A'' are constants. However, since $I_{ik}^{(A)}$ is trace free ς_A', $\varsigma_A'' = 0$, and thus

$$I_{ik}^{(A)} = \varsigma_A \eta_A^o[\partial_k \bar{u}_i^A + \partial_i \bar{u}_k^A - \frac{2}{3}\delta_{ik}\partial_j \bar{u}_j^A] \tag{2.83}$$

The dimensionless constant ζ_A may be conveniently eliminated in favour of a megascopic shear modulus μ_M (Hickey et al. 1995) through the definition

$$\mu_M = \eta_1^o\mu_1(1+\varsigma_1)+\eta_2^o\mu_2(1+\varsigma_2) \tag{2.84}$$

Thus the physical meaning of ζ_A is observed to be a measure of the difference between μ_M and the simple averaged value of $\eta_1^o\mu_1+\eta_2^o\mu_2$.

Thus $\eta_A^o\mu_A$ may be replaced by $\mu_M^A=\eta_A^o\mu_A(1+\varsigma_A)$ in Equation 2.69 representing the contribution of solid A to the megascopic shear modulus.

At the same time each elastic solid acquires a new term involving space derivatives of the other solid's velocity. This new term

$$-\eta_B^o\mu_A\left(\frac{\mu_M^A}{\eta_B^o\mu_B}-1\right)\frac{\partial}{\partial x_k}[\partial_i\bar{u}_k^B+\partial_k\bar{u}_i^B-\frac{2}{3}\delta_{ik}\,\partial_j\bar{u}_j^B]$$

arises from Equation 2.83 and fails to vanish unless $\zeta_B = 0$.

In analogy with the generalization of the shear modulus, the megascopic heat conductivities can be introduced as phenomenological parameters and are related to component heat conductivities κ_A according to

$$\kappa_M^A=\eta_A^o\kappa_A(1+b_A) \tag{2.85}$$

where the dimensionless constants b_A reflect the pore-scale behaviour through the assumed relation

$$\vec{J}^{(A)}=\frac{1}{V}\int_{A_{AB}}(T_A-T_o)\,d\vec{A}=\eta_A^o c_A\nabla\bar{T}_A-\eta_B^o c_B\nabla\bar{T}_B \tag{2.86}$$

Thus additional megascopic terms are obtained in the averaged heat equations (Equation 2.74).

The two integrals $J^{(A)}$ are equal and opposite and represent the heat transfer from one component to the other across the macroscopic interfaces. Hence, solid A acts as an additional heat source for solid B and vice versa. These heat exchange terms between components should vanish if and only if the megascopic component temperatures are equal ($\bar{T}_A=\bar{T}_B$). These terms may be represented by a first-order scalar proportional to ($\bar{T}_A-\bar{T}_B$) and therefore yield

$$J^{(A)}=\frac{1}{V}\int_{A_{AB}}\kappa_A\nabla T_A\cdot d\vec{A}=\gamma(\bar{T}_A-\bar{T}_B) \tag{2.87}$$

where γ is the surface coefficient of heat transfer between the solid phases, a positive empirical parameter. This parameter may be estimated by

$$\gamma = O|\kappa A/(VL)| \tag{2.88}$$

where κ is the effective conductivity between the solids, A is the interfacial surface area between the solids within the averaging volume V, and L is the characteristic pore-scale length.

Counting the number of variables and equations, one observes that an additional equation is needed for completeness when dilatational motions are considered. At this point note that Newton's second law has not been completely specified. When the medium is compressed the two solids may be compressed or deformed, each having its own bulk modulus and shear modulus, thus changing the porosity. Thus the relationship between $\nabla\cdot\vec{u}_A$, $\nabla\cdot\vec{u}_B$ and $\eta_A-\eta_A^o$ must be specified in order to completely describe a compression. Thus one obtains the relationship (assuming that locally the phases remain in thermal equilibrium)

$$\eta_A-\eta_A^o=\delta_A\nabla\cdot\vec{u}_A-\delta_B\nabla\cdot\vec{u}_B \tag{2.89}$$

where δ_A and δ_B are dimensionless parameters. A basic physical understanding of this relationship for a fluid and a solid has been presented in the context of dilatational experiments (Hickey et al. 1995). In the present case it is possible to obtain this equation through the following arguments. The volume fraction of phase A may be changed by altering the megascopic compressive stresses on the component phases or by altering the other forces such as body forces. Thus

$$\eta_A-\eta_A^o=a\bar{\sigma}_{jj}^A+b\bar{\sigma}_{jj}^B+B_o \tag{2.90}$$

where B_o represents the contribution from forces other than the stresses. For phenomena such as seismic wave propagation these forces may be set to 0. According to Equation 2.68 for $\bar{\sigma}_{jk}^A$

$$\sigma_{jj}^A=3K_A[\bar{u}_{jj}^A+\frac{\eta_A-\eta_A^o}{\eta_A^o}] \tag{2.91}$$

Thus,

$$[1+\frac{3aK_A}{\eta_A^o}-\frac{3bK_B}{\eta_B^o}](\eta_A-\eta_A^o)=3aK_A\bar{u}_{jj}^A-3bK_B\bar{u}_{jj}^B \tag{2.92}$$

This yields the porosity equation (Equation 2.89) which is independent of the equations of motion for the solids.

SUMMARY OF THE EQUATIONS FOR A HOMOGENEOUS MEDIUM (NEGLECTING THERMOMECHANICAL COUPLING)

Equations of motion (for the elastic phases):

$$\rho_A\frac{\partial^2}{\partial t^2}\vec{u}_A=K_A\nabla(\nabla\cdot\vec{u}_A)-\frac{K_A}{\eta_A^o}\nabla\eta_A$$

$$+\frac{\eta_B^o}{\eta_A^o}\mu_A(\frac{\mu_M^B}{\eta_B^o\mu_B}-1)[\nabla^2\frac{\partial\vec{u}_B}{\partial t}+\tfrac{1}{3}\nabla(\nabla\cdot\frac{\partial\vec{u}_B}{\partial t})] \tag{2.93}$$

$$+(-1)^A\frac{\rho_{12}}{\eta_A^o}\frac{\partial^2}{\partial t^2}(\vec{u}_A-\vec{u}_B)+\frac{\mu_M^A}{\eta_A^o}[\nabla^2\vec{u}_A+\tfrac{1}{3}\nabla(\nabla\cdot\vec{u}_A)]$$

Equation of motion (for porosity–internal momentum):

$$\eta_A-\eta_A^o=\delta_A\nabla\cdot\vec{u}_A-\delta_B\nabla\cdot\vec{u}_B \tag{2.94}$$

Equations of continuity:

$$\frac{\rho_A-\rho_A^o}{\rho_A^o}+\frac{\eta_A-\eta_A^o}{\eta_A^o}+\nabla\cdot\vec{u}_A=0 \tag{2.95}$$

Pressure equations:

$$\frac{1}{K_A}p_A=-\nabla\cdot\vec{u}_A-\frac{\eta_A-\eta_A^o}{\eta_A^o} \tag{2.96}$$

Aside from the (unperturbed) porosity, five other megascopic empirical parameters (ρ_{12}, μ_M^A, and δ_A) appear in these equations. In contrast, the parameters K_A, ρ_A^o, etc., are the pore-scale physical parameters specifying the constituents. If thermomechanical coupling is not ignored then Equations 2.92 and 2.95 are replaced by Equations 2.67 and 2.68 and the heat equation (Equation 2.73) must be included.

SEISMIC WAVE PROPAGATION

Consider a homogeneous acoustic plane wave propagating through the composite elastic medium due to the coupled compressional and shear motions of the two elastic solids. Dilatational waves may be quantified by taking the divergence of Equation 2.93 and eliminating porosity using Equation 2.93. Rotational waves are obtained by taking the curl of these equations. This process is called the *Helmholtz decomposition* and yields the following:

Substituting

$$\vec{u}_{s1}=\vec{\nabla}\phi_{s1}+\vec{\nabla}\times\vec{\psi}_{s1} \tag{2.97}$$

$$\vec{u}_{s2}=\vec{\nabla}\phi_{s2}+\vec{\nabla}\times\vec{\psi}_{s2} \tag{2.98}$$

into the equations of motion, two equations for the dilatational motions and two equations for the rotational motions are obtained. In each case, one of the potentials may be eliminated, yielding

$$\left(\nabla^2+\frac{\omega^2}{\alpha_1^2}\right)\left(\nabla^2+\frac{\omega^2}{\alpha_2^2}\right)\phi_{s1}=0 \tag{2.99}$$

and an identical result may be obtained for ϕ_{s2} by eliminating ϕ_{s1}. A potential for the first and second P waves may now be constructed (see Spanos 2002) such that

$$\left(\nabla^2 + \frac{\omega^2}{\alpha_1^2}\right)\phi_1 = 0 \tag{2.100}$$

and

$$\left(\nabla^2 + \frac{\omega^2}{\alpha_2^2}\right)\phi_2 = 0 \tag{2.101}$$

Similarly, potentials for the first and second S waves may be constructed, yielding

$$\left(\nabla^2 + \frac{\omega^2}{\beta_1^2}\right)\psi_1 = 0 \tag{2.102}$$

and

$$\left(\nabla^2 + \frac{\omega^2}{\beta_2^2}\right)\psi_2 = 0 \tag{2.103}$$

where

$$\alpha_1^2, \alpha_2^2 = \frac{1}{2\Delta D}\left[Tr(P^{\dagger}D) \pm \sqrt{Tr^2(P^{\dagger}D) - 4\Delta P \Delta D}\right] \tag{2.104}$$

$$\beta_1^2, \beta_2^2 = \frac{1}{2\Delta D}\left[Tr(S^{\dagger}D) \pm \sqrt{Tr^2(S^{\dagger}D) - 4\Delta S \Delta D}\right] \tag{2.105}$$

and

$$\Delta D = D_{11}D_{22} - D_{12}D_{21} \tag{2.106}$$

$$\Delta P = P_{11}P_{22} - P_{12}P_{21} \tag{2.107}$$

$$\Delta S = S_{11}S_{22} - S_{12}S_{21} \tag{2.108}$$

$$Tr(P^{\dagger}D) = D_{11}P_{22} + D_{22}P_{11} - D_{21}P_{12} - D_{12}P_{21} \tag{2.109}$$

$$Tr(S^{\dagger}D) = D_{11}S_{22} + D_{22}S_{11} - D_{21}S_{12} - D_{12}S_{21} \tag{2.110}$$

$$P_{11} = \eta_1^o K_1\left(1 - \frac{\delta_1}{\eta_1^o}\right) + \frac{4}{3}\mu_M^1 \quad P_{12} = K_1\delta_2 \tag{2.111}$$

$$P_{22} = \eta_2^o K_2 (1 - \frac{\delta_2}{\eta_2^o}) + \frac{4}{3}\mu_M^2 \quad P_{21} = K_2 \delta_1 \tag{2.112}$$

$$D_{11} = \eta_1^o \rho_1^o - \rho_{12} \quad D_{12} = \rho_{12} \tag{2.113}$$

$$D_{22} = \eta_2^o \rho_2^o - \rho_{12} \quad D_{21} = \rho_{12} \tag{2.114}$$

$$S_{11} = \mu_M^1 \; S_{12} = \eta_2^o \mu_1 (\frac{\mu_M^2}{\eta_2^o \mu_2} - 1) \tag{2.115}$$

$$S_{22} = \mu_M^2 \; S_{21} = \eta_1^o \mu_2 (\frac{\mu_M^1}{\eta_1^o \mu_1} - 1) \tag{2.116}$$

$$\phi_1 = c_1^\alpha \phi_{s1} - c_2^\alpha \phi_{s2} \tag{2.117}$$

$$\phi_2 = m_1^\alpha \phi_{s1} - m_2^\alpha \phi_{s2} \tag{2.118}$$

$$c_1^\alpha = \frac{[\Delta P - (D_{11} P_{22} - D_{12} P_{12}) \alpha_2^2]}{\Delta P (D_{12} P_{22} - D_{22} P_{12})(\alpha_1^2 - \alpha_2^2)} \tag{2.119}$$

$$c_2^\alpha = \frac{\alpha_2^2}{\Delta P(\alpha_1^2 - \alpha_2^2)} \tag{2.120}$$

$$m_1^\alpha = \frac{\alpha_1^2}{\Delta P(\alpha_2^2 - \alpha_1^2)} \tag{2.121}$$

$$m_2^\alpha = \frac{[\Delta P - (D_{11} P_{22} - D_{12} P_{12}) \alpha_1^2]}{\Delta P (D_{12} P_{22} - D_{22} P_{12})(\alpha_2^2 - \alpha_1^2)} \tag{2.122}$$

These plane waves may now be quantified as follows:

$$\Omega_{s1} = |\Omega_{os1}| \exp(i\phi_{s1}) \exp(i[kx - \omega t]) \tag{2.123}$$

$$\Omega_{s2} = |\Omega_{os2}| \exp(i\phi_{s2}) \exp(i[kx - \omega t]) \tag{2.124}$$

Here, $\frac{\omega}{k}$ is the phase velocity, $\frac{\partial \omega}{\partial k}$ is the group velocity, $\frac{|\Omega_{os1}|}{|\Omega_{os2}|}$ is the relative magnitude and $(\phi_{s1} - \phi_{s2})$ is the phase angle.

BOUNDARY CONDITIONS

The boundary conditions for a composite elastic medium are obtained from basic physical statements such as conservation of mass, Newton's third law and conservation of energy. From the continuity equation

$$\frac{\partial \rho}{\partial t} + \nabla \cdot (\rho V) \tag{2.125}$$

one obtains the natural definition of a medium velocity given by

$$\mathbf{v}^{(a)} = \frac{\eta_o^{(a)} \rho_1^{(a)} \mathbf{v}_1^{(a)} + (1-\eta_o^{(a)}) \rho_2^{(a)} \mathbf{v}_2^{(a)}}{\eta_o^{(a)} \rho_1^{(a)} + (1-\eta_o^{(a)}) \rho_2^{(a)}} \tag{2.126}$$

which in turn yields the boundary condition

$$\mathbf{v}^{(a)} \cdot \mathbf{n} = \mathbf{v}^{(b)} \cdot \mathbf{n} \tag{2.127}$$

The boundary conditions on the various stress tensors are as follows.

Any three of the following conditions specify Newton's third law

$$\eta_a \tau_{ik}^{(a)f} n_k = \tau_{ik}^{(b)f} n_k [\eta_a \eta_b - \alpha] + \tau_{ik}^{(b)s} n_k [\eta_a (1-\eta_b) - \alpha] \tag{2.128}$$

$$(1-\eta_a) \tau_{ik}^{(a)s} n_k = \tau_{ik}^{(b)f} n_k [(1-\eta_a)\eta_b - \alpha] + \tau_{ik}^{(b)s} n_k [(1-\eta_a)(1-\eta_b) - \alpha] \tag{2.129}$$

$$\eta_b \tau_{ik}^{(b)f} n_k = \tau_{ik}^{(a)f} n_k [\eta_a \eta_b - \alpha] + \tau_{ik}^{(a)s} n_k [(1-\eta_a)\eta_b - \alpha] \tag{2.130}$$

$$(1-\eta_b) \tau_{ik}^{(b)s} n_k = \tau_{ik}^{(a)f} n_k [\eta_a (1-\eta_b) - \alpha] + \tau_{ik}^{(a)s} n_k [(1-\eta_a)(1-\eta_b) - \alpha] \tag{2.131}$$

$$\eta_a \tau_{ik}^{(a)f} n_k + (1-\eta_a) \tau_{ik}^{(a)s} n_k = \eta_b \tau_{ik}^{(b)f} n_k + (1-\eta_b) \tau_{ik}^{(b)f} n_k \tag{2.132}$$

where

$$\alpha = \sqrt{\eta_a \eta_b (1-\eta_a)(1-\eta_b)} \tag{2.133}$$

For the final boundary condition, it is assumed that the tangential components of the velocities of momentum flux are continuous:

$$\vec{V}_t^{(a)} = \vec{V}_t^{(b)} \tag{2.134}$$

REFLECTION TRANSMISSION PROBLEMS

Ignoring gravitational body forces, the equation of motion of the medium is

$$\frac{\partial}{\partial t}(\rho V_i) = \partial_k \tau_{ik} \tag{2.135}$$

where

$$\tau_{ik} = \eta \tau_{ik}^{f} + (1-\eta)\tau_{ik}^{s} \tag{2.136}$$

is the stress tensor of the medium. Now, allowing for attenuation the (mean) energy flux vector is taken to be

$$J_i(\vec{X}) = -<\tau_{ik}(\vec{X},t)V_k(\vec{X},t)> \tag{2.137}$$

On account of the boundary conditions, the normal component of the energy flux vector is continuous at the boundary, that is,

$$\vec{J}^{(a)} \cdot \bar{n} = \vec{J}^{(b)} \cdot \vec{n} \tag{2.138}$$

For harmonic plane waves $\sim \exp\ i\,(\vec{k}\cdot\vec{x}\ \omega t)$, $w > 0$, it is straightforward to show that

$$\vec{J}\,(\vec{x}) = e^{-2\vec{k}_I \cdot \vec{X}}\vec{J}(0) \tag{2.139}$$

where $\vec{k}_I$ is the imaginary part of $\vec{k}$

$$\vec{k} \equiv \vec{k}_R + i\,\vec{k}_I \tag{2.140}$$

An identical $\vec{x}$-dependence to that specified in Equation 2.138 is obtained for any quantity quadratic in the displacements or velocities, for example, the mean kinetic energy

$$E(\vec{x}) = \left\langle \frac{1}{2}\rho v^2 \right\rangle \tag{2.141}$$

Along the direction of the energy flux vector $\vec{J}$, which in general differs from those of $\vec{k}_R$ and $\vec{k}_I$, one obtains, writing $\vec{x} = \hat{J}\,r$,

$$E(\hat{J}\,r) = e^{-2\vec{k}_I \cdot \hat{J}r}\ E(0) \tag{2.142}$$

After a distance of one 'wavelength along $\vec{J}$,

$$\vec{k}_R \cdot \vec{x} = \vec{k}_R \hat{J}\,r = 2\pi \tag{2.143}$$

E is attenuated by a factor of $e^{-2\pi/Q}$, where

$$Q^{-1} = 2(\vec{k}_I \cdot \vec{J})/(\vec{k}_R \cdot \vec{J}) \tag{2.144}$$

(Buchen 1971). Thus

$$E(\hat{J}r) = e - \frac{\vec{k}_R \cdot \hat{J}r}{Q} E(0) \tag{2.145}$$

It should be noted that the equations of motion determine for each mode the value of $\vec{k}_R \cdot \vec{k}_I$ as a function of ω, but not the angle between $\vec{k}_R$ and $\vec{k}_I$, which are in general not parallel. Hence it must be expected that except for normal incidence, where symmetry considerations may be adequate, the Q of each transmitted or reflected wave can only be determined as part of the complete solution of a reflection/transmission problem.

TWO ELASTIC COMPONENTS WITH THE VOLUME FRACTIONS VARYING WITH DEPTH

The importance of this discussion is to introduce a result that is observed in many forms throughout this book. The inverse problem associated with a single elastic solid whose properties are varied in the same way with depth was described by a single Sturm–Liouville equation (Henkin and Markushevich 1987; Markushevich 1987). This problem for a fluid-filled porous medium was constructed by Beals et al. (1995). They showed that the inverse problem associated with this multicomponent system is described by a matrix Sturm–Liouville equation. In general it is important to note that most of the generalizations in the context of non-linear field theory presented for multicomponent systems represent matrix formulations of previous linear physical theory.

The megascale equations presented in the section 'Summary of the Equations for a Homogeneous Medium' for two elastic solids coupled at the macroscale assumes that the medium is homogeneous and isotropic at the megascale. In this section, it is assumed that the phases' volume fractions vary exponentially with depth. The complete derivation of these equations is given by the following (Sahay et al. 2001):

Macroscopic continuity equations:

$$\eta_A^o \frac{(\rho_A - \rho_A^o)}{\rho_A^o} + \frac{\partial}{\partial x_j}(\eta_A^o u_j^A) + (\eta_A - \eta_A^o) = 0 \tag{2.146}$$

Macroscopic equations of motion:

$$\eta_A^o \rho_A^o \frac{\partial^2 u_j^A}{\partial t^2} = \tau_{jk,k}^A + (-1)^A I_j \tag{2.147}$$

Macroscopic stresses:

$$\tau_{jk}^{A} = \eta_{A}^{o}\sigma_{jk}^{A} = [K_{A}\eta_{A}^{o}\frac{\partial u_{i}^{A}}{\partial x_{i}} + (\eta_{A} - \eta_{A}^{o} - u_{i}^{A}\frac{\partial \eta_{A}^{o}}{\partial x_{i}})]\delta_{jk} \tag{2.148}$$

$$+2\eta_{A}^{o}\mu_{A}\mu_{jkmn}[u_{mn}^{A} + S_{mn}(\eta_{A} - \eta_{A}^{o} - u_{i}^{A}\frac{\partial \eta_{A}^{o}}{\partial x_{i}})]$$

Interfacial body force equation:

$$I_{j} = \rho_{jk}^{12}(\frac{\partial^{2} u_{k}^{1}}{\partial t^{2}} - \frac{\partial^{2} u_{k}^{2}}{\partial t^{2}}) + \rho_{jk}^{b}\frac{\partial^{2} u_{k}^{m}}{\partial t^{2}} \tag{2.149}$$

Here, ρ_{jk}^{12} is the apparent density associated with the relative acceleration of the components and ρ_{jk}^{b} is the tensorial parameter associated with a generalized buoyancy force acting between the components (see Sahay et al. 2001).

Porosity equation:

$$\eta_{1} - \eta_{1}^{o} - u_{j}^{1}\frac{\partial \eta_{1}^{o}}{\partial x_{i}} = a_{jk}^{1}\sigma_{jk}^{1} + a_{jk}^{2}\sigma_{jk}^{2} \tag{2.150}$$

This equation describes how the porosity changes in response to the stresses in the components.

SUMMARY

The equations of motion, boundary conditions, Raleigh wave, body waves and the reflection transmission problem for a composite medium composed of elastic components were constructed. It was shown that if the proportion of the components varies with depth then the surface waves may propagate as body waves. It was also shown that the interpretation of the wave numbers representing the direction of propagation and the attenuation of the waves in elasticity theory is not correct for composite media. The direction of propagation of the waves is determined by the direction of energy flux and the attenuation is specified by the change in amplitude of the energy after one wavelength along the direction of propagation of the wave. It was also noted that most of the generalizations in the context of non-linear field theory for multicomponent systems represent matrix formulations of previous linear physical theory.

REFERENCES

Beals, R., Henkin, G.M. and Novikova, 1995, The inverse boundary problem for the Rayleigh system, *J. Math. Phys.*, 36(12), 6688–6708.

Buchen, R.W., 1971, Plane waves in linear viscoelastic media. *Geophysics J. R. Astron. Soc.*, **23**, 531–542.

de la Cruz, V., Sahay, P.N., Spanos T.J.T., 1993, Thermodynamics of porous media, *Proc. Roy. Soc. Lond. A.*, 443, 247–255.

de la Cruz, V., Spanos T.J.T., 1989, Thermomechanical coupling during seismic wave propagation in a porous medium, *J. Geophys. Res.*, 94, 637–642.
Henkin, G.M., and Markushevich, V.M., 1987, Inverse Problems for the Surface Elastic Waves in a Layered Half-Space, Inverse Problems, 191–204.
Hickey, C.J., Spanos, T.J.T., de la Cruz, V., 1995, Deformation parameters of permeable media, *Geophys. J. Int.*, 121, 359–376.
Landau, L.D., Lifshitz, E.M., 1975, *Fluid Mechanics*, Pergammon, Toronto.
Markushevich, V.M., 1987, The Determination of Elastic Parameters of a Half-Space Using a Monochromatic Vibration Field at Surface, Wave Motion, 9, 37–49.
Sahay, P.N., Spanos, T.J.T., de la Cruz, V., 2001, Seismic wave propagation in inhomogeneous and anisotropic porous media, *Geophys. J. Int.*, 145, 209–223.
Spanos, T.J.T., 2002, *Thermophysics of Porous Media, Monographs and Surveys in Pure and Applied Mathematics*, Chapman and Hall, Boca Raton, FL.
Yilmaz, O., 2000, *Seismic Data Processing: Investigations in Geophysics*, vol. 2, Society of Exploration Geophysicists, Tulsa.

CHAPTER 3

Wave Propagation in Fluid-Filled Porous Media

OBJECTIVES OF THIS CHAPTER

Wave propagation in fluid-filled porous media involves the coupled motions of the fluid and solid at the pore scale. However, these motions are observed at a scale many orders of magnitude larger. In the case of compressional seismic waves, the fluid and solid have different bulk moduli and therefore compress by different amounts; this compression is also affected by the solid shear modulus and the structure of the solid matrix. As a result, the volume fraction of space (the porosity) occupied by the two phases is altered during the compression. This brings an additional degree of freedom into the description of the motions. Since the fluid and solid motions are coupled, such motions are described at the mega scale by a non-linear field theory. As well, the nature of the coupling depends on the process being observed. For example, static compressions involve no fluid flow. However, seismic compressions involve some fluid motions because all pores are not compressed by the same amount, and the amount of compression of the solid and fluid differs (e.g., in a water-filled silica matrix, the fluid bulk modulus and solid shear modulus are both about an order of magnitude less than the solid bulk modulus). Analogous results for material parameters are observed in the case of an adiabatic or isothermal bulk modulus. Other waves discussed in this chapter are the porosity–pressure waves, which are two highly non-linear coupled waves that involve elastic deformations of the matrix close to the incompressible limit of fluid motions coupled to flow of the fluid (Davidson et al. 1999; Dusseault et al. 1999; Spanos et al. 1999, 2003, 2004; Spanos 2002; Udey 2012). Applications of this wave process to enhanced oil recovery and environmental remediation have been patented in the U.S.A., U.K. and Canada (Davidson et al. 2001, 2002a, 2002b, 2005). These Patents are the property of Wavefront Technology Solutions along with 47 other patents at this time.

Here, the description of wave propagation in fluid-filled porous media is constructed in a similar fashion to that presented in Chapter 2 for composite elastic media. However, the presence of fluid phases introduces new physical processes associated with fluid motions at the pore scale. In particular, viscous dissipation introduces attenuation through the shear and bulk viscosities, and incompressible fluid flow coupled to elastic deformations of the matrix introduces a new type of wave (a porosity–pressure wave: here, the porosity and pressure waves are so strongly coupled, they will often be referred to as a single wave), which will be described in the section 'Porosity–Pressure Wave Propagation' and in Chapter 6. Initially, the porous matrix is assumed to be elastic, but when granular motion is allowed for in a stressed matrix it is observed that a wave may extract mechanical energy from the stress of the matrix. In the case of the porosity wave, the extraction of mechanical energy may be less than the viscous dissipation; then the wave is able to propagate to greater distances. In the case where the extraction of mechanical energy is as great as the viscous dissipation, a soliton is obtained, and in the case where the extraction of mechanical energy at some position is greater than the viscous dissipation, rock bursts or earthquakes may be generated.

PORE-SCALE EQUATIONS

Inside the elastic solid, the macroscopic equation of motion is given by

$$\frac{\partial^2}{\partial t^2}\left[\rho_s u_i^s\right]=\frac{\partial}{\partial x_k}\sigma_{ik}^s+B_i^s \tag{3.1}$$

where

$$\sigma_{ik}^s=-K_s\alpha_s(T_s-T_o)\delta_{ik}+K_s u_{jj}^s\delta_{ik}+2\mu_s\left(u_{ik}^s-\frac{2}{3}\delta_{ik}\partial_j u_j^s\right) \tag{3.2}$$

is the elastic stress tensor for the solid. Here, T_o is the ambient temperature, and $T_s(\vec{x},t)$ is the actual temperature. The temperature difference T_s–T_o is a first-order quantity associated with thermal expansion; ρ_s, u_i^s, K_s, μ_s and α_s are the mass density, displacement, bulk modulus, shear modulus and thermal expansion coefficient, respectively, for the solid matrix.

Inside the viscous fluid, the macroscopic equation of motion is given by

$$\frac{\partial^2}{\partial t^2}\left[\rho_f u_i^f\right]=\frac{\partial}{\partial x_k}\sigma_{ik}^f+B_i^f \tag{3.3}$$

where

$$\sigma_{ik}^f=\mu_f\left(v_{i,k}^f+v_{k,i}^f-\frac{2}{3}\delta_{ik}v_{j,j}^f\right)+\xi_f\delta_{ik}v_{j,j}^f-p_f\delta_{ik} \tag{3.4}$$

is the stress tensor for the fluid. The quantities ρ_f, u_i^f, v_i^f, μ_f and ξ_f are the mass density, displacement, velocity, shear viscosity and bulk viscosity, respectively, for the fluid. The body forces B_i^f and B_i^s will be assumed 0 in the following discussion.

The first-order equation of heat transfer in the solid medium is as follows (Landau and Lifshitz 1975).

$$\rho_s c_v^s\frac{\partial T_f}{\partial t}+\alpha_s K_s T_f\frac{\partial}{\partial t}\nabla\cdot\vec{u}_s-(\kappa_s\nabla^2 T_s)=0 \tag{3.5}$$

For the fluid, the first-order equation of heat transfer is given by the following (de la Cruz and Spanos 1989):

$$\rho_f c_p^f \frac{\partial T_f}{\partial t} + \alpha_f T_f \frac{\partial}{\partial t}\frac{\partial p_f}{\partial t} - \nabla\cdot(\kappa_f \nabla T_f) = 0 \tag{3.6}$$

The equations of continuity are given by

$$\frac{\partial \rho_s}{\partial t} + \nabla\cdot(\rho_s \vec{v}^s) = 0 \tag{3.7}$$

and

$$\frac{\partial \rho_f}{\partial t} + \nabla\cdot(\rho_f \vec{v}^f) = 0 \tag{3.8}$$

The mechanical boundary conditions at the fluid–solid interface are

$$v^f = \frac{\partial \vec{u}^s}{\partial t} \tag{3.9}$$

$$-p_f n_i + \sigma_{ik}^f n_k = \sigma_{ik}^s n_k \tag{3.10}$$

and the boundary condition on temperature is

$$\kappa_f \nabla T_f = \kappa_s \nabla T_s \tag{3.11}$$

EQUATIONS FOR A HOMOGENEOUS MEDIUM (WITH THERMOMECHANICAL COUPLING)

The megascopic continuum equations that describe wave propagation in a fluid-filled porous medium can be constructed by using volume averaging in conjunction with physical arguments. For the purpose of the present discussion, a porous medium is envisaged as an elastic matrix whose pores are interconnected and are completely filled with a viscous compressible fluid. The medium is also assumed to be megascopically homogeneous and isotropic. Thermomechanical coupling refers to the first-order heating of the phases from compression and the expansion/contraction of the phases due to heating and cooling.

Consider the pore-scale equation of continuity for a fluid:

$$\frac{\partial \rho_f}{\partial t} + \nabla\cdot\,(\rho_f \mathbf{v}^f) = 0 \tag{3.12}$$

Volume averaging yields

$$\frac{1}{V}\int_V \left[\frac{\partial \rho_f}{\partial t} + \nabla\cdot(\rho_f \mathbf{v}^f)\right] dV = 0 \tag{3.13}$$

which yields

$$\frac{\partial(\eta\bar{\rho}_f)}{\partial t}+\nabla\cdot(\overline{\eta\rho_f\mathbf{v}^f})=0 \tag{3.14}$$

The linearized version is obtained by writing

$$\eta=\eta_o+(\eta-\eta_o) \tag{3.15}$$

$$\bar{\rho}_f=\rho_f^o+(\bar{\rho}_f-\rho_f^o) \tag{3.16}$$

Keeping only first-order terms then yields

$$\frac{1}{\rho_f^o}\frac{\partial}{\partial t}\bar{\rho}_f+\frac{1}{\eta_o}\frac{\partial}{\partial t}\eta+\frac{1}{\eta_o}\nabla\cdot(\eta_o\bar{\mathbf{v}}^f)=0 \tag{3.17}$$

as the generalization of Equation 3.12, valid even if the porosity is non-uniform, $\eta_o=\eta_o(\vec{x})$. Equation 3.17 is the megascopic equation of continuity for the fluid.

The corresponding equation for the solid component may be constructed as follows. The fractional volume change in the interior of the elastic solid during deformation, $\nabla\cdot\mathbf{u}_s$, may be written as

$$\nabla\cdot\mathbf{u}^s=-\frac{(\rho_s-\rho_s^o)}{\rho_s^o} \tag{3.18}$$

Taking the volume average of each side yields

$$(1-\eta_o)\frac{(\bar{\rho}_s-\rho_s^o)}{\rho_s^o}=-\nabla\cdot\frac{1}{V}\int\mathbf{u}^s dV-\frac{1}{V}\int_{\mathbf{A}_{sf}}\mathbf{u}^s\cdot ds \tag{3.19}$$

Here, $\mathbf{u}_s\cdot\mathbf{ds}$ is the volume swept out by the displacement $\mathbf{u}_s$ of the boundary surface element

$$\frac{1}{V}\int_{\mathbf{A}_{sf}}\mathbf{u}^s\cdot ds=-(\eta-\eta_o) \tag{3.20}$$

Thus

$$(1-\eta_o)\frac{(\bar{\rho}_s-\rho_s^o)}{\rho_s^o}=-(1-\eta_o)\nabla\cdot\bar{\mathbf{u}}^s+(\eta-\eta_o) \tag{3.21}$$

Two megascopic pressure equations may now be constructed with the aid of these continuity equations.

Taking the volume average of the following pore-scale equation for the fluid (in the absence of thermal effects)

$$\frac{(\rho_f-\rho_f^o)}{\rho_f^o}=\frac{(p_f-p_f^o)}{K_f} \tag{3.22}$$

yields

$$\frac{(\bar{\rho}_f - \rho_f^o)}{\rho_f^o} = \frac{(\bar{p}_f - p_f)}{K_f} \tag{3.23}$$

Combining the megascopic continuity equation for the fluid with this yields a megascopic pressure equation

$$\frac{1}{K_f}\frac{\partial}{\partial t}\bar{p}_f = -\nabla \cdot \bar{\mathbf{v}}^f - \frac{1}{\eta_o}\frac{\partial}{\partial t}\eta \tag{3.24}$$

Similarly, taking the volume average of the following pore-scale equation for the solid (in the absence of thermomechanical coupling)

$$\frac{(\rho_s - \rho_s^o)}{\rho_s^o} = \frac{(p_s - p_o)}{K_s} \tag{3.25}$$

yields

$$\frac{(\bar{\rho}_s - \rho_s^o)}{\rho_s^o} = \frac{(\bar{p}_s - p_o)}{K_s} \tag{3.26}$$

Combining the megascopic continuity equation for the fluid with this equation yields

$$\frac{1}{K_s}(\bar{p}_s - p_o) = -\nabla \cdot \bar{\mathbf{u}}^s + \frac{(\eta - \eta_o)}{(1 - \eta_o)} \tag{3.27}$$

If thermomechanical coupling is included, then Equations 3.22 and 3.25 become

$$\frac{(\rho_f - \rho_f^o)}{\rho_f^o} = \frac{(p_f - p_f)}{K_f} + \alpha_f \frac{\partial}{\partial t}T_f \tag{3.28}$$

and

$$\frac{(\rho_s - \rho_f^o)}{\rho_s^o} = \frac{(p_s - p_o)}{K_s} + \alpha_s (T_s - T_o) \tag{3.29}$$

Taking the volume average of these equations yields

$$\frac{1}{K_f}\frac{\partial}{\partial t}\bar{p}_f = -\nabla \cdot \bar{\mathbf{v}}^f - \frac{1}{\eta_o}\frac{\partial}{\partial t}\eta + \alpha_f \frac{\partial}{\partial t}\bar{T}_f \tag{3.30}$$

and

$$\frac{1}{K_s}(\bar{p}_s - p_o) = -\nabla \cdot \bar{\mathbf{u}}^s + \frac{(\eta - \eta_o)}{(1 - \eta_o)} + \alpha_s (\bar{T}_s - T_o) \tag{3.31}$$

The volume average of the equations of motion yields

$$(1-\eta_o)\rho_s\frac{\partial^2\bar{\mathbf{u}}_i^s}{\partial t^2}=(1-\eta_o)K_s\,\partial_i(\nabla\cdot\bar{\mathbf{u}}^s)-K_s\,\nabla\eta+(1-\eta_o)\mu_s[\nabla^2\bar{\mathbf{u}}_i^s+{}^1\!/_3\,\partial_i(\nabla\cdot\bar{\mathbf{u}}^s)] \\ -\ (1-\eta_o)K_s\alpha_s\,\partial_i\bar{T}_s-\mathbf{I}_i^{(3)}+\mu_s\,\partial_k I_{ik}^{(4)} \tag{3.32}$$

$$\eta_o\rho_f\frac{\partial}{\partial t}\bar{\mathbf{v}}_i^f=-\eta_o\,\partial_i\bar{\mathrm{p}}_f+\eta_o[\mu_f\nabla^2\bar{\mathbf{v}}_i^f+{}^1\!/_3\mu_f\,\partial_i(\nabla\cdot\mathbf{v}_f)]-\mathbf{I}_i^{(1)}+\mu_f\,\partial_k I_{ik}^{(2)} \\ +\ \eta_o\xi_f\,\partial_i(\nabla\cdot\mathbf{v}_f)+\xi_f\,\partial_i\frac{\partial_n}{\partial t} \tag{3.33}$$

The integrals (denoted by I) over the fluid–solid interface in the above equations represent the coupling between the constituents, and representative expressions in terms of megascopic observables may be uniquely obtained through physical arguments. These expressions introduce the majority of the megascopic parameters in the theory, except for porosity. An interpretation of the introduced megascopic parameters in such expressions may depend on the type of process taking place. These area integrals are not all independent but are related due to the pore-scale boundary conditions. The area integrals given by

$$\mathbf{I}_i^{(1)}=\frac{1}{\mathrm{V}}\int_{A_{fs}}[(p_f-p_o)\delta_{ik}-\tau_{ik}^f]\mathrm{n}_k\mathrm{dA} \tag{3.34}$$

and

$$\mathbf{I}_i^{(3)}=\frac{1}{\mathrm{V}}\int_{A_{fs}}(\tau_{ik}^f+p_o\delta_{ik})\mathrm{n}_k\mathrm{dA} \tag{3.35}$$

are related, due to the continuity of stress at the pore-scale interface (Newton's third law), as

$$\mathbf{I}_i^{(3)}=-\mathbf{I}_i^{(1)} \tag{3.36}$$

The area integrals given by

$$I_{ik}^{(2)}=\frac{1}{V}\int_{A_{fs}}(\mathrm{v}_k^f\mathrm{n}_i+\mathrm{v}_i^f\mathrm{n}_k-\frac{2}{3}\mathrm{v}_j^f\mathrm{n}_j\,\delta_{ik})\mathrm{dA} \tag{3.37}$$

and

$$I_{ik}^{(4)}=\frac{1}{V}\int_{A_{sf}}(\mathrm{u}_k^s\mathrm{n}_i+\mathrm{u}_i^s\mathrm{n}_k-\frac{2}{3}\mathrm{u}_j^s\mathrm{n}_j\,\delta_{ik})\,\mathrm{dA} \tag{3.38}$$

are related by

$$\frac{\partial}{\partial t}I_{ik}^{(4)}=-I_{ik}^{(2)} \tag{3.39}$$

Taking the volume average of the heat equations for the fluid and solid yields

$$(1-\eta_o)\,\rho_s c_v^s \frac{\partial \bar{T}_s}{\partial t} - T_o K_s \alpha_s \left[\frac{\partial \eta}{\partial t} - (1-\eta_o)\frac{\partial}{\partial t}\nabla\cdot\bar{\mathbf{u}}^s\right]$$
$$-(1-\eta_o)\kappa_s \nabla^2 \bar{T}_s - \nabla\cdot\kappa_s \mathbf{I}^{(7)} - I^{(8)} = 0 \tag{3.40}$$

$$\eta_o \rho_f c_p^f \frac{\partial \bar{T}_f}{\partial t} - \eta_o T_o \alpha_f \frac{\partial \bar{p}_f}{\partial t} - \eta_o \kappa_f \nabla^2 \bar{T}_f - \nabla\cdot\kappa_f \mathbf{I}^{(5)} - I^{(6)} = 0 \tag{3.41}$$

where the two area integrals

$$\mathbf{I}^{(5)} = \frac{1}{V}\int_{A_{fs}} (\mathrm{T}_f - \mathrm{T}_o)\,\mathrm{d}\,\mathbf{A} \tag{3.42}$$

and

$$\mathbf{I}^{(7)} = \frac{1}{V}\int_{A_{sf}} (\mathrm{T}_s - \mathrm{T}_o)\,\mathrm{d}\,\mathbf{A} \tag{3.43}$$

are related by

$$\mathbf{I}^{(5)} = -\mathbf{I}^{(7)} \tag{3.44}$$

due to continuity of temperatures of the two components on the interface. The two area integrals

$$\mathbf{I}^{(6)} = \frac{1}{V}\int_{A_{fs}} \kappa_f \nabla T_f \cdot d\mathbf{A} \tag{3.45}$$

and

$$I^{(8)} = \frac{1}{V}\int_{A_{sf}} \kappa_s \nabla T_s \cdot d\mathbf{A} \tag{3.46}$$

are related by

$$I^{(6)} = -I^{(8)} \tag{3.47}$$

due to the continuity of heat flux.

The integral $I_{ik}^{(1)}(=-I_{ik}^{(3)})$ is the force (per unit volume) exerted on the fluid by the solid matrix across the interface due to motion. From the point of view of the megascopic continuum equations, it is a body force. For steady flow, this is the term responsible for the Darcian resistance (a Galilean invariant) $\frac{\mu_f \eta_o^2}{K}(\bar{\mathrm{v}}^f - \bar{\mathrm{v}}^s)$, where K is the permeability. For non-steady flow, an additional term, proportional to the relative acceleration, may also be presented (Landau and Lifshitz 1975; Johnson 1980; de la Cruz and Spanos 1989): $\rho_{12}\frac{\partial}{\partial t}(\bar{\mathrm{v}}^f - \bar{\mathrm{v}}^s)$.

Now note that if gravity is included in the discussion the equations in their current form do not satisfy the principle of equivalence. In the presence of gravity, there will be an induced buoyancy force acting on the solid by the fluid, say $-\rho^b g_i$. However, a uniformly accelerating frame can simulate gravity. Since relative acceleration is invariant, another linear combination of the accelerations is needed. Including gravity, this additional term is of the form $\rho^b\left(\frac{\partial \bar{v}_i^m}{\partial t} - g_i\right)$, where $\frac{\partial \bar{v}_i^m}{\partial t}$ is the acceleration of the poro-continuum

$$\frac{\partial \bar{v}_i^m}{\partial t} = \frac{(1-\eta_o)\rho_o^s}{\rho_o^m}\frac{\partial \bar{v}_i^s}{\partial t} + \frac{\eta_o\rho_o^f}{\rho_o^m}\frac{\partial \bar{v}_i^f}{\partial t} \tag{3.48}$$

and $\rho_o^m = (1-\eta_o)\rho_o^s + \eta_o\rho_o^f$ is the mass density of the poro-continuum. When gravity is switched off, this term still arises and has the form $\rho^b \frac{\partial \bar{v}_i^m}{\partial t}$.

The area integral $I_{ik}^{(4)}$ may be expressed in megascopic form in the present case as follows: According to Equation 3.31, $\mu_s I_{ik}^{(4)}$ is the piece needed to fully determine the megascopic solid stress tensor (which we will denote by τ_{ij}^s). It can be shown quite generally (de la Cruz et al. 1993; also see Chapter 4) that the dependence of τ_{ij}^s on the deformation $\bar{u}_{ij}^s$ and $\phi - \phi_o$ (here $\phi = 1 - \eta$ and $\phi_o = 1 - \eta_o$) occurs only through the combination

$$\bar{u}_{jj}'^s = \bar{u}_{ij}^s + \frac{1}{3}\delta_{ij}(\phi-\phi_o)/\phi_o \tag{3.49}$$

where

$$\bar{u}_{ij}^s = \frac{1}{2}(\partial_i \bar{u}_j^s + \partial_j \bar{u}_i^s) \tag{3.50}$$

Here, the symmetric tensor $I_{ik}^{(4)}$ has the general form

$$I_{ik}^{(4)} = c\phi_o\left[\partial_k \bar{u}_i^s + \partial_i \bar{u}_k^s - \frac{2}{3}\delta_{ik}\,\partial_j \bar{u}_j^s\right] + c'\delta_{ik}\bar{u}_{ik}^s + c''\delta_{ik}(\phi-\phi_o) \tag{3.51}$$

where c, c' and c'' are constants. However, since $I_{ik}^{(4)}$ is trace free, $c', c'' = 0$, and thus

$$I_{ik}^{(4)} = c\phi_o[\partial_k \bar{u}_i^s + \partial_i \bar{u}_k^s - \frac{2}{3}\delta_{ik}\,\partial_j \bar{u}_j^s] \tag{3.52}$$

The dimensionless constant c may be conveniently eliminated in favour of a megascopic shear modulus μ_M (Hickey et al. 1995) through the definition

$$\mu_M = \phi_o\mu_s(1+c) \tag{3.53}$$

Thus the physical meaning of c in Equation 3.52 is observed to be a measure of the difference between μ_M and the simple volume-averaged value $\phi_o\,\mu_s$.

Thus $(1-\eta_o)\mu_s$ may be replaced by μ_M in Equation 3.32. At the same time, Equation 3.33, the fluid equation of motion, acquires a new term involving space derivatives of the solid velocity. The new term

$$-(1-\eta_o)\mu_f\left(\frac{\mu_M}{(1-\eta_o)\mu_s}-1\right)\partial_k[\bar{v}^s_{i,k}+\bar{v}^s_{k,i}-\frac{2}{3}\bar{v}^s_{j,j}\delta_{i,k}] \tag{3.54}$$

arises from Equation 3.52, which fails to vanish unless c = 0.

In analogy with the generalization of the shear modulus, the megascopic heat conductivities can be introduced as phenomenological parameters and are related to component heat conductivities κ_f, κ_s according to

$$\kappa^f_M=\eta_o\kappa_f\,(1+c_f) \tag{3.55}$$

and

$$\kappa^s_M=(1-\eta_o)\kappa_s(1+c_s) \tag{3.56}$$

where the dimensionless constants c_s, c_f reflect the pore-scale behaviour through the assumed relation

$$\mathbf{I}^{(5)}=\frac{1}{V}\int_{A_{fs}}(T_f-T_o)d\vec{A}=\eta_o c_f\nabla\bar{T}_f-(1-\eta_o)c_s\nabla\bar{T}_s \tag{3.57}$$

so that from Equation 3.44, continuity of temperature

$$\mathbf{I}^{(7)}=\frac{1}{V}\int_{A_{sf}}(T_s-T_o)d\vec{A}=(1-\eta_o)c_{ss}\nabla\bar{T}_s-\eta_o c_f\nabla\bar{T}_f \tag{3.58}$$

Thus additional megascopic terms in the averaged heat equations 3.40 and 3.41 are obtained, and two additional parameters are introduced. Nozad et al. (1985) also constructed the basis for a two-equation model of transient heat conduction in porous media using volume averaging. However, the area integrals were not evaluated using physical arguments. Given that the two additional megascopic parameters, defined by Equations 3.55 and 3.56, come about due to the effect of pore structure on heat conduction through the porous medium, the two megascopic heat conductivities might be related.

In most studies of heat transfer in porous media filled with a static fluid (Woodside and Messmer 1961a, 1961b; Huang 1971; Zimmerman 1989; Verma et al. 1991; among others), a one-equation model is used and only one effective thermal conductivity parameter, usually referred to as the *stagnant effective thermal conductivity* (Huang 1971; Hsu and Cheng 1990), is required. This model is based on the assumption that a single temperature characterizes the energy transport process (Nozad et al. 1985). This assumption is referred to as *local thermal equilibrium* (Zarotti and Carbonell 1984; Nozad et al. 1985).

By reducing their two-equation model to a one-equation model, Nozad et al. (1985) showed that

$$\nabla^2[\eta_o \kappa_f + (1-\eta_o)\kappa_s]\bar{T}$$

$$+\nabla \cdot \frac{1}{V}\int_{A_{fs}} \kappa_f (T_f - T_o) - \kappa_s (T_s - T_o) dA = \nabla \cdot (\kappa_d \nabla \bar{T}) \tag{3.59}$$

where $\bar{T}$ is the characteristic single temperature and κ_d is the stagnant effective thermal conductivity. Applying such an approach to the present analysis would lead to only one degree of freedom, which would be related to the stagnant effective thermal conductivity as

$$\eta_o c_f - (1-\eta_o) c_s = \frac{\kappa_d - [\eta_o \kappa_f + (1-\eta_o)\kappa_s]}{\kappa_f - \kappa_s} \tag{3.60}$$

The generalization to include convection has been addressed by Yoshida et al. (1990) and Hsu and Cheng (1990). This generalization also contains one effective thermal conductivity. In thermal processes, stagnant effective thermal conductivities are usually considered adequate to describe the conductive heat transfer (Huang 1971).

The two integrals, $I^{(6)}$ and $I^{(8)}$, are equal and opposite and represent the heat transfer from one component to the other across the macroscopic interfaces. Hence, the fluid component acts as an additional heat source for the solid and vice versa. These heat exchange terms between components should vanish if and only if the megascopic component temperatures are equal ($\bar{T}_f = \bar{T}_s$). These terms may be represented by a first-order scalar proportional to ($\bar{T}_s - \bar{T}_f$), and therefore they yield

$$I^{(6)} \int_{A_{fs}} \kappa_f \nabla T_f \cdot d\vec{A} = \gamma(\bar{T}_s - \bar{T}_f) \tag{3.61}$$

where γ is the positive empirical parameter. The heat transfer between components represented by this term should contribute to the attenuation of the dilatational waves.

The surface coefficient of heat transfer, γ, may be estimated by

$$\gamma = O|\kappa A/(VL)| \tag{3.62}$$

where κ is the effective conductivity between the fluid and solid, A is the interfacial surface area between the fluid and the solid within the volume, V is the averaging volume, and L is the characteristic pore-scale length.

Counting the number of variables and equations, it is observed that an additional equation is needed for completeness when dilatational motions are considered. At this point note that Newton's second law has not been completely specified. When the medium is compressed, the solid may be compressed or deformed, the fluid may be compressed or may flow, and the relative proportions of the two phases inside of a volume element may change, thus changing the porosity. Furthermore, for a static compression both phases are simply compressed; however, for a seismic deformation, fluid flow is an integral part of a compression. Thus the relationship between $\nabla \cdot \mathbf{v}_s$, $\nabla \cdot \mathbf{v}_f$ and $\frac{\partial n}{\partial t}$ must be specified in order to completely describe a compression. Thus the following relationship is obtained (assuming that locally the phases remain in thermal equilibrium)

$$\frac{\partial \eta}{\partial t} = \delta_s \nabla \cdot v_s - \delta_f \nabla \cdot v_f \tag{3.63}$$

where δ_s and δ_f are dimensionless parameters. Equation 3.63 is required to complete Newton's second law; thus it is an equation of motion that specifies the interactions of the phases at the macroscale (pore scale). A basic physical understanding of this relationship has been presented in the context of dilational experiments (Hickey et al. 1995). Its thermodynamic basis will be presented in Chapter 4.

For now it is possible to obtain this equation through the following arguments (Spanos 2002). The porosity may be changed, quasi-statically by altering the megascopic compressive stresses on the component phases or by altering other forces such as body forces. Thus

$$\eta - \eta_o = a\bar{\sigma}^s_{jj} + b\bar{\sigma}^f_{jj} + B \tag{3.64}$$

where B denotes the contribution from forces other than the stresses. For phenomena such as seismic wave propagation, these forces are set to 0. According to $\bar{\sigma}^s_{jk}$, the following relation is obtained:

$$\bar{\sigma}^s_{jj} = 3K_s \left[\bar{u}^s_{ii} - \frac{\eta - \eta_o}{1 - \eta_o} \right] \tag{3.65}$$

For fluids, it is sufficient to take $\bar{\sigma}^f_{jj} = -3\bar{p}_f$. Thus

$$\left[1 + \frac{3\,a\,K_s}{1 - \eta_o} \right] (\eta - \eta_o) = 3\,\mathrm{a}\,\mathrm{K}_s\,\bar{u}^s_{jj} - 3\,b\,\bar{p}_f \tag{3.66}$$

Differentiating both sides with respect to time and making use of the pressure equation (Equation 3.16) yields the quasi-static porosity equation

$$\frac{\partial \eta}{\partial t} = \delta_s \nabla \cdot v_s - \delta_f \nabla \cdot v_f \tag{3.67}$$

SUMMARY OF THE EQUATIONS (NEGLECTING THERMOMECHANICAL COUPLING AND THE PRINCIPLE OF EQUIVALENCE)

The megascopic equations describing deformations of elastic matrix with a Newtonian fluid saturating the pore spaces are given by de la Cruz and Spanos (1989), Hickey et al. (1995) and Spanos (2002).

Equations of motion:

$$\rho_s \frac{\partial}{\partial t}\vec{v}_s = K_s \vec{\nabla}(\vec{\nabla}\cdot\vec{u}_s) - \frac{K_s}{(1-\eta_o)}\vec{\nabla}\eta + \frac{\mu_f \eta_o^2}{K(1-\eta_o)}(\vec{v}_f - \vec{v}_s)$$
$$- \frac{\rho_{12}}{(1-\eta_o)}\frac{\partial}{\partial t}(\vec{v}_f - \vec{v}_s) + \frac{\mu_M}{(1-\eta_o)}\left[\nabla^2 u_s + \tfrac{1}{3}\vec{\nabla}(\vec{\nabla}\cdot\vec{u}_s)\right] \tag{3.68}$$

and

$$\rho_f \frac{\partial}{\partial t}\vec{v}_f = -\vec{\nabla}p_f + \left[\mu_f \nabla^2 \vec{v}_f + (\xi_f + \tfrac{1}{3}\mu_f)\vec{\nabla}(\vec{\nabla}\cdot\vec{v}_f)\right] + \frac{\xi_f}{\eta_o}\vec{\nabla}\frac{\partial\eta}{\partial t}$$
$$+ \frac{(1-\eta_o)}{\eta_o}\mu_f\left(\frac{\mu_M}{(1-\eta_o)\mu_s} - 1\right)\left[\nabla^2 \frac{\partial \vec{u}_s}{\partial t} + \tfrac{1}{3}\vec{\nabla}(\vec{\nabla}\cdot\frac{\partial \vec{u}_s}{\partial t})\right] \tag{3.69}$$
$$- \frac{\mu_f \eta_o}{K}(\vec{v}_f - \vec{v}_s) + \frac{\rho_{12}}{\eta_o}\frac{\partial}{\partial t}(\vec{v}_f - \vec{v}_s)$$

Equations of continuity:

$$\frac{(\rho_s - \rho_s^o)}{\rho_s^o} - \frac{(\eta - \eta_o)}{(1-\eta_o)} + \nabla\cdot\mathbf{u}_s = 0 \tag{3.70}$$

$$\frac{1}{\rho_f^o}\frac{\partial}{\partial t}\rho_f + \frac{1}{\eta_o}\frac{\partial}{\partial t}\eta + \nabla\cdot\mathbf{v}_f = 0 \tag{3.71}$$

Pressure equations:

$$\frac{1}{K_s}p_s = -\nabla\cdot\mathbf{u}_s + \frac{(\eta-\eta_o)}{(1-\eta_o)} \tag{3.72}$$

$$\frac{1}{K_f}\frac{\partial}{\partial t}p_f = -\nabla\cdot\mathbf{v}_f - \frac{1}{\eta_o}\frac{\partial}{\partial t}\eta \tag{3.73}$$

Porosity equation:

$$\frac{\partial\eta}{\partial t} = \delta_s \vec{\nabla}\cdot\vec{\mathbf{v}}_s - \delta_f \vec{\nabla}\cdot\vec{\mathbf{v}}_f \tag{3.74}$$

Here, $\mathbf{v}_f$, for example, stands for $\bar{\mathbf{v}}_f$, the averaged velocity over the fluid portion within an averaging volume element. The subscript or superscript *o* refers to unperturbed quantities.

Aside from the (unperturbed) porosity η_o, five other 'megascopic' empirical parameters (K, ρ_{12}, μ_M, δ_s and δ_f) appear in these equations. In contrast, the parameters K_s, μ_f, etc., are the pore-scale physical parameters specifying the constituents.

In the case of static compressions, the dimensionless parameters δ_s and δ_f have the following values (de la Cruz et al. 1992):

$$\delta_s = (1-\eta_o)\frac{(K_s - K_{ud})(K_s - K_d)}{(K_s - K_f)[K_s - (1-\eta_o)K_d]} \tag{3.75}$$

$$\delta_f = \eta_o \frac{K_s(K_{ud} - K_f) - (1-\eta_o)K_d(K_s - K_f)}{(K_s - K_f)[K_s - (1-\eta_o)K_d]} \tag{3.76}$$

However, seismic compressions involve local fluid motions due to the difference in compressibility of the fluid and solid as well as the pore structure. These effects are much smaller than the zeroth order terms (for 1/K) given by Equations 3.75 and 3.76. These equations now become

$$\delta_s = (1-\eta_o)\frac{(K_s - K_{ud})(K_s - K_d)}{(K_s - K_f)[K_s - (1-\eta_o)K_d]} + \frac{\Xi(\mu_s, \mu_f, \xi_f)}{[K_s - (1-\eta_o)K_d]} \tag{3.77}$$

$$\delta_f = \eta_o \frac{K_s(K_{ud} - K_f) - (1-\eta_o)K_d(K_s - K_f)}{(K_s - K_f)[K_s - (1-\eta_o)K_d]} + \frac{\Theta(\mu_s, \mu_f, \xi_f)}{[K_s - (1-\eta_o)K_d]} \tag{3.78}$$

where Ξ and Θ are experimentally determined parameters that depend on the pore structure and describe a linear combination of pore-scale parameters, μ_s, μ_f, ξ_f.

CONSTRUCTION OF THE WAVE EQUATIONS

Here, $\mathbf{u}_s$ is the megascopically averaged solid displacement. Here, wave motions where the time dependence is given by $e^{-i\omega t}$ are considered; thus the megascopically averaged 'fluid displacement vector' is defined by

$$\mathbf{u}^f = \frac{1}{-i\omega} v^f \tag{3.79}$$

Here, μ_m is the megascopic shear modulus of the solid component (see Hickey et al. 1995). The parameters δ_s and δ_f are process dependent and thus may have different values for wave propagation than for static compressions (see de la Cruz et al. 1993). In the present analysis, the term $\rho^b \frac{\partial v_i^{(m)}}{\partial t}$ will be ignored, but it is an indication of what is to come in later chapters.

Assuming time harmonic fields

$$u^s = U^s e^{-i\omega t}, U^f e^{-i\omega t} \tag{3.80}$$

and adopting the notation

$$\delta_\mu = (1-\eta_o)\mu_f\left(\frac{\mu_m}{(1-\eta_o)\mu_s} - 1\right) \tag{3.81}$$

yields

$$\left[\eta_o\rho_f^o\omega^2+i\omega\frac{\eta_o^2\mu_f}{K}-\rho_{12}\omega^2\right]\vec{U}^f-\left[i\omega\frac{\eta_o^2\mu_f}{K}-\rho_{12}\omega^2\right]\vec{U}^s$$

$$+\eta_o\left[K_f(1-\frac{\delta_f}{\eta_o})-\mathrm{i}\,\omega\xi(1\frac{\delta_f}{\eta_o})-\frac{4}{3}\mathrm{i}\omega\,\mu_f\right]\vec{\nabla}(\vec{\nabla}\cdot\vec{U}^f) \tag{3.82}$$

$$+\left[(K_f-\mathrm{i}\omega\xi)\delta_s-\frac{4}{3}i\,\omega\,\delta\mu\right]\vec{\nabla}(\vec{\nabla}\cdot\vec{U}^s)$$

$$-\mathrm{i}\omega\,\delta_\mu\vec{\nabla}\times\vec{\nabla}\times\vec{U}^s+\mathrm{i}\omega\eta_o\mu_f\vec{\nabla}\times\vec{\nabla}\times\vec{U}^f=0$$

$$-\left[i\omega\frac{\eta_o^2\mu_f}{K}-\rho_{12}\omega^2\right]\vec{U}^f+\left[(1-\eta_o)\rho_s^o\omega^2+i\omega\frac{\eta_o^2\mu_f}{K}-\rho_{12}\omega^2\right]\vec{U}^s$$

$$+K_s\delta_f\vec{\nabla}(\vec{\nabla}\cdot\vec{U}^f)+\left[(1-\eta_o)\mathrm{K}_s(1-\frac{\delta_s}{(1-\eta_o)})+\frac{4}{3}\mu_M\right]\vec{\nabla}(\vec{\nabla}\cdot\vec{U}^s) \tag{3.83}$$

$$-\mu_M\vec{\nabla}\times\vec{\nabla}\times\vec{U}^s=0$$

These equations may be written as

$$P_{11}\vec{\nabla}(\vec{\nabla}\cdot\vec{U}^s)+\mathrm{P}_{12}\,\vec{\nabla}(\vec{\nabla}\cdot\vec{U}f)-\mathrm{S}_{11}\,\vec{\nabla}\times\vec{\nabla}\times\vec{U}^s$$

$$+\omega^2[\mathrm{D}_{11}\vec{\mathrm{U}}^s+\mathrm{D}_{12}\;\vec{\mathrm{U}}^f]=0 \tag{3.84}$$

$$P_{21}\,\vec{\nabla}(\vec{\nabla}\cdot\vec{U}^s)+P_{22}\vec{\nabla}(\vec{\nabla}\cdot\vec{U}^f)-S_{21}\vec{\nabla}\times\vec{\nabla}\times\vec{U}^s-S_{22}\vec{\nabla}\times\vec{\nabla}\times\vec{U}^f)$$

$$+\omega^2[\mathrm{D}_{21}\vec{U}^s+\mathrm{D}_{22}\vec{U}^f]=0 \tag{3.85}$$

where

$$P_{11}=(1-\eta_o)\mathrm{K}_s(1-\frac{\delta_s}{1-\eta_o})+\frac{4}{3}\mu_M \qquad P_{12}=K_s\delta_f \tag{3.86}$$

$$P_{21}=(K_f-\mathrm{i}\omega\xi)\delta_s-\frac{4}{3}i\,\omega\,\delta_\mu \tag{3.87}$$

$$P_{22}=\eta_o\left[K_f\left(1-\frac{\delta_f}{\eta_o}\right)-i\,\omega\xi\left(1-\frac{\delta_f}{\eta_o}\right)-\frac{4}{3}i\,\omega\,\mu_f\right] \tag{3.88}$$

$$\mathrm{S}_{11}=\mu_M\;\mathrm{S}_{21}=\mathrm{i}\,\omega\,\delta_\mu\;\mathrm{S}_{22}=\mathrm{i}\,\omega\,\eta_o\mu_f \tag{3.89}$$

$$D_{11} = (1-\eta_o)\rho_s^o + i\frac{\eta_o^2\mu_f}{\omega K} - \rho_{12} \quad D_{12} = \rho_{12} - i\frac{\eta_o^2\mu_f}{\omega K} \tag{3.90}$$

$$D_{21} = \rho_{12} - \frac{i\,\eta_o^2\mu_f}{\omega K} \quad D_{22} = \eta_o\,\rho_f^o + i\frac{\eta_o^2\mu_f}{\omega K} - \rho_{12} \tag{3.91}$$

The scalar and vector potentials

$$\vec{U}^s = \vec{\nabla}\,\phi_s + \vec{\nabla}\times\vec{\psi}_s \tag{3.92}$$

$$\vec{U}^f = \vec{\nabla}\,\phi_f + \vec{\nabla}\times\vec{\psi}_f \tag{3.93}$$

are now introduced, where $\vec{\nabla}\cdot\vec{\psi}_s = \vec{\nabla}\cdot\vec{\psi}_f = 0$.

Substituting the potentials 3.92 and 3.93 into Equations 3.84 and 3.85 yields

$$\vec{\nabla}\left(P_{11}\,\nabla^2\phi_s + P_{12}\nabla^2\phi_f + \omega^2[D_{11}\,\phi_s + D_{12}\,\phi_f]\right)$$

$$\vec{\nabla}\times\left(S_{11}\,\nabla^2\vec{\psi}_s + \omega^2[D_{11}\,\vec{\psi}_s + D_{12}\vec{\psi}_f]\right) = 0 \tag{3.94}$$

$$\vec{\nabla}\left(P_{21}\,\nabla^2\phi_s + P_{22}\nabla^2\phi_f + \omega^2[D_{21}\,\phi_s + D_{22}\phi_f]\right)$$

$$\vec{\nabla}\times\left(S_{21}\,\nabla^2\vec{\psi}_s + S_{22}\nabla^2\vec{\psi}_f + \omega^2[D_{21}\,\vec{\psi}_s + D_{22}\vec{\psi}_f]\right) = 0 \tag{3.95}$$

For compressional motions one obtains the system of equations

$$\begin{pmatrix} P_{11} & P_{12} \\ P_{21} & P_{22} \end{pmatrix}\begin{pmatrix} \nabla^2\,\phi_s \\ \nabla^2\,\phi_f \end{pmatrix} + \omega^2\begin{pmatrix} D_{11} & D_{12} \\ D_{21} & D_{22} \end{pmatrix}\begin{pmatrix} \phi_s \\ \phi_f \end{pmatrix} = \begin{pmatrix} 0 \\ 0 \end{pmatrix} \tag{3.96}$$

which may be rewritten as

$$\begin{pmatrix} \nabla^2\,\phi_s \\ \nabla^2\,\phi_f \end{pmatrix} + \frac{\omega^2}{\Delta P}\begin{pmatrix} P_{22} & -P_{12} \\ -P_{21} & P_{11} \end{pmatrix}\begin{pmatrix} D_{11} & D_{12} \\ D_{21} & D_{22} \end{pmatrix}\begin{pmatrix} \phi_s \\ \phi_f \end{pmatrix} = \begin{pmatrix} 0 \\ 0 \end{pmatrix} \tag{3.97}$$

where $\Delta P = P_{11}\,P_{22} - P_{12}\,P_{21}$.

Writing out the two equations given in Equation 3.97 and eliminating ϕ_f yields

$$\left(\nabla^2 + \frac{\omega^2}{\alpha_1^2}\right)\left(\nabla^2 + \frac{\omega^2}{\alpha_2^2}\right)\phi_s = 0 \tag{3.98}$$

where

$$\alpha_1^2, \alpha_2^2 = \frac{1}{2\Delta D}Tr(P\dagger D) \pm \sqrt{Tr^2(P\dagger D) - 4\Delta P\,\Delta D} \tag{3.99}$$

and

$$\Delta D = D_{11}D_{22} - D_{12}D_{21} \tag{3.100}$$

$$\mathrm{Tr}(P^{\dagger}D) = D_{11}P_{22} + D_{22}P_{11} - D_{21}P_{12} - D_{12}P_{21} \tag{3.101}$$

An identical result can be obtained by eliminating ϕ_s; thus

$$\phi_s = v_{s1}^{\alpha}\ \phi_1 + v_{s2}^{\alpha}\phi_2 \tag{3.102}$$

and

$$\phi_f = v_{f1}^{\alpha}\ \phi_1 + v_{f2}^{\alpha}\ \phi_2 \tag{3.103}$$

such that

$$\left(\nabla^2 + \frac{\omega^2}{\alpha_1^2}\right)\phi_1 = 0 \tag{3.104}$$

and

$$\left(\nabla^2 + \frac{\omega^2}{\alpha_2^2}\right)\phi_2 = 0 \tag{3.105}$$

Here, a suitable choice of v_{si}^{α} and $v_{fi}^{\alpha}(i = 1, 2)$ is given by

$$v_{s1}^{\alpha} = a(D_{12}P_{22} - D_{22}P_{12})\alpha_1^2 \tag{3.106}$$

$$v_{s2}^{\alpha} = a(D_{12}P_{22} - D_{22}P_{12})\alpha_2^2 \tag{3.107}$$

$$v_{f1}^{\alpha} = d[\Delta P\ - (D_{11}P_{22} - D_{21}P_{12})\alpha_1^2] \tag{3.108}$$

$$v_{f2}^{\alpha} = d[\Delta P\ - (D_{11}P_{22} - D_{21}P_{12})\alpha_2^2] \tag{3.109}$$

where a/d is an arbitrary constant that may be incorporated into the potentials (i.e. without loss of generality, let a = d = 1).

For rotational motions

$$\begin{pmatrix} S_{11} & 0 \\ S_{21} & S_{22} \end{pmatrix}\begin{pmatrix} \nabla^2 & \vec{\psi}_s \\ \nabla^2 & \vec{\psi}_f \end{pmatrix} + \omega^2\begin{pmatrix} D_{11} & D_{12} \\ D_{21} & D_{22} \end{pmatrix}\begin{pmatrix} \vec{\psi}_s \\ \vec{\psi}_f \end{pmatrix} = \begin{pmatrix} 0 \\ 0 \end{pmatrix} \tag{3.110}$$

which may be rewritten as

$$\begin{pmatrix} \nabla^2 & \vec{\psi}_s \\ \nabla^2 & \vec{\psi}_f \end{pmatrix} + \frac{\omega^2}{\Delta S}\begin{pmatrix} S_{22} & 0 \\ -S_{21} & S_{11} \end{pmatrix}\begin{pmatrix} D_{11} & D_{12} \\ D_{21} & D_{22} \end{pmatrix}\begin{pmatrix} \vec{\psi}_s \\ \vec{\psi}_f \end{pmatrix} = \begin{pmatrix} 0 \\ 0 \end{pmatrix} \tag{3.111}$$

where $\Delta S = S_{11}S_{22}$. Eliminating $\vec{\psi}_f$ yields

$$\left(\nabla^2 + \frac{\omega^2}{\beta_1^2}\right)\left(\nabla^2 + \frac{\omega^2}{\beta_2^2}\right)\vec{\psi}_s = 0 \tag{3.112}$$

where

$$\beta_1^2, \beta_2^2 = \frac{1}{2\Delta D} Tr(S\dagger D) \pm \sqrt{Tr^2(S\dagger D) - 4\Delta S\ \Delta D} \tag{3.113}$$

An identical result can be obtained by eliminating $\vec{\psi}_s$; thus

$$\vec{\psi}_s = v_{s1}^{\beta}\ \vec{\psi}_1 + v_{s2}^{\beta}\ \vec{\psi}_2 \tag{3.114}$$

$$\vec{\psi}_f = v_{f1}^{\beta}\ \vec{\psi}_1 + v_{f2}^{\beta}\ \vec{\psi}_2 \tag{3.115}$$

such that

$$\left(\nabla^2 + \frac{\omega^2}{\beta_1^2}\right)\vec{\psi}_1 = 0 \tag{3.116}$$

and

$$\left(\nabla^2 + \frac{\omega^2}{\beta_2^2}\right)\vec{\psi}_2 = 0 \tag{3.117}$$

Here, a suitable choice of v_{si}^{β} and v_{fi}^{β} (i = 1, 2) is given by

$$v_{s1}^{\beta} = a\ D_{12}S_{22}\ \beta_1^2 \tag{3.118}$$

$$v_{s2}^{\beta} = a\ D_{12}S_{22}\ \beta_2^2 \tag{3.119}$$

$$v_{f1}^{\beta} = d\left[\Delta S - D_{11}S_{22}\ \beta_1^2\right] \tag{3.120}$$

$$v_{f2}^{\beta} = d\left[\Delta S - D_{11}S_{22}\ \beta_2^2\right] \tag{3.121}$$

where a/d is an arbitrary constant that may be incorporated into the potentials (i.e. without loss of generality, let a = d = 1).

Here, multiple seismic waves are introduced, along with a natural attenuation mechanism described by the fluid motions. As well, with the introduction of a fluid, a new mechanism for moving fluid through a porous medium is obtained, which is described by a porosity–pressure wave. In the limiting case where this wave becomes overdamped, a porosity–pressure diffusion process is obtained. In the intermediate regime a porosity–pressure wave is obtained, which if generated periodically may retain energy from previous waves to propagate further with each consecutive pulse. It will be observed that this wave may extract mechanical energy from the stress field of the matrix.

BOUNDARY CONDITIONS

Let V be an arbitrary volume element that encloses two fluid-filled porous media. The total mass in V is denote by M. The boundary surface is defined to be the surface such that the volumes V_a and V_b ascribed to a and b, respectively, satisfy

$$V_a\rho^{(a)} + V_b\rho^{(b)} = M \tag{3.122}$$

where

$$\rho^{(a)} = \eta_a\rho_f^{(a)} + (1-\eta_a)\rho_s^{(a)} \tag{3.123}$$

is the mass density in medium a, and similarly for b.

The megascopic equations of continuity for the two components of a fluid-filled porous medium are

$$\frac{\partial}{\partial t}\left(\frac{\rho_f - \rho_f^o}{\rho_f^o} + \frac{\eta - \eta_o}{\eta_o}\right) + \vec{\nabla}\cdot\vec{v}_f = 0 \tag{3.124}$$

$$(1-\eta_o)(\rho_s - \rho_s^o)/\rho_s^o = -(1-\eta_o)\vec{\nabla}\cdot\vec{u}_s + (\eta - \eta_o) \tag{3.125}$$

where subscript or superscript *o* denotes unperturbed values.

Combining the previous two equations yields

$$\frac{\partial}{\partial t}(\eta\rho_f + (1-\eta)\rho_s) = -\vec{\nabla}\cdot(\eta_o\rho_f^o\vec{v}_f + (1-\eta_o)\rho_s^o\vec{v}_s) \tag{3.126}$$

Here

$$\rho = \eta\rho_f + (1-\eta)\rho_s \tag{3.127}$$

is the mass density of the medium; thus Equation 3.126 is the equation of continuity, which has already been derived using volume averaging beyond the first-order construction presented here. Thus Equation 3.126 may be written as

$$\frac{\partial}{\partial t}\rho + \vec{\nabla}\cdot(\rho\vec{v}) = 0 \tag{3.128}$$

where

$$\vec{v} = (\eta\rho_f\vec{v}_f + (1-\eta)\rho_s\vec{v}_s)/\rho \tag{3.129}$$

is the velocity of momentum flux, and conservation of mass requires that the velocity of momentum flux be continuous across the boundary. Of course, this condition must also hold at every position in the interior of each medium, which becomes an important constraint in determining how the equations of motion are modified when considering a heterogeneous porous medium.

Thus

$$\vec{v}^{(a)} \cdot \vec{n} = \vec{v}^{(b)} \cdot \vec{n} \tag{3.130}$$

where $\vec{n}$ is the unit normal

$$\vec{v}^{(a)} = (\eta_a\rho_f^a\vec{v}_f^a + (1-\eta_a)\rho_s^a\vec{v}_s^a)/(\eta_a\rho_f^a + (1-\eta_a)\rho_s^a) \tag{3.131}$$

and similarly for b.

The other boundary conditions are specified by Newton's third law. These are also statements that must hold at every point in space and also become important statements constraining the equations of motion both at a boundary and when heterogeneity is introduced.

In terms of $\tau_{jk}^{(a)}$ and $\tau_{jk}^{(b)}$, where

$$\tau_{jk} = \eta\tau_{jk}^f + (1-\eta)\tau_{jk}^s \tag{3.132}$$

is the total stress tensor and the superscripts a and b specify the medium, Newton's third law requires

$$\tau_{jk}^{(a)}n_k = \tau_{jk}^{(b)}n_k \tag{3.133}$$

When written in terms of the component phases, one obtains

$$\eta_a\tau_{ik}^{(a)f}n_k = \tau_{ik}^{(b)f}n_k[\eta_a\eta_b - \alpha] + \tau_{ik}^{(b)s}n_k[\eta_a(1-\eta_b) - \alpha] \tag{3.134}$$

$$(1-\eta_a)\tau_{ik}^{(a)s}n_k = \tau_{ik}^{(b)f}n_k[(1-\eta_a)\eta_b - \alpha] + \tau_{ik}^{(b)s}n_k[(1-\eta_a)(1-\eta_b) - \alpha] \tag{3.135}$$

$$\eta_b\tau_{ik}^{(b)f}n_k = \tau_{ik}^{(a)f}n_k[\eta_a\eta_b - \alpha] + \tau_{ik}^{(a)s}n_k[(1-\eta_a)\eta_b - \alpha] \tag{3.136}$$

$$(1-\eta_b)\tau_{ik}^{(b)f}n_k = \tau_{ik}^{(a)f}n_k[\eta_a(1-\eta_b) - \alpha] + \tau_{ik}^{(a)s}n_k[(1-\eta_a)(1-\eta_b) - \alpha] \tag{3.137}$$

where

$$\alpha = \sqrt{\eta_a \eta_b (1-\eta_a)(1-\eta_b)} \tag{3.138}$$

Here, any three of Equations 3.133 through 3.137 completely specify Newton's third law at the boundary.

In Figure 3.1, the standard case of a solid–solid boundary without a critical angle is shown for comparison with Figures 3.2 and 3.3. In Figure 3.2, medium I is given a

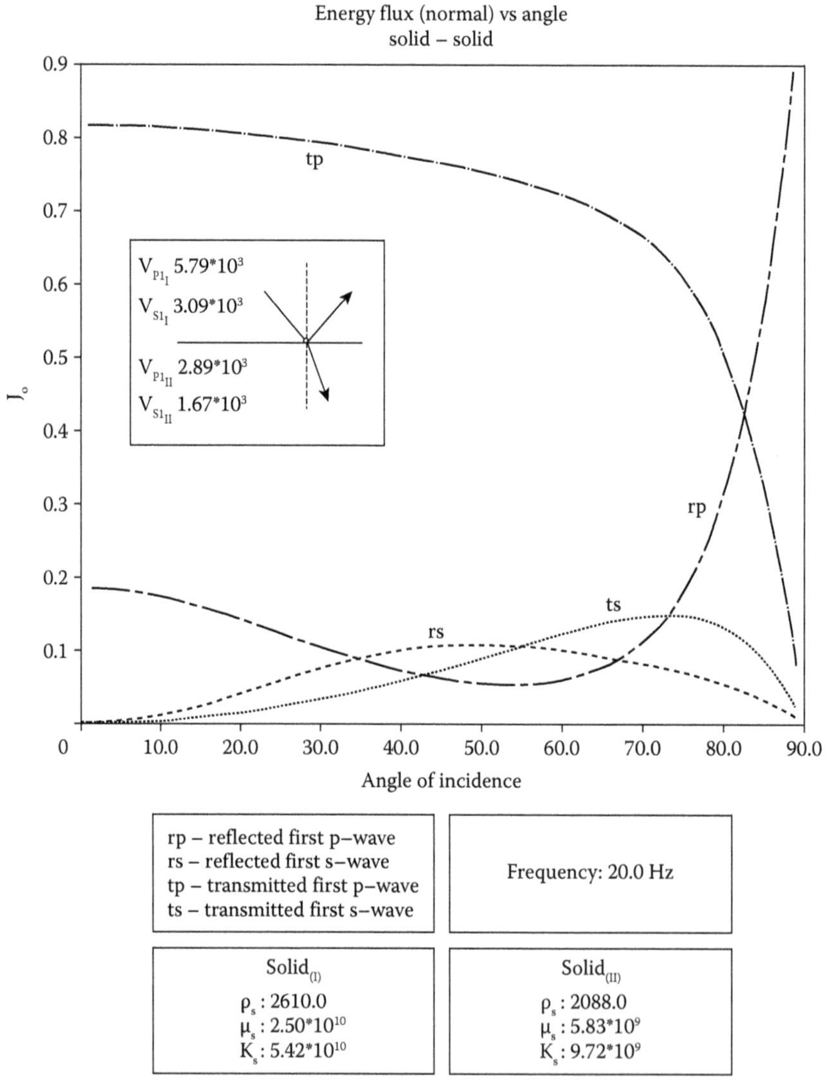

FIGURE 3.1 Reflected and transmitted waves between two elastic media. The speed in the incident medium is greater than the transmitted medium. (Adapted from Spanos, T. J. T., et al., *Fundamental Thermodynamic Requirements for Porous Media Description*, pp. 513–521, 2004.)

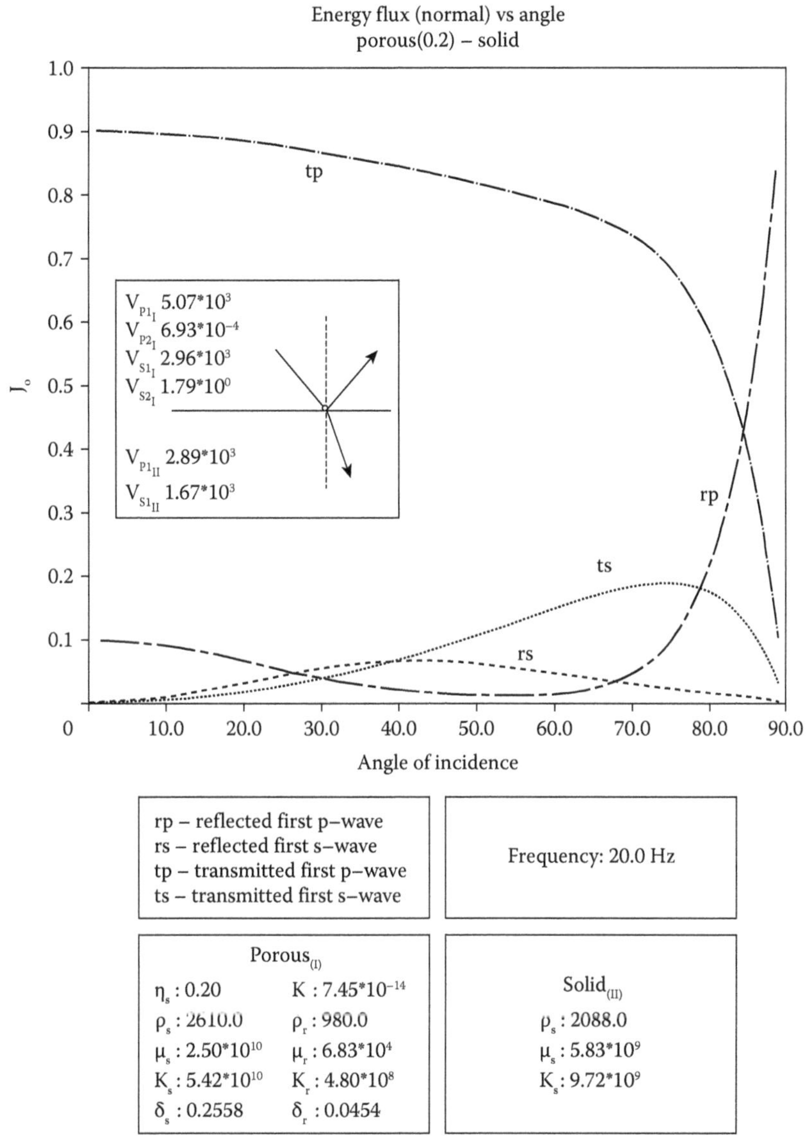

FIGURE 3.2 Here, the incident medium in Figure 3.1 is given porosity of 0.20 and pores are filled with a highly viscous fluid with the properties of bitumen. (Adapted from Spanos, T. J. T., et al., *Fundamental Thermodynamic Requirements for Porous Media Description*, pp. 513–521, 2004.)

porosity of 0.2 and a permeability of 7.45×10^{-14} m^2 and the pores are filled with bitumen. It is interesting to note an extremely small reflected P wave is predicted at intermediate angle, which is consistent with the experimental observations of Kanasewich (1991). In Figure 3.3, the fluid is taken to be water and the variation in energy flux with angle is more pronounced. In Figure 3.4, the standard case of a solid–solid boundary with a critical angle is given to compare with Figures 3.5 and 3.6, involving porous media. In Figure 3.5, medium II is given a porosity of 0.2 and a permeability of 7.45×10^{-14} m^2

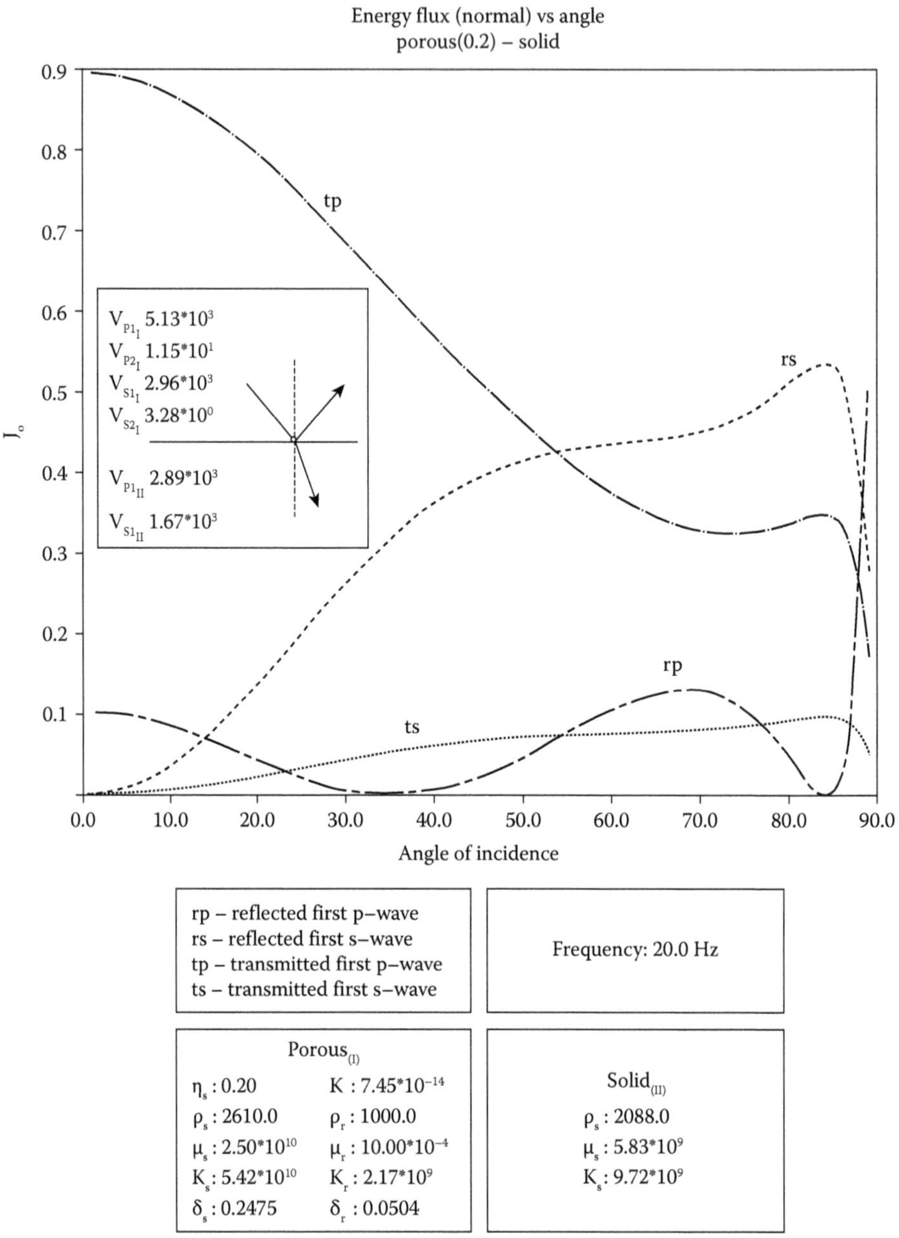

FIGURE 3.3 The fluid in Figure 3.2 is changed to water. (Adapted from Spanos, T. J. T., et al., *Fundamental Thermodynamic Requirements for Porous Media Description*, pp. 513–521, 2004.)

and is filled with water. In Figure 3.6, both media are given a porosity of 0.2, medium II is once again filled with water and medium I is filled with bitumen. In Figure 3.7, the mode conversions that occur at a water–solid boundary are shown in order to compare with Figure 3.8, which shows the mode conversions at a water–porous medium boundary. In Figure 3.8, the fluid is water, the porosity is 0.35 and the permeability is once again 7.45×10^{-14} m^2. In Figure 3.8, a critical angle appears, but beyond the critical

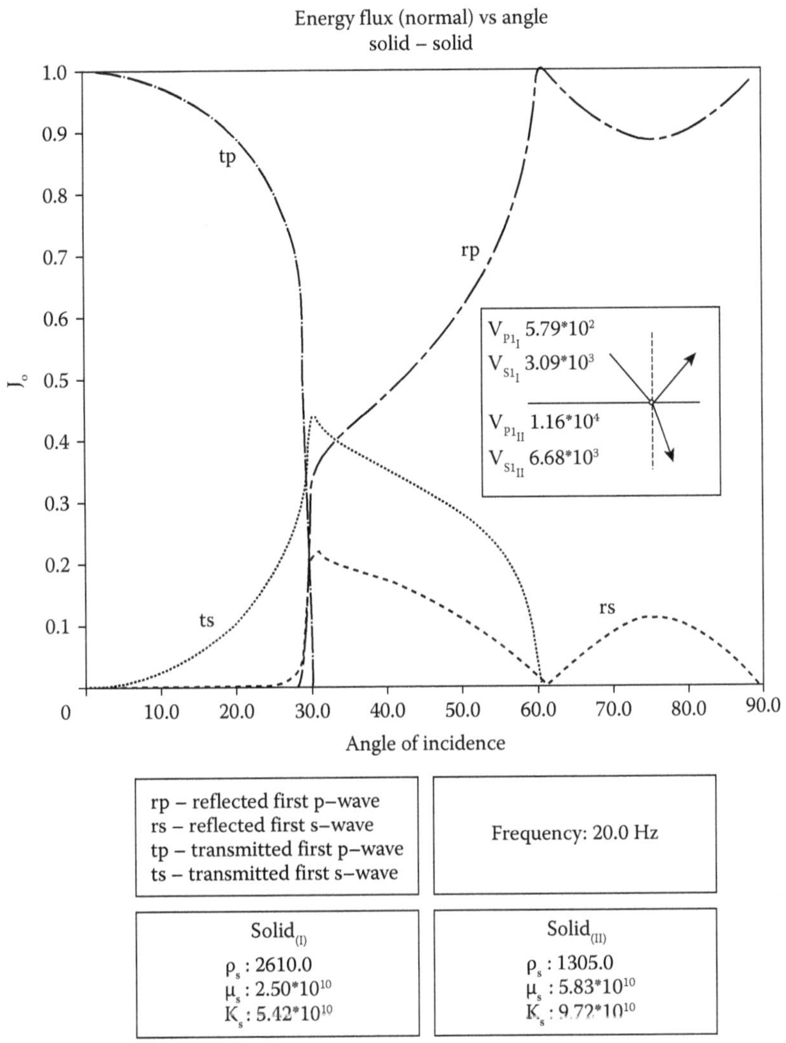

FIGURE 3.4 Reflected and transmitted waves between two elastic media. (Adapted from Spanos, T. J. T., et al., *Fundamental Thermodynamic Requirements for Porous Media Description*, pp. 513–521, 2004.)

angle an energy flux once again occurs in the P and S waves transmitted into the porous medium. In the limit, as the porosity goes to 0, the component perpendicular to the surface vanishes and these waves become pure surface waves. Figure 3.9 shows that for angles greater than the critical angle the attenuation of these transmitted P and S waves into the porous medium increases sharply.

THE EFFECT OF HETEROGENEITY

When the pore-scale equations are volume averaged and the most basic laws of physics (conservation of mass, Netwton's second law, Newton's third law, etc.) are required to hold at all scales, then it is observed that it is not parameters that change but the structure and

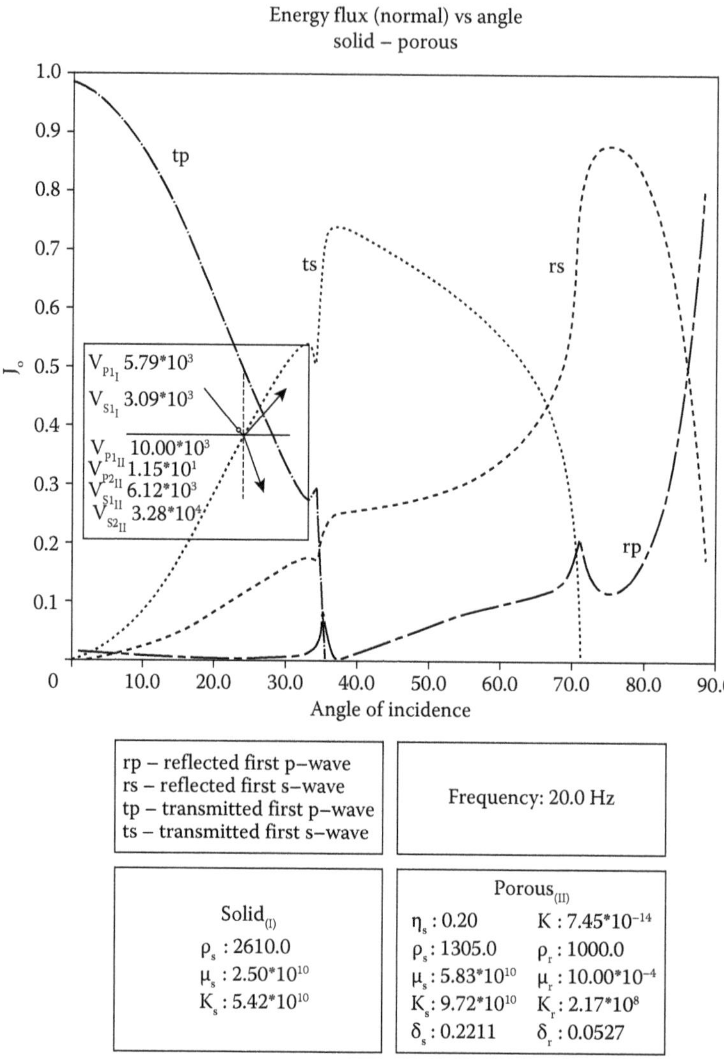

FIGURE 3.5 Medium II in Figure 3.4 is given a porosity of 0.2 and filled with water. (Adapted from Spanos, T. J. T., et al., *Fundamental Thermodynamic Requirements for Porous Media Description*, pp. 513–521, 2004.)

form of the megascopic equations are altered by heterogeneity. A similar result will be observed in Chapter 5 for single-phase flow in a heterogeneous, porous medium.

As an example, consider an inhomogeneous porous medium whose components are each homogeneous when unperturbed. Allow the porosity to vary exponentially with depth. The equations describing an inhomogeneous porous medium of this type are given by Sahay et al. (2001).

Macroscopic continuity equations:

$$\eta_o \frac{\partial \rho_f}{\partial t} + \rho_f^o \frac{\partial \eta}{\partial t} + \rho_f^o \frac{\partial}{\partial x_j}(\eta_o v_j^f) = 0 \tag{3.139}$$

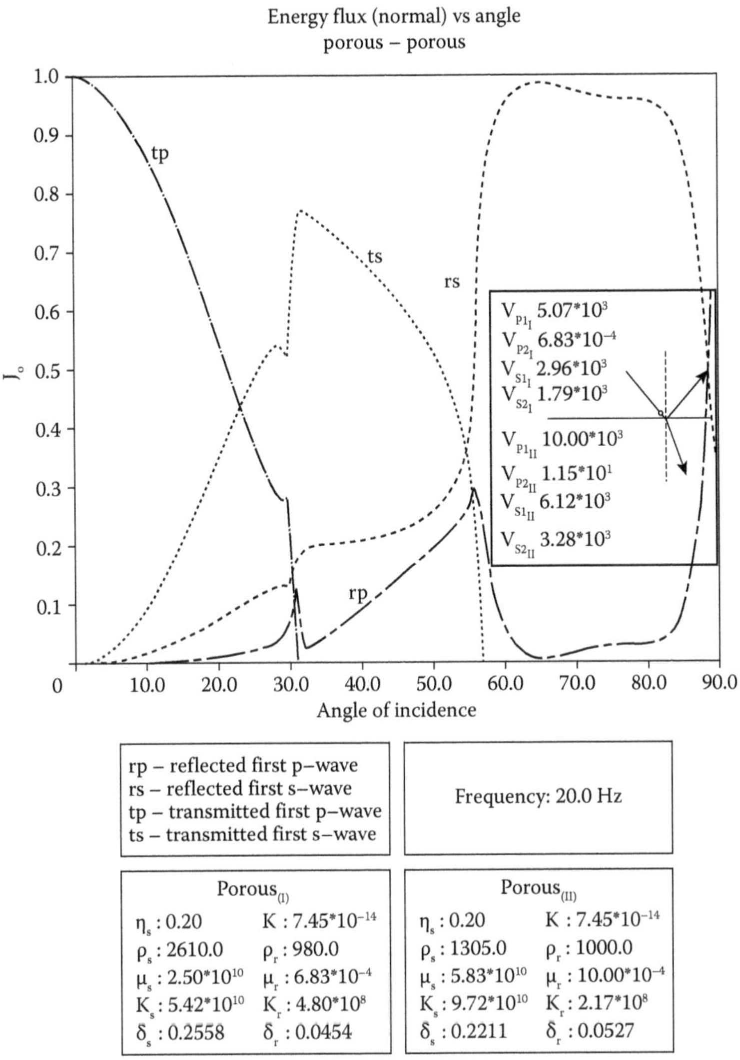

FIGURE 3.6 Medium I in Figure 3.4 is also given a porosity of 0.2 and filled with bitumen. (Adapted from Spanos, T. J. T., et al., *Fundamental Thermodynamic Requirements for Porous Media Description*, pp. 513–521, 2004.)

$$(1-\eta_o)\frac{(\rho_s-\rho_s^o)}{\rho_s^o}+\frac{\partial}{\partial x_j}[(1-\eta_o)u_j^s]-(\eta-\eta_o)=0 \tag{3.140}$$

Macroscopic equations of motion

$$\eta_o\rho_f^o\frac{\partial}{\partial t}v_i^f=\partial_k\tau_{ik}^f-F_i \tag{3.141}$$

$$\phi_o\rho_s^o\frac{\partial}{\partial t}v_i^S=\partial_k\tau_{ik}^s+F_i \tag{3.142}$$

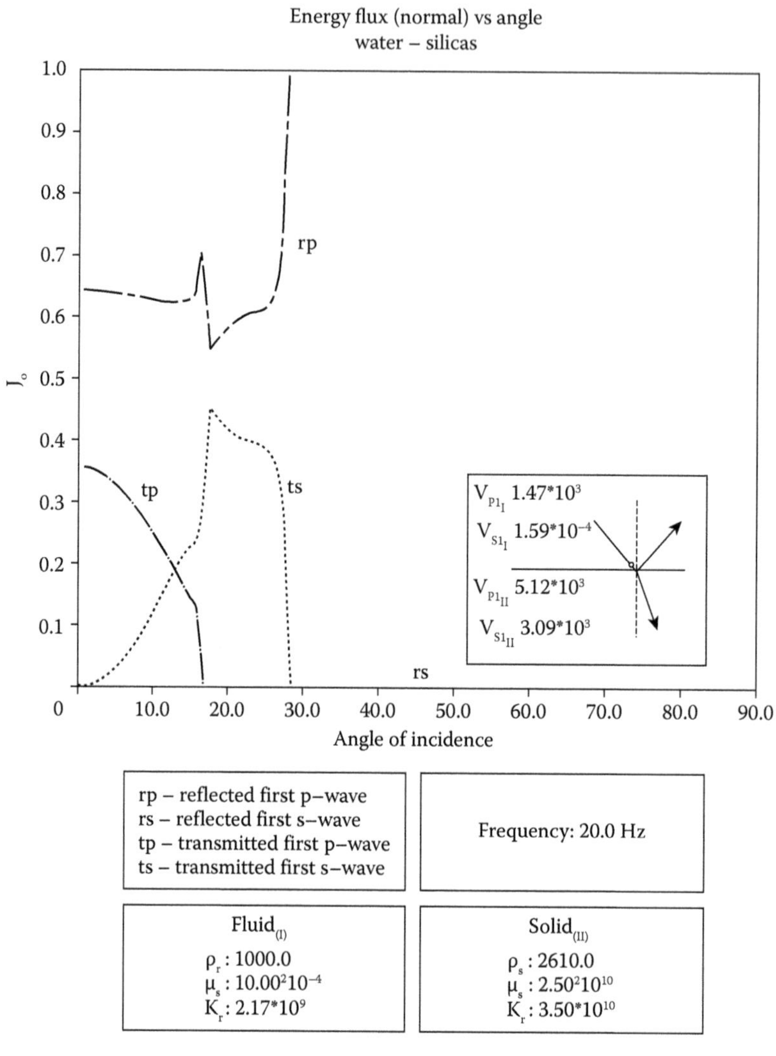

FIGURE 3.7 An elastic solid and a viscous fluid. This incident P wave is in the fluid assumed to be water. (Adapted from Spanos, T. J. T., et al., *Fundamental Thermodynamic Requirements for Porous Media Description*, pp. 513–521, 2004.)

Macroscopic stresses:

$$\tau^f_{jk} = -\eta_o p_f \delta_{jk} + \xi_f \left[\frac{\partial}{\partial x_i}(\eta_o v^f_i) + \frac{\partial \eta}{\partial t} \right] \delta_{jk} + \mu_f \eta_o \left(\frac{\partial v^f_j}{\partial x_k} + \frac{\partial v^f_k}{\partial x_j} - \frac{2}{3}\delta_{jk}\frac{\partial v^f_i}{\partial x_i} \right)$$

$$-\mu_f (1-\eta_o) c_{jkmn} \left(\frac{\partial v^s_m}{\partial x_n} + \frac{\partial v^s_n}{\partial x_m} - \frac{2}{3}\delta_{mn}\frac{\partial v^s_i}{\partial x_i} \right) \qquad (3.143)$$

$$+\mu_f \left[(v^f_j - v^s_j)\frac{\partial \eta_o}{\partial x_k} + (v^f_k - v^s_k)\frac{\partial \eta_o}{\partial x_j} - \frac{2}{3}\delta_{jk}(v^f_i - v^s_i)\frac{\partial \eta_o}{\partial x_i} \right]$$

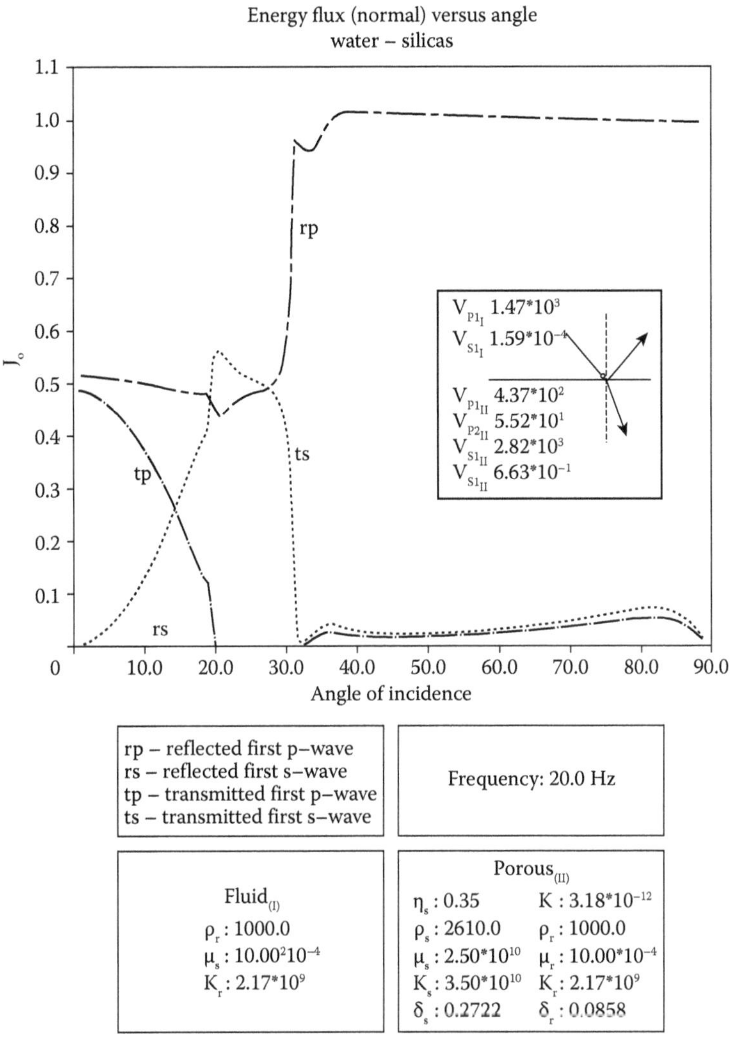

FIGURE 3.8 The elastic medium in Figure 3.7 is now a porous medium with a porosity of 0.3. The pore spaces are filled with water. (Adapted from Spanos, T. J. T., et al., *Fundamental Thermodynamic Requirements for Porous Media Description*, pp. 513–521, 2004.)

$$\tau^s_{jk} = K_s\left[\frac{\partial}{\partial x_j}[(1-\eta_o)u^s_j]-(\eta-\eta_o)\right]\delta_{jk}+\mu_{jkmn}\left(\frac{\partial v^s_m}{\partial x_n}+\frac{\partial v^s_n}{\partial x_m}-\frac{2}{3}\delta_{mn}\frac{\partial v^s_i}{\partial x_i}\right) \quad (3.144)$$

Macroscopic fluid pressure equation:

$$\eta_o\frac{\partial p_f}{\partial t} = -K_f\left[\frac{\partial(\eta_o v^f_j)}{\partial x_j}+\frac{\partial \eta}{\partial t}\right] \quad (3.145)$$

Body force equation:

$$F_j = -p_f\frac{\partial \eta_o}{\partial x_j}+Q_{jk}(v^f_k - v^s_k)-\rho_{jk}\left(\frac{\partial v^f_k}{\partial t}-\frac{\partial v^s_k}{\partial t}\right)+\rho^b_{ik}\frac{\partial v^m_k}{\partial t} \quad (3.146)$$

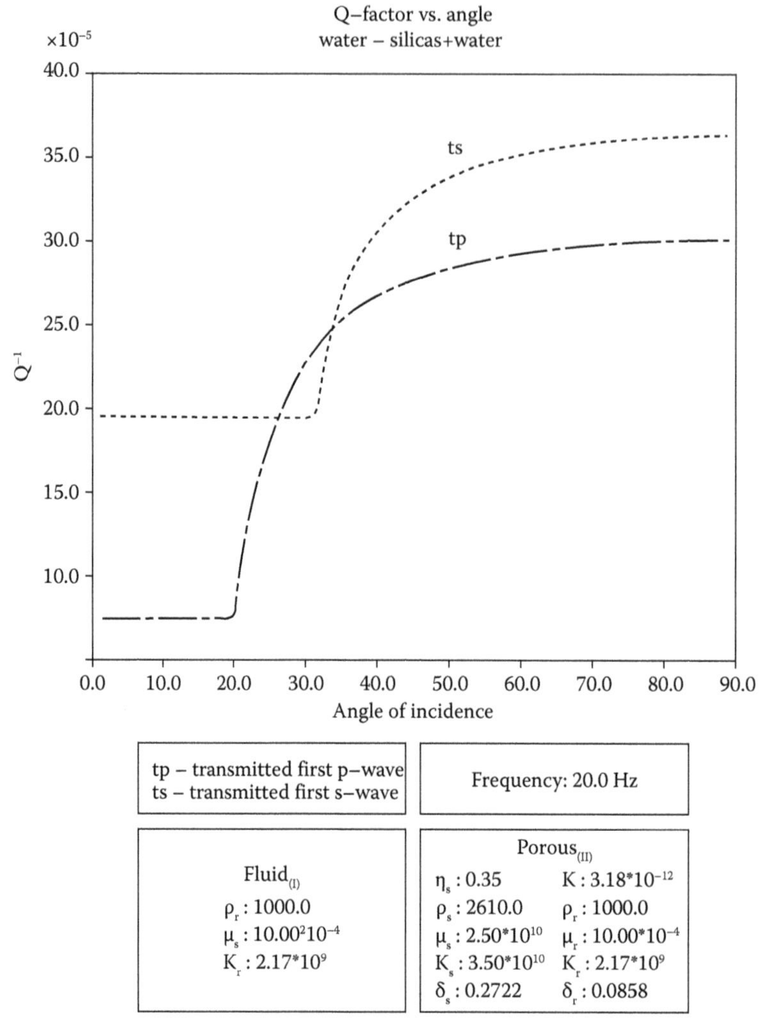

FIGURE 3.9 The inverse of Q for the transmitted P and S waves is plotted as a function of the angle of incidence. (Adapted from Spanos, T. J. T., et al., *Fundamental Thermodynamic Requirements for Porous Media Description*, pp. 513–521, 2004.)

Porosity equation:

$$\frac{\partial \eta}{\partial t} + v_j^s \frac{\partial \eta_o}{\partial x_j} = \Delta_{jk}^s v_{jk}^s - \delta_f v_{jk}^f \tag{3.147}$$

Here, it is important to note that additional terms appear in these equations that depend on the porosity gradient. Note that without these terms the description of a laboratory experiment would be different for someone in the lab and someone walking past the lab. In the homogeneous limit, these additional terms vanish. This is true for most heterogeneous media. The form of the equations of motion depends on the structure of the medium. When that structure is removed, the form of the equations of motion returns to the form of the equations constructed for a homogeneous medium. As in the case of a homogeneous medium, it is assumed that one has the time harmonic fields

$$\mathbf{u}^s = \mathbf{U}^s\, \mathbf{e}^{-i\omega t},\ \mathbf{u}^f = \mathbf{U}^f\, \mathrm{e}^{-i\omega t} \tag{3.148}$$

Again, substituting the scalar and vector potentials

$$\vec{\mathrm{U}}^s = \vec{\nabla}\phi_s + \vec{\nabla}\times\vec{\psi}_s \tag{3.149}$$

$$\vec{\mathrm{U}}^f = \vec{\nabla}\phi_f + \vec{\nabla}\times\vec{\psi}_f \tag{3.150}$$

where

$$\vec{\nabla}\cdot\vec{\psi}_s = \vec{\nabla}\cdot\vec{\psi}_f = 0 \tag{3.151}$$

into the equations of motion.

If a particular direction of propagation is chosen, one may write out a separate wave equation for the first and second P waves using the same procedure as was presented for a homogeneous medium. In the case of an S wave one must choose both a direction of propagation and the orientation of the transverse displacements. The form of the wave equations obtained is

$$\left(\nabla^2 + \omega\, \vec{\mathbf{A}}_1\cdot\vec{\nabla} + \frac{\omega^2}{\alpha_1^2}\right)\phi_1 = 0 \tag{3.152}$$

$$\left(\nabla^2 + \omega\, \vec{\mathbf{A}}_2\cdot\vec{\nabla} + \frac{\omega^2}{\alpha_2^2}\right)\phi_2 = 0 \tag{3.153}$$

$$\left(\nabla^2 + \omega\, \vec{\mathbf{B}}_1\cdot\vec{\nabla} + \frac{\omega^2}{\beta_1^2}\right)\vec{\psi}_1 = 0 \tag{3.154}$$

$$\left(\nabla^2 + \omega\, \vec{\mathbf{B}}_2\cdot\vec{\nabla} + \frac{\omega^2}{\beta_2^2}\right)\vec{\psi}_2 = 0 \tag{3.155}$$

Note that the additional term in each of the above equations acts as a propagating source (sink) through which the various modes are able to exchange energy. This occurs because of interaction of the waves with the inhomogeneity and anisotropy of the medium. For an isotropic, homogenous medium, such interactions only occur at the boundary of the medium.

POROSITY DIFFUSION

Deformable fluid-saturated porous media are described by a set of equations described in the section 'Summary of the Equations' and originally constructed by de la Cruz and Spanos (1985, 1989). These equations were obtained by applying volume averaging to the macroscopic (pore-scale) equations and then specifying the unique form of the megascopic equations through the constraints that the basic laws of physics (conservation of mass, Newton's third law, the principle of equivalence, etc.) must hold at all scales. In order to achieve this result, an additional equation not obtained from volume averaging but from basic physical arguments was required. This and other such equations to be discussed later introduce additional physical processes at the megascale that result from the interaction of the component materials. The first of these processes, porosity–pressure diffusion, will be discussed in this section. The set of equations given in the section 'Summary of the Equations' can be written in a simplified form after neglecting all inertial terms (second derivatives of displacements with respect to time). Furthermore, the fluid may be considered almost incompressible if the velocity of flow is much less than the velocity of sound in the fluid (Landau and Lifshitz 1975). Obviously, this condition is satisfied for slow fluid filtration, which has been described by de la Cruz and Spanos (1983). As a first step in considering this diffusion process, the divergence of the equations of motion for the fluid and solid is taken. Then, substituting the porosity and pressure equations into the resulting equations yields

$$\frac{\rho_{\eta f}}{\eta_o}\left(\frac{\partial^2\eta}{\partial t^2}+2a_{\eta f}\frac{\partial\eta}{\partial t}-2b_{\eta f}\nabla^2\frac{\partial\eta}{\partial t}\right)$$

$$=\frac{\rho_{pf}}{K_f}\left(\frac{\partial^2 p_f}{\partial t^2}+2a_{pf}\frac{\partial p_f}{\partial t}-2b_{pf}\nabla^2\frac{\partial p_f}{\partial t}-v_{pf}^2\nabla^2 p_f\right) \tag{3.156}$$

$$\frac{\alpha_1\rho_{\eta s}}{\eta_o}\left(\frac{\partial^2\eta}{\partial t^2}+2a_{\eta s}\frac{\partial\eta}{\partial t}-v_{\eta s}^2\nabla^2\eta\right)=\frac{\alpha_2\rho_{ps}}{K_f}\left(\frac{\partial^2 p_f}{\partial t^2}+2a_{ps}\frac{\partial p_f}{\partial t}-v_{ps}^2\nabla^2 p_f\right) \tag{3.157}$$

where

$$\sigma_M=(1-\eta_o)\mu_f\left(\frac{\mu_M}{(1-\eta_o)\mu_s}-1\right)\qquad a_{\eta f}=\frac{1}{2}\frac{\mu_f}{K}\frac{\eta_o(\alpha_1+1)}{\rho_{\eta f}} \tag{3.158}$$

$$\alpha_1=\frac{\eta_o-\delta_f}{\delta_s}\qquad b_{\eta f}=\frac{2}{3}\frac{\mu_f-\frac{\sigma_M}{\eta_o}\alpha_1}{\rho_{\eta f}} \tag{3.159}$$

$$\alpha_2=\frac{\delta_f}{\delta_s}\qquad a_{pf}=\frac{1}{2}\frac{\mu_f}{K}\eta_o\frac{(\alpha_2-1)}{\rho_{pf}} \tag{3.160}$$

$$\rho_{\eta f}=\rho_f-\rho_{12}(\alpha_1+1)\qquad b_{pf}=\frac{1}{2\rho_{\eta f}}\left(\xi_f+\frac{4}{3}\left(\mu_f-\frac{\sigma_M}{\eta_o}\alpha_2\right)\right) \tag{3.161}$$

$$v_{pf}^2=\frac{K_f}{\rho_{pf}}\qquad a_{\eta s}=\frac{1}{2}\frac{\mu_f}{K}\frac{\eta_o^2(\alpha_1+1)}{\rho_{\eta s}\alpha_1(1-\eta_o)} \tag{3.162}$$

$$\rho_{\eta s}=\rho_s-\frac{\rho_{12}(\alpha_1+1)}{\alpha_1(1-\eta_o)}\qquad v_{\eta s}^2=\frac{K_\eta}{\rho_{\eta s}} \tag{3.163}$$

$$\rho_{ps}=\rho_s-\frac{\rho_{12}(\alpha_2-1)}{\alpha_2(1-\eta_o)}\qquad a_{ps}=\frac{1}{2}\frac{\mu_f}{K}\frac{\eta_o(\alpha_2-1)}{\rho_{ps}\alpha_2(1-\eta_o)} \tag{3.164}$$

$$K_\eta = \left(1 - \frac{\eta_o}{(1-\eta_o)\alpha_1}\right) \qquad K_s + \frac{4}{3}\frac{\mu_M}{(1-\eta_o)} \cdot v_{ps}^2 = \frac{K_M}{\rho_{ps}} \tag{3.165}$$

$$K_M = K_s + \frac{4}{3}\frac{\mu_M}{(1-\eta_o)} \qquad \rho_{pf} = \rho_f - \frac{\rho_{12}}{\eta_o}(\alpha_2 - 1) \tag{3.166}$$

In the limit where the inertial terms vanish, one obtains two coupled diffusion equations for pressure and porosity:

$$\frac{\partial p_f}{\partial t} - D_f \nabla^2 p_f = C\frac{\partial \eta}{\partial t} \tag{3.167}$$

$$\frac{\partial \eta}{\partial t} - D_s \nabla^2 \eta = B\frac{\partial p_f}{\partial t} \tag{3.168}$$

where

$$D_s = \frac{v_{\eta M}^2}{a_{\eta M}} \qquad D_f = \frac{v_{pf}^2}{a_{pf}} \tag{3.169}$$

$$B = 2\frac{\alpha_2 \rho_{pM} \eta_o a_{pM}}{\alpha_1 \rho_{\eta M} K_f} \qquad C = \frac{\rho_{\eta f} K_f a_{\eta f}}{\eta_o \rho_{pf} a_{pf}} \tag{3.170}$$

For $K \approx 10^{-8}\,\mathrm{cm}^2$, $K_s \approx 10^{10}\,\mathrm{dyne/cm}^2$, $\mu_f \approx 10^{-2}\,\mathrm{g/(cm\times s)}$ and $\eta \approx 10^{-1}$, the two diffusion constants are approximately 10^5 cm^2/s, which is straightforward to measure in laboratory conditions and is the value of the diffusion constant obtained for diffusional propagation of microseismic activity following the loading of reservoirs in the earth.

In the case of static compressions, the dimensionless parameters δ_s and δ_f have the following values (de la Cruz et al. 1992):

$$\delta_s = (1-\eta_o)\frac{(K_s - K_{ud})(K_s - K_d)}{(K_s - K_f)[K_s - (1-\eta_o)K_d]} \tag{3.171}$$

$$\delta_f = \eta_o \frac{K_s(K_{ud} - K_f) - (1-\eta_o)K_d(K_s - K_f)}{(K_s - K_f)[K_s - (1-\eta_o)K_d]} \tag{3.172}$$

It is now important to note that, for the motions being considered here, flow of the fluid and elastic deformations of the matrix occur, with the fluid phase behaving in an almost incompressible fashion. In this limit, the above values for δ_s and δ_f no longer depend on the fluid compressibility or the undrained bulk modulus, since only elastic deformations associated with the drained bulk modulus and associated fluid flow occur. Thus:

$$\delta_s = (1-\eta_o)\frac{(K_s - K_d)}{[K_s - (1-\eta_o)K_d]} + \frac{A(\mu_s, \mu_f)}{K_d} \tag{3.173}$$

$$\delta_f = \eta_o \frac{(1-\eta_o)K_d}{[K_s - (1-\eta_o)K_d]} + \frac{B(\mu_s, \mu_f)}{K_d} \tag{3.174}$$

where A and B are experimentally determined parameters that depend on the pore structure and specify a linear combination of the macroscopic parameters μ_s, μ_f.

POROSITY–PRESSURE WAVE PROPAGATION

Now, keeping all of the inertial terms in Equations 3.156 and 3.157, these equations of motion now yield two coupled wave equations.

If the values of δ are chosen to be the values obtained for seismic wave propagation, then they describe the two P waves described in the section 'Construction of the Wave Equations'. However, if the values of δ are chosen to be the values obtained in the section 'Porosity Diffusion', then they describe a porosity pressure wave more than an order of magnitude slower, which propagates thorough the medium by moving fluid coupled to the elastic deformations of the matrix. In both cases the attenuation of the wave occurs due to viscous dissipation in the fluid.

Here, the criteria for a fluid to behave in an incompressible fashion are as follows (Landau and Lifshitz 1975):

1. The velocity of the fluid is small compared to the velocity of sound in the fluid $v << c$.
2. The time over which the fluid undergoes significant changes is large compared to the distance over which those changes occur, divided by the speed of sound $t >> L/c$.

When both of these criteria are met, the fluid is able to avoid compression through flow and thus can be regarded as incompressible. In the present case of a porous medium, the solid is unable to flow but may still deform elastically. Here, porosity describes how the fluid and solid interact in the context of specific processes. In the case of a porosity wave, incompressible fluid flow is coupled to elastic deformations of the solid matrix. These motions occur orders of magnitude slower than seismic P waves and orders of magnitude faster than steady-state fluid flow.

Here, the starting point in the description of the porosity–pressure wave will be experimental observation of this process followed by a complete theoretical description using the physical results presented above.

Laboratory experiments on flow enhancement by fluid pulsing began at the University of Waterloo and the University of Alberta in 1997. The experiments were done in tightly packed and stressed silica sand packs in Plexiglas and steel cells. One of the first observations of these experiments was that if any rubber or soft material was placed in contact with the porous matrix all of the energy of the pulse would be absorbed by the rubber and no flow enhancement would occur. At this time a number of graduate theses were written quantifying these experiments. These sand packs were generally 1-darcy permeability and 30% porosity. Some of the observations were as follows:

1. The tighter the sand pack and the more highly it was stressed, the better the flow enhancement.
2. The pulses created wave motion in the porous medium moving about 40 times slower than a P wave.

3. The waves attenuated due to viscous dissipation. The higher the viscosity of the fluid, the more quickly the waves attenuated.
4. The higher the fluid viscosity, the slower the waves propagated.
5. The waves could be generated by either a sudden increase or a sudden decrease in pressure.
6. The effectiveness of the pressure pulses could be tuned, resulting in the most effective pulses having a rise time of about 1/10th of a second, followed by a relaxation time back to the original pressure of about 1 second.
7. The most effective laboratory pulser was two solenoids, one producing the sudden pressure increase and the other venting back to the original pressure.

Figure 3.10 shows the experiment where the porosity wave was measured for the first time. Here, a silica sand pack of 33% porosity is saturated with glycerine. The high viscosity of the glycerine causes the wave to be an order of magnitude slower than it would be with water. The diffusion of pressure is also slowed down so it can be easily observed at the laboratory scale (Figure 3.11). This pressure build up and decay is shown graphically in Figures 3.12, and 3.13.

Many other laboratory cells were used, including steel cells similar to the Lucite cell illustrated above. One cell that has been very useful in giving insight is a visual cell shown in Figure 3.14. Figure 3.15 illustrates the solenoids that delivered the pulse and then vented the pressure. The response when pulsing water is shown in Figure 3.16.

Now, returning to Equations 3.156 and 3.157 and allowing δ_s and δ_f to have the values given by Equations 3.173 and 3.174, it is observed that wave equations are obtained in which the fluid and solid behave almost incompressibly. Thus elastic deformations of the matrix are obtained that are coupled to a net flux of the fluid. Coupled porosity pressure waves are described that move fluid through the porous matrix. These wave equations have

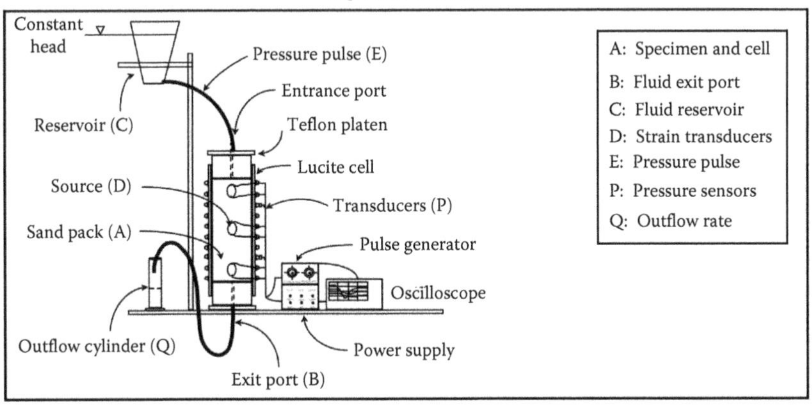

FIGURE 3.10 Experimental set-up for measuring the speed of the porosity wave in a laboratory experiment.

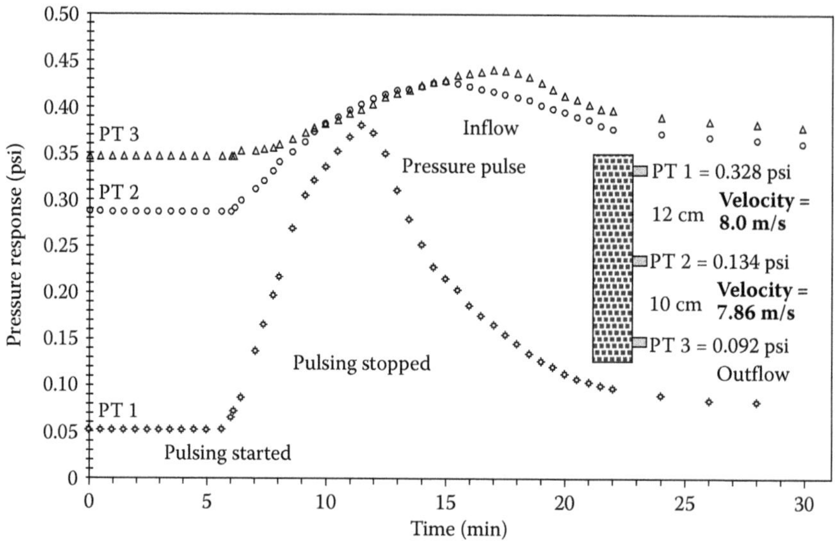

FIGURE 3.11 Wave speed and pressure changes at the three pressure transducers as the porosity–pressure waves are introduced to the medium and then stopped. Here, the waves move at under 10 m/sec, the pressure builds with successive pulses and then pressure diffusion is observed when pulsing stops. The pressure changes at Transducers 2 and 3 have been moved to make them easier to observe the relative changes in pressure.

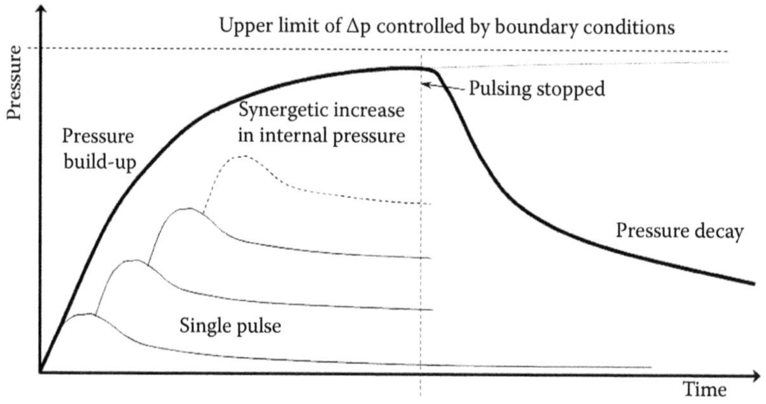

FIGURE 3.12 Illustration of the pressure waves, the pressure build-up and pressure diffusion following each wave.

been studied numerically for the past 16 years at Wavefront Technology Solutions; a summary of this work is given in the following discussion.

First, consider an approximation for the incompressible limit of Equations 3.155 and 3.156; in a computer program, this can be achieved by setting the compressibility of the fluid to 0 ($c_f = 0$). In this case, the solid wave equation becomes

$$\frac{\alpha_1 \rho_{\eta s}}{\eta_o}\left(\frac{\partial^2 \eta}{\partial t^2} + 2a_{\eta s}\frac{\partial \eta}{\partial t} - v_{\eta s}^2 \nabla^2 \eta\right) = 0 \qquad (3.175)$$

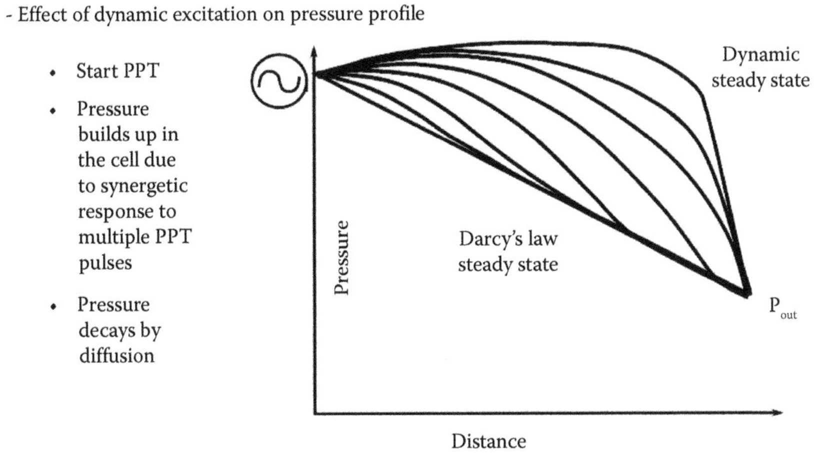
- Effect of dynamic excitation on pressure profile
• Start PPT
• Pressure builds up in the cell due to synergetic response to multiple PPT pulses
• Pressure decays by diffusion
Dynamic steady state
Darcy's law steady state
Pressure
P out
Distance

FIGURE 3.13 Illustration of the pressure changes that occur within the laboratory experiment during pulsing.

FIGURE 3.14 A pressurized tank containing the injection fluid passes through a two solenoid system to a porous medium in a Plexiglas cell. The first solenoid causes a pressure increase in 1/10th of a second; the second solenoid then causes a pressure decrease to the original pressure over 1 second (this is accomplished by adjusting a valve).

or more simply

$$\frac{\partial^2 \eta}{\partial t^2} + 2a_{\eta s}\frac{\partial \eta}{\partial t} - v_{\eta s}^2 \nabla^2 \eta = 0 \tag{3.176}$$

This equation describes a damped porosity wave. Without attenuation, it would travel with speed $v_{\eta s}$. Under conditions where the inertial term (the second time derivative) is very small, it reduces to a porosity diffusion equation.

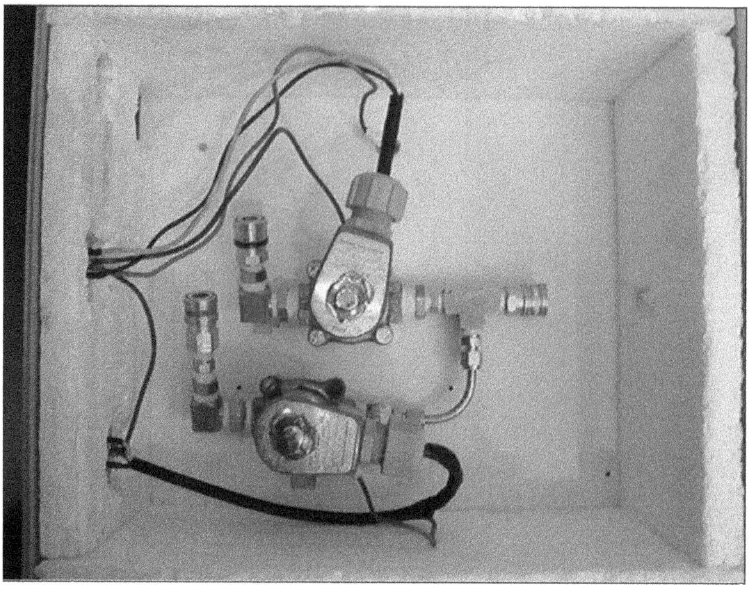

FIGURE 3.15 The solenoids that delivered the pulse and then vented the pressure.

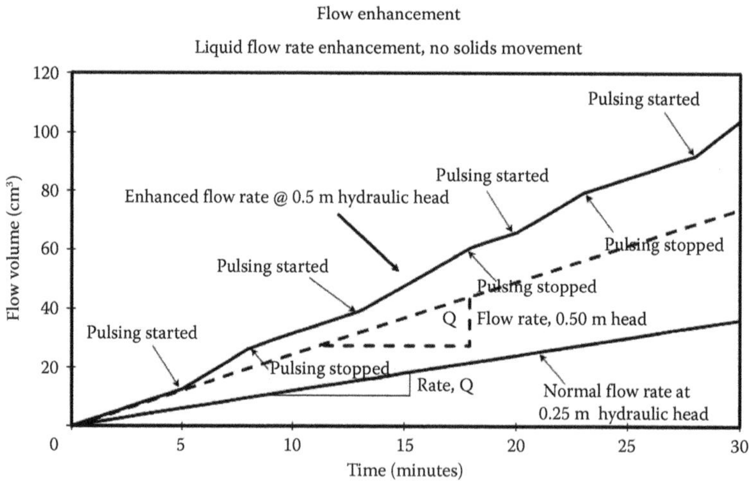

FIGURE 3.16 When the fluid is water and pulsing stops, the flow returns the same slope (Darcy flow) almost immediately, unlike the previous case of glycerine where the diffusion process could be observed.

The fluid wave equation becomes

$$\nabla^2 p_f = \frac{\rho_{\eta f}}{\eta_o}\left(\frac{\partial^2 \eta}{\partial t^2} + 2a_{\eta f}\frac{\partial \eta}{\partial t} - 2b_{\eta f}\nabla^2\frac{\partial \eta}{\partial t}\right) \tag{3.177}$$

This equation shows how the fluid pressure responds to the porosity wave. We see that as the wave propagates (either as a wave or diffusively), it loses energy to the porous medium, and the pressure (typically increases) in response to this energy loss. Of course, in more complicated scenarios like earthquakes or pressure pulsing, the appropriate source terms need to be included into these equations.

The choice of independent parameters used to solve these equations may be (δ_s, δ_f) or (α_1,α_2). Experience has shown that (δ_s, δ_f) is a bad choice since the solutions are extremely sensitive to small changes in (δ_s, δ_f) and the computation using these quantities is prone to numerical error.

Computation using the parameters (α_1, α_2) is more numerically stable and since they appear in the equations of motion they are 'closest' to physically measurable parameters for the case of a porosity wave. With (α_1, α_2) specified, the values of (δ_s, δ_f) can easily be computed in a numerically stable way:

$$\delta_s = \frac{\eta_o}{\alpha_1 + \alpha_2} \tag{3.178}$$

and

$$\delta_f = \frac{\eta_o \alpha_2}{\alpha_1 + \alpha_2} \tag{3.179}$$

Here, large changes in α_1 or α_2 cause relatively small changes in δ_s and δ_f.

Now consider the wave speed $v_{\eta s}$. It is defined by

$$v_{\eta s}^2 = \frac{K_\eta}{\rho_{\eta s}} \tag{3.180}$$

Substituting for the values on the right-hand side

$$v_{\eta s}^2 = \frac{\left(1 - \dfrac{\eta_o}{(1-\eta_o)\alpha_1}\right) K_s + \dfrac{4}{3}\dfrac{\mu_M}{(1-\eta_o)}}{\rho_s - \dfrac{\rho_{12}(\alpha_1 + 1)}{\alpha_1(1-\eta_o)}} \tag{3.181}$$

Solving this equation for α_1 yields

$$\alpha_1 = \frac{K_s \dfrac{\eta_o}{(1-\eta_o)} - v_{\eta s}^2 \dfrac{\rho_{12}}{(1-\eta_o)}}{K_s + \dfrac{4}{3}\dfrac{\mu_M}{(1-\eta_o)} - v_{\eta s}^2 \left(\rho_s - \dfrac{\rho_{12}}{(1-\eta_o)}\right)} \tag{3.182}$$

Thus, given a measured value or estimate of $v_{\eta s}$, the value of α_1 can be computed.

In the first part of the present discussion, the fluid was assumed completely incompressible. That is almost true and thus this calculation yields an accurate approximation. Now assume that the porosity changes are negligible while the pressure is allowed to change. Here

$$\frac{\rho_{pf}}{K_f}\left(\frac{\partial^2 p_f}{\partial t^2} + 2a_{pf}\frac{\partial p_f}{\partial t} - 2b_{pf}\nabla^2\frac{\partial p_f}{\partial t} - v_{pf}^2\nabla^2 p_f\right) = 0 \tag{3.183}$$

or more simply

$$\frac{\partial^2 p_f}{\partial t^2} + 2a_{pf}\frac{\partial p_f}{\partial t} - 2b_{pf}\nabla^2\frac{\partial p_f}{\partial t} - \mathrm{v}_{pf}^2\nabla^2 p_f = 0 \tag{3.184}$$

This equation describes a damped pressure wave that without attenuation would propagate at speed v_{pf}, which is defined by

$$\mathrm{v}_{pf}^2 = \frac{K_f}{\rho_{pf}} \tag{3.185}$$

Substituting into the right-hand side yields

$$\mathrm{v}_{pf}^2 = \frac{K_f}{\rho_f - \frac{\rho_{12}}{\eta_o}(\alpha_2 - 1)} \tag{3.186}$$

which may be solved for α_2 to yield

$$\alpha_2 = \frac{\eta_o}{\rho_{12}}\left(\rho_f - \frac{K_f}{\mathrm{v}_{pf}^2}\right) + 1 \tag{3.187}$$

Thus, given a measured value or estimate of v_{pf}, the value of α_2 can be computed.

Now, note that the results presented here do not appear consistent with the results of the previous section, where two coupled diffusion equations were obtained when the inertial terms were ignored. However, note first that the two diffusion constants in the previous section were approximately equal.

In reality, the fluid is almost incompressible near the incompressible limit and thus two coupled wave equations are obtained. These wave equations describe a porosity wave strongly coupled to a pressure wave moving at almost the same speed near the incompressible limit of fluid motions. The previous approximations allowed these processes to be decoupled so that the porosity or pressure could be calculated in isolation and then impose an external constraint on the pressure response or porosity response. The accuracy of this approximation may now be considered. As will be seen in more complex systems, the usefulness of such computational approximations becomes very important if they are reasonably accurate and capture the essential physical processes. In this case, wave speed, attenuation and pressure change are the important quantities to determine.

The actual porosity and pressure waves must be calculated from Equations 3.156 and 3.157. These equations may be rewritten as a wave equation for porosity coupled to changes in pressure and a wave equation for pressure coupled to changes in porosity. However, now another observation about these equations is that they are not symmetric. That is because viscous dissipation in the fluid is allowed for but no irreversible motions of the solid is described in these equations. Later, this will be allowed and shown to incorporate coupling with the stress field of the solid matrix and introduce a propagating source term into the equations of motion. (That is, the coupled porosity pressure wave may then extract mechanical energy from the stress field of the matrix. This theory then predicts soliton waves that carry fluids great distances deep in the earth, the interaction of earthquakes and other dynamic processes.)

A numerical study of these waves is presented in Chapter 6.

SUMMARY

The equations that describe seismic wave propagation in fluid-filled porous media were constructed. It was observed that the volume fraction of space (the porosity) occupied by the two phases is altered during compression. This brings an additional degree of freedom into the description of the motions. It was shown that the boundary conditions are once again statements of conservation of mass, Newton's third law and continuity in the transverse velocity of momentum flux. A number of reflection transmission problems were presented showing the differences between elastic media and porous media, as well as how different the results are when different fluids are present. As well, it was shown how the introduction of heterogeneity changes the form of the equations of motion.

It was shown that the equations of motion constructed for seismic wave propagation in porous media predict additional processes. The first process analysed was porosity–pressure diffusion. Here, the equations predict that coupled porosity pressure diffusion occurs at a rate much greater than normal pressure diffusion. It was shown that these processes may be calculated theoretically and measured experimentally. It was then shown that porosity–pressure waves are also predicted by the equations of motion; these are two highly non-linear coupled waves that involve elastic deformations of the matrix coupled to flow of the fluid through the porous medium. It was shown that these wave processes are quite easily demonstrated in the lab and a complete theoretical description of these waves was constructed.

Subject to the assumption of the fluid behaving in an incompressible fashion, a very simple description of the porosity wave was obtained. Under these conditions, straightforward relations for the dilatational parameters were obtained.

REFERENCES

Davidson, B.C., Dusseault, M.B., Geilikman, M.B., Hayes, K.W., and Spanos, T.J.T., June 5, 2001, United States Patent #6,241,019.

Davidson, B.C., Dusseault, M.B., Geilikman, M.B., Hayes, K.W., and Spanos, T.J.T., June 18, 2002a, United States Patent #6,405,797.

Davidson, B.C., Dusseault, M.B., Geilikman, M.B., Hayes, K.W., and Spanos, T.J.T., January 2, 2002b, United Kingdom Patent#GB232819.

Davidson, B.C., Dusseault, M.B., Geilikman, M.B., Hayes, K.W., and Spanos, T.J.T., June 7, 2005, Canadian Patent #CA2232948.

Davidson, B.C., Spanos, T.J.T., and Dusseault, M.B., 1999, Laboratory experiments on pressure pulse flow enhancement in porous media, Proceedings of the CIM Regina Technical Meeting, Regina, Saskatchewan, October.

de la Cruz, V., Hube, J., and Spanos, T.J.T., 1992, Reflection and transmission of seismic waves at the boundaries of porous media, *Wave Motion*, **16**, 1–16.

de la Cruz, V., Sahay, P.N., and Spanos, T.J.T., 1993, Thermodynamics of porous media, *Proc. Roy. Soc. Lond. A*, **443**, 247–255.

de la Cruz, V., and Spanos, T.J.T., 1983. Mobilization of Oil Ganglia, AIChE J., 29 (7), 854–858.

de la Cruz, V., and Spanos, T.J.T., 1985, Seismic wave propagation in a porous medium, *Geophysics*, **50**(10), 1556–1565.

de la Cruz, V., and Spanos, T.J.T., 1989, Thermomechanical coupling during seismic wave propagation in a porous medium, *J. Geophys. Res.*, **94**, 637–642.

Dusseault, M.B., Spanos, T.J.T., and Davidson, B.C., 1999, A dynamic pulsing workover technique for wells, Proceedings of the CIM Regina Technical Meeting. Regina, Saskatchewan.

Hickey, C.J., Spanos, T.J.T., and de la Cruz, V., 1995, Deformation parameters of permeable media, *Geophys. J. Int.*, **121**, 359–370.

Huang, J.H., 1971, Effective thermal conductivity of porous rocks, *J. Geophys. Res.*, **76**(26), 6420–6427.

Hsu, C.T., and Cheng, P., 1990, Thermal dispersion in a porous medium, *Int. J. Heat Mass Transfer*, **33(8)**, 1587–1597.

Johnson, D.L., 1980, Equivalence between fourth sound in helium at low temperatures and the Biot slow wave in consolidated porous media, *Appl. Phys. Lett.*, **37**(12), 1065–1067. Kanasewich, E.R., Personal Communication, 1991.

Landau, L.D., and Lifshitz, E.M., 1975, *Fluid Mechanics*, Pergamon, Toronto.

Nozad, I., Carbonell, R.G., and Whitaker, S., 1985, Heat conduction in multiphase systems—I, *Chem. Eng. Sci.*, **40**(5), 843–855.

Sahay, P.N., Spanos, T.J.T., and de la Cruz, V., 2001, Seismic wave propagation in inhomogeneous and anisotropic porous media, *Geophys. J. Int.*, **145**, 209–223.

Spanos, T.J.T., 2002, The Thermophysics of Porous Media, Chapman & Hall/CRC Press, Monographs and Surveys in Pure and Applied Mathematics series. pp. 212.

Spanos, T.J.T., Dusseault, M.B., and Davidson, B.C., 1999, Pressure Pulsing at the Reservoir Scale a new IOR approach, Proceedings of the CSPG and Petroleum Society, Calgary, June 14–18, paper 99-11, 11 p.

Spanos, T.J.T., Dusseault, M.B., Davidson, B.C., Shand, D., and Samarroo, M., 2003, Pressure Pulsing at the Reservoir Scale a new IOR approach, *J. Can. Petrol. Technol.*, February, 16–28, 2003.

Spanos, T.J.T., Dusseault, M.B., and Udey, N., 2004, *Fundamental Thermodynamic Requirements for Porous Media Description*, Elsevier Geo-Engineering Book Series Volume 2, Coupled Thermo-Hydro-Mechanical-Chemical Processes in Geo-Systems Fundamentals, Modelling, Experiments and Applications Edited by Ove Stephanson, Netherlands, pp. 513–521.

Udey, N., 2012, Coupled porosity and saturation waves in porous media, In *Mathematical and Numerical Modeling in Porous Media: Applications in Geosciences*, edited by M.A. Diaz Viera, P. Sahay, M. Coronado, and A.O. Tapia, CRC Press, Boca Raton, FL., 313–343.

Verma, L.S., Shrotriya, A.K., Singh, R., and Chaudhary, D.R., 1991, Prediction and measurement of effective thermal conductivity of three-phase system, *J. Phys. D Appl. Phys.*, **24**, 1515–1526.

Woodside, W., and Messmer, J.H., 1961a, Thermal conductivity of porous media. I. Unconsolidated sands, *J. Appl. Phys.*, **3**(9), 1688–1699.

Woodside, W., and Messmer, J.H., 1961b, Thermal conductivity of porous media. II. Consolidated rocks, *J. Appl. Phys.*, **32**(9), 1699–1706.

Yoshida, H., Yun, J.H., Echigo, R. and Tomimura, T., 1990. Transient characteristics of combined conduction, convection, and radiation heat transfer in porous media, Int. J. Heat Mass Transfer, 33(5), 847–857.

Zarotti, F., and Carbonell, R.G., 1984. Development of transport equations for multiphase systems II, Chem. Engng Sci., 39, 263–278.

Zimmerman, R.W., 1989, Thermal conductivity of fluid-saturated rocks, *J. Petrol. Sci. Eng.*, **3**, 219–227.

CHAPTER 4

Equilibrium Thermodynamics of Porous Media

OBJECTIVES OF THIS CHAPTER

Megascopic systems such as porous media contain additional dynamic variables not present in macroscopic systems such as elastic solids and Newtonian fluids (e.g. volume fractions such as porosity and saturation). In the present chapter, equilibrium thermodynamics will be presented. In later chapters, in the context of non-equilibrium thermodynamics, it is observed that these new dynamic variables allow for additional physical processes to occur (e.g. porosity waves, saturation waves, soliton waves, cold melting—the transition from an intact porous medium to a fluid containing solid suspensions, etc.). The theory yields predictions that are very specific in describing these new processes, with details of how they occur and how they interact. These predictions make this non-linear classical field theory subject to experimental scrutiny. Thermodynamics places strong constraints on the interactions of the various processes, which further clarifies experimental predictions. As well, it is important not to hold too firmly to the results obtained from equilibrium thermodynamics in this chapter, because in the context of non-equilibrium thermodynamics the description contains a lot of new information.

Thermodynamics in general deals with the average properties of a system and describes transformations between work and energy in various forms. The present discussion of thermodynamics will be expanded from that of a chemically homogeneous system based on mass fractions to include the new dynamic variables porosity and saturation. Through these variables, macroscopic mixtures as well as microscopic mixtures are introduced into the averaging (volume fractions as well as mass fractions). As a result, transformations between the structure (or order) of a medium and mechanical work are observed. When chemical reactions are introduced in a heterogeneous fashion, at various scales, the new dynamical variables are described in terms of mass fractions of the component phases mixed at the various scales. This causes the concept of a chemical potential to be substantially generalized at the megascale, requiring that the structure (or order) potentials associated with porosity and saturation be incorporated into this megascopic general potential, which contains the order potentials (associated with volume fractions originating at the macroscale) and chemical potentials (associated with mass fractions originating at the molecular scale) as special cases.

This discussion begins with the construction of the equilibrium thermodynamics of megascopic systems. The limitation of equilibrium thermodynamics is that it describes static processes. It is also used to describe reversible dynamic processes by assuming quasi-static changes in a system. This theory is also commonly referred to as *thermostatics*, since it is the limit that non-equilibrium thermodynamics reduces to in the absence of motion. Following this discussion, the megascopic non-equilibrium thermodynamics of components mixed at the macroscale will be described.

The most fundamental physical quantity that must be addressed in thermodynamics is the entropy. For molecular mixtures, entropy is the Boltzmann constant times the number of ways in which the components of a system may be rearranged, keeping the total energy constant. If we now allow for macroscopic components to be rearranged and then go up in scale by orders of magnitude, the fashion in which the system may be rearranged now has new degrees of freedom originating at the macroscale. In the present discussion, it is observed that entropy may be constructed as a combination of heat and temperature; stress and strain; a couple of general potentials (order potentials and chemical potential); and mass and volume fractions. As an introduction to the general potential, consider a system in total statistical equilibrium. Here, a closed system is observed over a time period that is long in comparison to its relaxation time. Now, divide the system up into a large number of subsystems composed of different macroscopic parts. Let w_n be the distribution function for one of these subsystems; w_n may be expressed as a function of energy, $w_n = w(E_n)$. The general potentials describe both how the subsystems are distributed when observed from a much larger scale and how molecular scale mixing occurs.

A general potential approaches the definition of the macroscopic chemical potential in the limit as macroscopic mixing vanishes. Also, in the limit as chemical reactions vanish in the presence of macroscopic mixing, they approach the order potentials presented in the following discussion. At this stage it should be noted that the general potentials (and thus the order potentials and chemical potential), heat, stress, energy and momentum are fundamental quantities. Thus, definitions of these quantities in terms of less fundamental quantities such as the Gibbs free energy, the Helmholtz free energy and the enthalpy are only relations that hold in the context of a specific thermodynamic system. When speaking in a more general context, it is the more fundamental quantities that define the free energies and the enthalpy.

Here, the megascopic relations for equilibrium thermodynamics are formulated on a firm basis. As a starting point, the equilibrium thermostatics associated with the mechanical motions of an elastic porous matrix saturated with a single compressible viscous fluid is considered. In Chapter 9, this formulation is generalized to include the non-equilibrium thermodynamics of such a medium and next to generalize this formulation to include multiple fluid and solid phases. Initially it is assumed that the pores are well connected. Volume-averaged equations are used to provide the linkage to the pore-scale thermodynamics rather than employing what might be called an 'axiomatic' approach. Part of the motivation is to find first for porosity and subsequently for saturation and megascopic concentrations, the new purely megascopic variables, their natural 'thermodynamic' roles. It will be shown that aside from their bookkeeping role (keeping track of proportions of the phases by volume and mass), the porosity, saturation and megascopic concentrations also appear in the work terms. For example, porosity is found to play a dynamic role independent of temperature, thus yielding a theory of porodynamics that has analogies with thermodynamics or more properly may be considered as an extension of the thermodynamics of molecular mixtures. A thermodynamic role for saturation in the case of compressible multiphase fluid motions is established in an analogous fashion. Furthermore, when the segregation of the phases by their mass fractions is considered, then the relevant thermodynamic variable becomes the megascopic concentration. The importance of the thermodynamic role that the above megascopic thermodynamic variables play in each case occurs due to the relation between the dilational motions of the component phases and the change of these megascopic variables. This relation and how it is process dependent can be clearly seen through the description of the fluid and solid components and their interactions as described in the following discussion.

De la Cruz et al. (1993) constructed megascopic equilibrium thermostatic relations for the fluid phase from the well-understood pore-scale description. The corresponding relations for the solid component were also treated. The internal energy for the porous medium was discussed in the context of a system consisting of two superposed continua. The effect of a spatially varying porosity on the thermodynamic relations was obtained (Spanos 2002).

It has been common to describe porous media in terms of two spatially superposed interacting media (e.g. Gassmann 1951; Biot 1956; Slattery 1967; Whitaker 1967; Keller 1977; Sanchez-Palencia 1980; Burridge and Keller 1981; de la Cruz and Spanos 1983, 1985, and 1989; Bear and Bachmat 1990; Hickey et al. 1995). The question then arises of how to express thermodynamic ideas using only quantities that are meaningful in this description. Biot's (1956) work on poroelasticity was founded on a 'megascopic thermodynamics' in which the relations were equivalent in form to single-component equilibrium thermostatics. In spite of the severe limitations imposed by the initial assumptions, this work became the foundation of descriptions of poromechanics for several decades. However, when the equilibrium thermostatics of porous media were constructed by de la Cruz et al. (1993), it was observed that multicomponent thermostatics required porosity to be treated as a variable. They observed that the definition of strain in the Biot theory could be redefined to include the porosity variation associated with a reversible quasi-static compression, making the two descriptions equivalent in the case of statics. The Biot theory, however, cannot be used to construct a physical description of dynamic processes because of the thermostatic foundations on which it is based. When dynamic processes are introduced, non-linear field theory and non-equilibrium thermodynamics are required for physical consistency at all scales. Many other formulations have also appeared in the literature (e.g. Scheidegger 1974; Marle 1982; Gurtin 1988; Koch and Brady, 1988; Gurtin and Struthers 1990; Garcia-Conlin and Uribe 1991; del Rio and Lopez de Haro 1992; Dullien 1992; Detournay, 1993). Many of these formulations are based on non-equilibrium thermodynamics associated with multicomponent systems. However, the system presented here is distinctly different than that of the non-equilibrium multicomponent systems discussed by Prirogine (1954) and de Groot and Mazur (1962) in that the laws of thermomechanics and thermodynamics are firmly established in the individual component phases at an intermediate scale in the present analysis. Thus, it appears that the non-equilibrium theory of de Groot and Mazur (1962) cannot be considered as applicable to porous media, since it has a different equilibrium limit than the theory constructed in this chapter and does not allow for the important dynamic role that porosity must play.

SUMMARY OF THE EQUATIONS THAT DESCRIBE THERMODYNAMIC DEFORMATIONS

Solid Properties

Internal energy:

$$dU_s = \tau_{ik}^s \, du_{ik}'^s + T_s \, dS_s \tag{4.1}$$

Solid stress:

$$\tau_{ik}^s = \phi_o K_s \alpha_s \left(T^s - T_o\right)\delta_{ik} + 2\mu_M \left[u_{ik}'^s - \tfrac{1}{3}\delta_{ik} u_{jj}'^s\right] + \phi_o K_s \delta_{ik} u_{jj}'^s \tag{4.2}$$

where $\phi = 1 - \eta$ and the solid strain is given by

$$u_{ik}'^s = u_{ik}^s + \frac{1}{3}\frac{\phi - \phi_o}{\phi_o}\delta_{ik} \tag{4.3}$$

Solid internal energy

$$\begin{aligned} U(S, u_{ik}'^s) &= \phi_o U_o^s(S_s) - \frac{T_o^s K_s \alpha_s}{c_v^s}(S_s - S_o) u_{jj}'^s \\ &+ \mu_m \left(u_{ik}'^s - \frac{1}{3}\delta_{ik} u_{jj}'^s\right)^2 \frac{1}{2}\phi_o K_{ad}^s u_{jj}'^{s2} \end{aligned} \tag{4.4}$$

In equilibrium thermodynamics, entropy balance of the solid component must describe how the energy of the system is arranged at all scales. Thus

$$S_o^s\left(T^s\right) = -\phi_o \frac{dF_o^s\left(T^s\right)}{dT^s} \tag{4.5}$$

the entropy balance of the solid component is given by (Landau and Lifshitz 1975)

$$S^s - S_o^s = \phi_o c_v^s \left(T^s - T_o\right) / T_o + \phi_o K_s \alpha_s u_{jj}'^s \tag{4.6}$$

and the heat capacity by

$$c_v^s = -T_o \frac{d^2 F_o^s(T_o)}{dT_o^2} \tag{4.7}$$

In the current formulation, the Helmholtz free energy for the solid represents the total energy required to create the solid part of the system minus the heat obtained from the environment at temperature T^s. In deformations discussed later, granular motions will be allowed for, where it will be observed that changes in porosity can also cause changes in entropy. For reversible motions of the matrix involving only elastic deformations of the matrix and thus only mechanical energy and heat, the Helmholtz free energy for the solid may be written as

$$dF_s = \tau_{ik}^s du_{ik}'^s - S_s dT^s \tag{4.8}$$

$$\begin{aligned} F_s(T^s, u_{ik}'^s) &= \phi_o F_o^s(T^s) - \phi_o K_s \alpha_s (T^s - T_o) u_{jj}'^s \\ &+ \mu_M \left(u_{ik}'^s - \frac{1}{3}\delta_{ik} u_{jj}'^s \right)^2 + \frac{1}{2}\phi_o K_s u_{jj}'^{s2} \end{aligned} \tag{4.9}$$

The enthalpy of the solid represents the solid internal energy of the system plus the work required to make room for the solid in an environment with constant pressure

$$dH_s = -u_{ik}'^s d\tau_{ik}^s + T^s dS_s \tag{4.10}$$

$$\begin{aligned} H\left(S_s, u_{ik}'^s\right) &= \phi_o U_o^s(S_s) - (T_o K_s \alpha_s / c_v)\left(S_s - S_o^s\right) u_{jj}'^s - \mu_M \left(u_{ik}'^s - \frac{1}{3}\delta_{ik} u_{jj}'^s \right)^2 \\ &+ \phi_o \frac{1}{2}\left[K_{s(ad)} - K_s\right] u_{jj}'^{s2} \end{aligned} \tag{4.11}$$

The Gibbs free energy for the solid represents the solid internal energy of the system plus the work required to make room for it in an environment with constant pressure minus the heat obtained from an environment at temperature T^s.

$$dG_s = -u_{ik}'^s d\tau_{ik}^s - S_s dT^s \tag{4.12}$$

$$G_s\left(T^s, u_{ik}'^s\right) = \phi_o F_o^s\left(T^s\right) - \mu_M \left(u_{ik}'^s - \frac{1}{3}\delta_{ik} u_{jj}'^s \right)^2 - \frac{1}{2}\phi_o K_s u_{jj}'^{s2} \tag{4.13}$$

Fluid Properties

The equivalent potentials for the fluid are given by

Internal energy:

$$dU_f = -\eta p_f d\, u'^f_{kk} + T^f dS_f \tag{4.14}$$

$$U_f\left(S_f, u'^f_{jj}\right) = \phi_o U^f_o\left(S_f\right) - \left(T_o \alpha_f / c^f_p\right)(S_f - S_o) p_f + \frac{1}{2}\eta_o K_{f(ad)} u'^{f2}_{jj} \tag{4.15}$$

Fluid compressive strain:

$$u'^f_{jj} = \nabla \cdot \mathbf{u}_f + \frac{1}{3}\frac{\eta - \eta_o}{\eta_o} \tag{4.16}$$

Ambient temperature:

$$T_o = \eta_o \frac{dU^f_o(S_o)}{dS_o} \tag{4.17}$$

Helmholtz free energy:

$$dF_f = -\eta p_f\, du'^f_{jj} - S_f\, dT^f \tag{4.18}$$

$$F_f\left(T^f, u'^f_{ik}\right) = \phi_o F^f_o\left(T^f\right) + \eta_o K_f \alpha_f \left(T^f - T_o\right) u'^f_{jj} - \frac{1}{2}\eta_o K_f\, u'^{f2}_{jj} \tag{4.19}$$

Thermostatic fluid entropy balance:

$$S_f - S^f_o = \eta_o c^f_p (T^f - T_o)/T_o - \eta_o \alpha_f P_f \tag{4.20}$$

$$S^f_o\left(T^f\right) = -\eta_o \frac{dF^f_o\left(T^f_o\right)}{dT^f_o} \tag{4.21}$$

Enthalpy:

$$dH_f = u'^f_{jj}\, d\left(\eta p_f\right) + T^f dS_f \tag{4.22}$$

$$H_f\left(S_f, u'^f_{jj}\right) = \eta_o U^f_o\left(S_f\right) - \left(T_o \alpha_f / c^f_p\right)\left(S_f - S_o\right) p_f + \frac{1}{2}\eta_o K_{f(ad)}\, u'^{f2}_{jj} + \eta_o p_f u'^s_{jj} \tag{4.23}$$

Gibbs free energy:

$$dG_f = u'^f_{jj}\, d\left(\eta p_f\right) - S_f\, dT^f \tag{4.24}$$

RECIPROCITY

Reciprocity is the constraint imposed by the Onsager's relations, which require that the entropy production must be positive or, in the present case of equilibrium thermostatics, that it must be 0.

The Onsager's relations are constructed as follows: The empirical flux laws are expressed by the following relations.

Temperature (heat):

$$\vec{J}_Q = -\kappa \nabla T \tag{4.25}$$

Porosity variations (order) associated with quasi-static deformations enter the equilibrium thermostatics through the strains as described in Equations 4.3 and 4.16.

$$\vec{J}_\Pi = -D_s \nabla \eta \tag{4.26}$$

Pressure (momentum):

$$\vec{J}_M = -D_f \nabla p \tag{4.27}$$

These quantities may also be expressed as

$$\vec{J}_Q = L_{QQ}\left(-\frac{\nabla T}{T}\right) - L_{Q\Pi}\nabla\eta - L_{QM}\nabla p \tag{4.28}$$

$$\vec{J}_\Pi = L_{\Pi Q}\left(-\frac{\nabla T}{T}\right) - L_{\Pi\Pi}\nabla\eta - L_{\Pi M}\nabla p \tag{4.29}$$

$$\vec{J}_M = L_{MQ}\left(-\frac{\nabla T}{T}\right) - L_{M\Pi}\nabla\eta - L_{MM}\nabla p \tag{4.30}$$

Now define

$$\Gamma_Q = -\frac{\nabla T}{T} \tag{4.31}$$

$$\Gamma_\Pi = -\nabla \eta \tag{4.32}$$

$$\Gamma_M = -\nabla p \tag{4.33}$$

and the coefficients in Equations 5.28 through 5.30 are given by

$$L_{QQ} = \frac{\partial \vec{J}_Q}{\partial \Gamma_Q} L_{Q\Pi} = \frac{\partial \vec{J}_Q}{\partial \Gamma_\Pi} L_{QM} = \frac{\partial \vec{J}_Q}{\partial \Gamma_M} L_{Qq} = \frac{\partial \vec{J}_Q}{\partial \Gamma_q} \tag{4.34}$$

$$L_{\Pi Q} = \frac{\partial \vec{J}_\Pi}{\partial \Gamma_Q} L_{\Pi\Pi} = \frac{\partial \vec{J}_\Pi}{\partial \Gamma_\Pi} L_{\Pi M} = \frac{\partial \vec{J}_\Pi}{\partial \Gamma_M} L_{\Pi q} = \frac{\partial \vec{J}_\Pi}{\partial \Gamma_q} \tag{4.35}$$

$$L_{MQ} = \frac{\partial \vec{J}_M}{\partial \Gamma_Q} L_{M\Pi} = \frac{\partial \vec{J}_M}{\partial \Gamma_\Pi} L_{MM} = \frac{\partial \vec{J}_M}{\partial \Gamma_M} L_{Mq} = \frac{\partial \vec{J}_M}{\partial \Gamma_q} \tag{4.36}$$

$$L_{qQ} = \frac{\partial \vec{J}_q}{\partial \Gamma_Q} L_{q\Pi} = \frac{\partial \vec{J}_q}{\partial \Gamma_\Pi} L_{qM} = \frac{\partial \vec{J}_q}{\partial \Gamma_M} L_{qq} = \frac{\partial \vec{J}_q}{\partial \Gamma_q} \tag{4.37}$$

According to Onsager's reciprocity relations

$$L_{M\Pi} = L_{\Pi M} \tag{4.38}$$

thus

$$\frac{\partial \vec{J}_M}{\partial \Gamma_\Pi} = \frac{\partial \vec{J}_\Pi}{\partial \Gamma_M} \tag{4.39}$$

which may be written as

$$\frac{\partial(-D_f \nabla p)}{\partial(-\nabla \eta)} = \frac{\partial(-D_s \nabla \eta)}{\partial(-\nabla p)} \tag{4.40}$$

or

$$D_f \frac{\partial(\nabla p)}{\partial(\nabla \eta)} = D_s \frac{\partial(\nabla \eta)}{\partial(\nabla p)} \tag{4.41}$$

where

$$D_f = \frac{v_{pf}^2}{2a_{pf}} \tag{4.42}$$

$$D_s = \frac{v_{\eta M}^2}{2a_{\eta M}} \tag{4.43}$$

$$a_{\eta f} = \frac{1}{2}\frac{\mu_f}{K}\frac{\eta_o(\alpha_1 + 1)}{\rho_{\eta f}} \tag{4.44}$$

$$a_{\eta s} = \frac{1}{2}\frac{\mu_f}{K}\frac{\eta_o^2(\alpha_1 + 1)}{\rho_{\eta s}\alpha_1(1 - \eta_o)} \tag{4.45}$$

$$v_{pf}^2 = \frac{K_f}{\rho_{pf}} \tag{4.46}$$

$$v_{\eta M}^2 = \frac{K_\eta}{\rho_{\eta M}} \tag{4.47}$$

$$a_{pf} = \frac{1}{2}\frac{\mu f}{K}\frac{\eta_o(\alpha_2 - 1)}{\rho_{pf}} \tag{4.48}$$

$$a_{\eta M} = \frac{\rho_{\eta s}}{\rho_{\eta M}} a_{\eta s} + \frac{\rho_{\eta f}}{\rho_{\eta M}}\frac{\alpha_2 K_M}{\alpha_1 K_f} a_\eta f \tag{4.49}$$

$$K_\eta = \left(1 - \frac{\eta_o}{(1-\eta_o)\alpha_1}\right) K_s + \frac{4}{3}\frac{\mu_M}{(1-\eta_o)} \tag{4.50}$$

$$\rho_{pf} = \left(\rho_f - \frac{\rho_{12}}{\eta_o}(\alpha_2 - 1)\right) \tag{4.51}$$

$$\rho_{\eta M} = \rho_{\eta s} + \rho_{\eta f}\frac{\alpha_2 K_M}{\alpha_1 K_f} \tag{4.52}$$

$$\rho_{\eta s} = \left(\rho_s - \frac{\rho_{12}(\alpha_1 + 1)}{\alpha_1(1-\eta_o)}\right) \tag{4.53}$$

$$\rho\eta_f = [\rho_f - \rho_{12}(\alpha_1 + 1)] \tag{4.54}$$

$$K_M = K_s + \frac{4}{3}\frac{\mu_M}{(1-\eta_o)} \tag{4.55}$$

$$\alpha_1 = \frac{\eta_o - \delta_f}{\delta_s} \tag{4.56}$$

$$\alpha_2 = \frac{\delta_f}{\delta_s} \tag{4.57}$$

Under quasi-static conditions, the equations of motion yield

$$D_f \nabla^2 p = -C D_s \nabla^2 \eta \tag{4.58}$$

and

$$D_s \nabla^2 \eta = -B D_f \nabla^2 p \tag{4.59}$$

Upon substituting the relations

$$D_f \frac{\partial(\nabla p)}{\partial(\nabla \eta)} = -C D_s \tag{4.60}$$

$$D_s \frac{\partial(\nabla \eta)}{\partial(\nabla p)} = -B D_f \tag{4.61}$$

into Onsager's relations, one obtains

$$C D_s = B D_f \tag{4.62}$$

where

$$B = \frac{\alpha_2 \rho_{pM} \eta_o a_{pM}}{\alpha_1 \rho_{\eta M} a_{\eta M} K_f} \tag{4.63}$$

and

$$C = \frac{\rho_{\eta f} K_f a_{\eta f}}{\eta_o \rho_{pf} a_{pf}} \tag{4.64}$$

Therefore

$$\left[\left(K_s + \frac{4}{3}\mu_M\right)\alpha_1 - \eta_o K_s\right]\frac{(\alpha_1 + 1)}{2} = \eta_o^2(\alpha_2 - 1)\left[1 - \frac{\left(K_s + \frac{4}{3}\mu_M\right)}{K_f}\alpha_2\right] \tag{4.65}$$

This relation places constraints on the alphas for equilibrium processes. Thus Onsager's relation places a constraint on α_2 to be in the range

$$\frac{K_f}{\left(K_s + \frac{4}{3}\mu_M\right)} \le \alpha_2 \le 1 \tag{4.66}$$

and α_1 is in the range

$$\frac{\eta_o K_s}{\left(K_s + \frac{4}{3}\mu_M\right)} \le \alpha_1 \le 1 \tag{4.67}$$

Here, it was assumed that $K_s > K_f$ and $K_s > \mu_M$.

Now, note that going through the same discussion assuming two elastic solids rather than a solid and a fluid, then Equations 4.66 and 4.67 become

$$\frac{\left(K_2 + \frac{4}{3}\mu_M^2\right)}{\left(K_1 + \frac{4}{3}\mu_M^1\right)} \le \alpha_2 \le 1 \text{ and } \frac{\eta_o K_1}{\left(K_1 + \frac{4}{3}\mu_M^1\right)} \le \alpha_1 \le 1 \tag{4.68}$$

Here, it was assumed that $K_1 \ge K_2$, $K_1 > \mu_M^1$ and $K_2 > \mu_M^2$.

Now note that in the limit where the solid properties become equal, $\alpha_2 = 1$, which from Equation 4.68 yields $\delta_1 = \delta_2$, the same result obtained in Chapter 2 from the equations of motion.

SUMMARY

The discussion of thermodynamics was expanded from that of a chemically homogeneous system based on mass fractions to include the new dynamic variable porosity. This variable allows macroscopic mixtures as well as microscopic mixtures to be introduced into the megascopic description (volume fractions as well as mass fractions). This introduces transformations between the structure (or order) of a medium and mechanical work. Allowing for macroscopic components to be rearranged and then going up in scale by orders of magnitude, the fashion in which the system may be rearranged at the megascale now has new degrees of freedom originating at the macroscale. Megascopic relations for equilibrium thermodynamics are formulated on a firm basis.

Reciprocity is the constraint imposed by the Onsager's relations, which require that the entropy production must be positive or in the present case of equilibrium thermostatics that it must be 0. These relations were shown to place constraints on the dilatational parameters that describe porosity changes that occur during dilatational motions.

REFERENCES

Bear, J., and Bachmat, Y., 1990, *Theory and Applications of Transport in Porous Media: Introduction to Modeling of Transport Phenomena in Porous Media*, Kluwer Academic, Dordrecht.

Biot, M.A., 1956, Theory of propagation of elastic waves in a fluid saturated porous solid, I, Low-frequency range, *J. Acoust. Soc. Am.*, 28, 168–178.

Burridge, R., and Keller, J.B., 1981, Poroelasticity equations derived from microstructure, *J. Acoust. Soc. Am.*, 70, 1140–1146.

De Groot, S.R., and Mazur, 1962, *Non-Equilibrium Thermodynamics*, Interscience, New York.

de la Cruz, V., and Spanos, T.J.T., 1983, Mobilization of oil ganglia, *AIChE J.*, 29(7), 854–858.

Del Rio, J.A., and Lopez De Haro, M., 1992, Extended irreversible thermodynamics as a framework for transport in porous media, *Transport in Porous Media*, 9, 207–221.

de la Cruz, V., and Spanos, T.J.T., 1985, Seismic wave propagation in a porous medium, *Geophysics*, 50(10), 1556–1565.

de la Cruz, V., and Spanos, T.J.T., 1989, Thermo-mechanical coupling during seismic wave propagation in a porous medium, *J. Geophys. Res.*, 94, 637–642.

de la Cruz, V., Sahay, P.N., and Spanos, T.J.T., 1993, Thermodynamics of Porous Media, *Proceedings of the Royal Society of London (rapid communication)*, 443(1917) 247–255.

Detournay, E., 1993. Constitutive equations: Overview and theoretical background, in A Short Course in Poroelasticity in Rock Mechanics, The 34th U.S. Symposium on Rock Mechanics, June 27–30, University of Wisconsin-Madison. Madison, Wisconsin.

Dullien, F.A.L., 1992, *Porous Media: Fluid Transport and Pore Structure*, Academic Press, San Diego, CA.

Garcia-Conlin, L.S., and Uribe, F.J., 1991, Extended irreversible thermodynamics beyond the linear regime: A critical overview, *J. Non Equilib. Thermodyn.*, 16, 89–128.

Gassmann, F., 1951, Uber die elastizitat poroser medien, *Ver der Natur. Gesellschaft*, 96, 1–23.

Gurtin, M.E., 1988. Multiphase thermodynamics with interfacial structure 1. Heat conduction and the capillary balance law, *Arch. Ration. Mech. Anal.*, 104, 185–221.

Gurtin, M.E., and Struthers, A., 1990, Multiphase thermodynamics with interfacial structure 3. Evolving phase boundaries in the presence of bulk deformation, *Arch. Ration. Mech. Anal.*, 112, 97–160.

Hickey, C.J., Spanos T.J.T., and de la Cruz, V., 1995, Deformation parameters of permeable media, *Geophys. J. Int.*, 121, 359–370.

Keller, J.B., 1977, *Statistical Mechanics and Statistical Methods in Theory and Application*, Plenum Press, New York.

Koch, D.L., and Brady, J.F., 1988, Anomalous diffusion in heterogeneous porous media, *Phys. Fluids*, 31, 965.

Landau, L.D., and Lifshitz, E.M., 1975. *Theory of Elasticity*, Pergamon Press, Toronto.

Marle, C.M., 1982, On macroscopic equations governing multiphase flow with diffusion and chemical reactions in porous media, *Int. J. Eng. Sci.*, 20(5), 643–662.

Prigogine, I., 1954, *Treatise on Thermodynamics*, Longmans, London.

Sanchez-Palencia, E., 1980, *Non-Homogeneous Media and Vibration Theory*, Lecture Notes in Physics 127, Springer-Verlag, New York.

Scheidegger, A.E., 1974, *The Physics of Flow Through Porous Media*, University of Toronto Press, Toronto.

Slattery, J.C., 1967, Flow of viscoelastic fluids through porous media, *AIChE J.*, 13, 1066–1071.

Spanos, T.J.T., 2002, The Thermophysics of Porous Media, Chapman & Hall/CRC Press, Monographs and Surveys in Pure and Applied Mathematics series, pp. 212.

Whitaker, S., 1967, Diffusion and dispersion in porous media, *AIChE J.*, 13, 420–427.

CHAPTER 5

Immiscible Fluid Flow in Porous Media

OBJECTIVES OF THIS CHAPTER

A system of megascopic equations for the flow of incompressible, immiscible fluid phases in porous media is presented (de la Cruz and Spanos 1983). The complete system of equations for compressible fluid flow is then constructed and the incompressible limit of these equations is considered. It is observed that a complete system of dynamical equations is obtained that are consistent with the previous equations in the steady-state limit.

The primary complication that must be addressed in this chapter is the complex interfacial phenomena occurring between fluid phases at the pore scale and how this information enters the megascopic description (Eastwood and Spanos, 1991). The effect of phase transitions on multiphase flow is also described. It has been observed that phase transitions can have an important stabilizing effect on displacement processes (Krueger 1982a, 1982b; de la Cruz et al. 1985).

The concept of capillary pressure in porous media has been reviewed by a number of authors (e.g. Scheidegger 1974; Barenblatt et al. 1990; Bear and Bachmat 1990; Dullien 1992; Lenormand et al. 1983; de Gennes 1983). The megascopic pressure difference between phases, however, depends on the megascopic variables, and thus its connection to the pore-scale capillary pressure is sometimes difficult to delineate (see Barenblatt et al. 1990; Bear and Bachmat 1990; Bentsen 1994). In the present discussion, the megascopic pressure difference is described by considering the incompressible limit of the equations of compressible fluid flow through porous media. Here, the equations for compressible fluid flow through porous media have been constructed from the well-understood equations and boundary conditions at the pore scale (Hickey 1994; Hickey et. al. 1995). It is also observed that Onsager's relations place constraints on new variables that arise to account for changes in volume fractions and saturation during multiphase flow (Equation 5.47). Furthermore, the thermodynamic understanding (de la Cruz et al. 1993) of the parameters and variables described in Chapter 4 may be used when considering the incompressible limit.

SINGLE-PHASE FLOW

Steady-state single-phase flow in porous media was quantified by Darcy's law in 1856:

$$\frac{\mu}{K}\vec{q} = -\vec{\nabla}p + \rho g\hat{z} \tag{5.1}$$

This relation is based on Darcy's experiment, which measures the volume of fluid that flows through a homogeneous porous medium of constant cross-sectional area, out an outlet in a fixed period of time and at a constant pressure gradient (Figure 5.1).

Consider a uniformly packed porous medium of length h and cross-sectional area A in a cylindrical jacket. Place a reservoir at each end with an inflow at one end and an outflow at the other. This flow may be established by connecting these reservoirs to two large reservoirs at fixed but different heights. This difference in height is referred to as the *head* and maintains a constant pressure (the constant force per unit area exerted on the surface of the porous medium, including both solid and fluid, by the fluid in the reservoir bounding the porous medium) at the two ends of the porous medium. This pressure may be monitored at the end points or any points in between by attaching tubes to the porous medium and observing the height of the fluid level in the tubes.

In Darcy's experiment, it is observed that the volumetric flow rate Q is proportional to the cross-sectional area of the porous medium, A, times the change in height in the tubes, $h_2 - h_1$, divided by the length of the porous medium h

$$Q = \text{Const}\frac{A(h_2 - h_1)}{h} \tag{5.2}$$

By varying the fluid viscosity, one observes that this relationship becomes

$$Q = \frac{K'}{\mu}\frac{A(h_2 - h_1)}{h} \tag{5.3}$$

where μ is the fluid viscosity and the constant K′ is a property of the particular porous medium being studied. Here, K′ is independent of μ, A, h, $(h_2 - h_1)$ and Q. This makes K′ a physical parameter, meaning that like viscosity it may be determined by an infinite number of different experiments that always yield the same value for K′. In contrast, Q, A, h and $(h_2 - h_1)$ are quantities that may be varied in the experiment and are therefore called variables. Darcy's law results from this experiment and is generally stated as the result that the flow rate (Darcy velocity) $q = \frac{Q}{A}$ is linearly related to the pressure gradient

$$-\nabla P = \frac{\rho g(h_2 - h_1)}{h} \tag{5.4}$$

yielding the relation called *Darcy's equation* (which may also include body forces such as gravity)

$$\frac{\mu}{K}q = -\nabla P \tag{5.5}$$

where $K = \frac{K'}{\rho g}$ is called the permeability of the porous medium and K is independent of ρ.

In order to obtain a theoretical understanding of this relationship, start with the well-understood macroscopic equations that describe fluid flow at the pore scale and obtain a megascopic description at the scale of thousands of pores. Megascopic equations governing fluid flow and deformations of the porous matrix reflect the properties and behaviour of the constituent materials and their interaction. Hence, in theoretical attempts to construct these equations it is important to maintain a strong connection with the well-established equations governing the behaviour of the macroscopically segregated components, which interact at the numerous interfaces in accordance with suitable boundary conditions. From this point of view some averaging scheme appears to be unavoidable. Equation 5.1 was derived theoretically, using volume averaging, by a number of authors (Anderson and Jackson 1967; Slattery 1967, 1969; Whitaker 1967, 1969; Newman 1977). In these derivations the inertial terms have been neglected. However, they must be included to describe such effects as the evolution to Darcy flow or fluid flow–associated seismic or porosity wave propagation. A summary of the complete set of equations including inertial effects is given in the following chapters.

In the description of megascopic systems, structure often introduces heterogeneity. This heterogeneity then affects the form of the equations of motion. If structure occurs at all scales and is indeterminate, then the megascopic equations of motion are indeterminate. As an illustration of this principle, the flow of a single fluid phase in a homogenous porous medium will be considered first. In the construction of these megascopic equations, explicit use is made of the homogeneous structure of the medium. When the porous medium is allowed to be composed of different homogeneous structures at various scales and of various shapes, the form of the megascopic equations is examined. Even when the two components are macroscopically homogeneous, the porous medium is megascopically inhomogenous if the unperturbed porosity varies with position.

HETEROGENEITY

Now that steady-state (independent of time) Darcy flow has been established in a homogeneous (uniform) isotropic (independent of the direction of flow) porous medium, the next question to ask is, what if these restrictions are relaxed? As a first venture to inhomogeneous porous media set up Darcy experiments, with porous media having different permeabilities, in parallel and series. First consider five porous media in parallel, as shown in Figure 5.2.

For each of the porous media

$$\frac{\mu}{K_i}q_i = -\nabla P \tag{5.6}$$

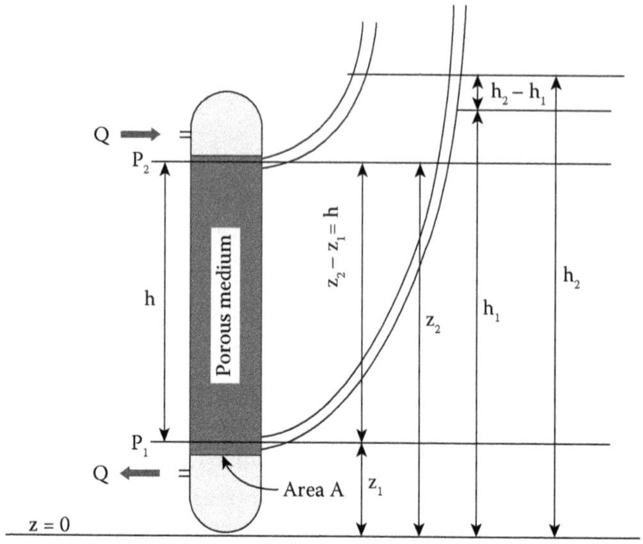

FIGURE 5.1 This figure illustrates Darcy's experiment, in which a difference in height between the inlet and outlet reservoirs causes a pressure drop and a volumetric flow rate Q through the porous medium. Here, the porous medium is homogeneous with a length h and cross-sectional area A. This experiment determines the permeability of the porous medium, which is independent of fluid viscosity, the length and cross-sectional area of the porous medium and the difference in height between the two reservoirs.

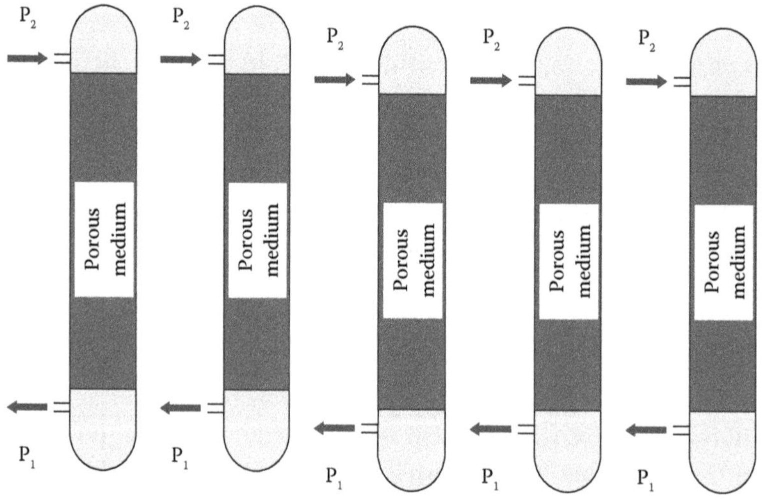

FIGURE 5.2 This figure shows five porous media hooked up in parallel. Each porous medium is connected to the same inlet reservoir and the same outlet reservoir. Therefore each reservoir has an identical head and thus the same pressure drop between the inlet and outlet.

for $i = 1$ to 5. Thus

$$q_T = -\frac{K_T}{\mu}\nabla P \tag{5.7}$$

where

$$q_T = \sum_{i=1}^{5} q_i \tag{5.8}$$

and

$$K_T = \sum_{i=1}^{5} K_i \tag{5.9}$$

Here, Equation 5.7 is Darcy's equation. Now consider five porous media in series as shown in Figure 5.3.

For each of the porous media

$$\frac{\mu}{K_i} q_i = -\nabla P_i \tag{5.10}$$

for $i = 1$ to 5. Thus

$$\frac{\mu}{K_T} q = \nabla P_T \tag{5.11}$$

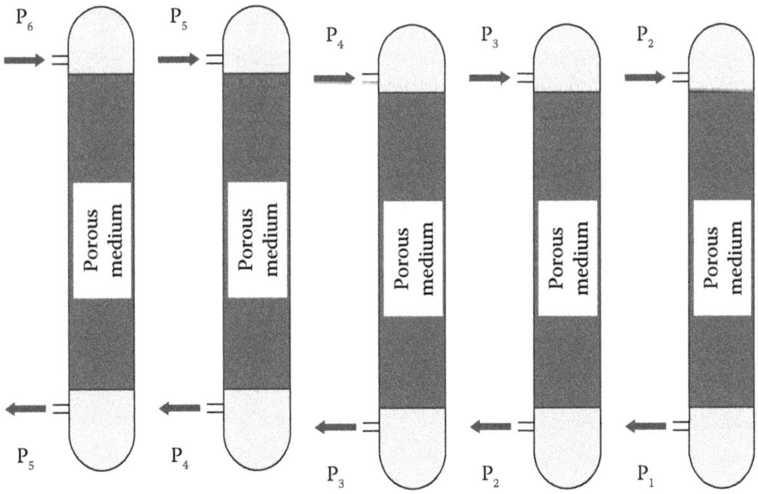

FIGURE 5.3 This figure shows five porous media hooked up in series. Thus the outlet pressure of the first porous medium is the inlet pressure for the second and so on. So each porous medium causes a pressure drop in succession with the same constant volumetric flow rate in each.

where

$$P_T = P_6 - P_1 \tag{5.12}$$

and

$$\frac{1}{K_T} = \sum_{i=1}^{5} \frac{1}{K_i} \tag{5.13}$$

So in both of these examples Darcy's equation describes the fluid motions and the permeability of the overall system may be calculated.

Now consider a porous medium constructed by combining different homogenous pieces of various sizes and shapes (Figure 5.4). Construct the equations of motion for each homogeneous piece. Then going up in scale the equations of motion for the composite medium may be constructed. This demonstrates that the form of the equations of motion depends on the properties and structure of the heterogeneities. Here, first restrict this discussion to single-phase steady flow. Then the megascopic incompressible flow of a fluid through a homogeneous porous medium is given by

$$\frac{\mu}{K}\vec{q} = -\vec{\nabla}p \tag{5.14}$$

$$\vec{\nabla}\cdot\vec{q} = 0 \tag{5.15}$$

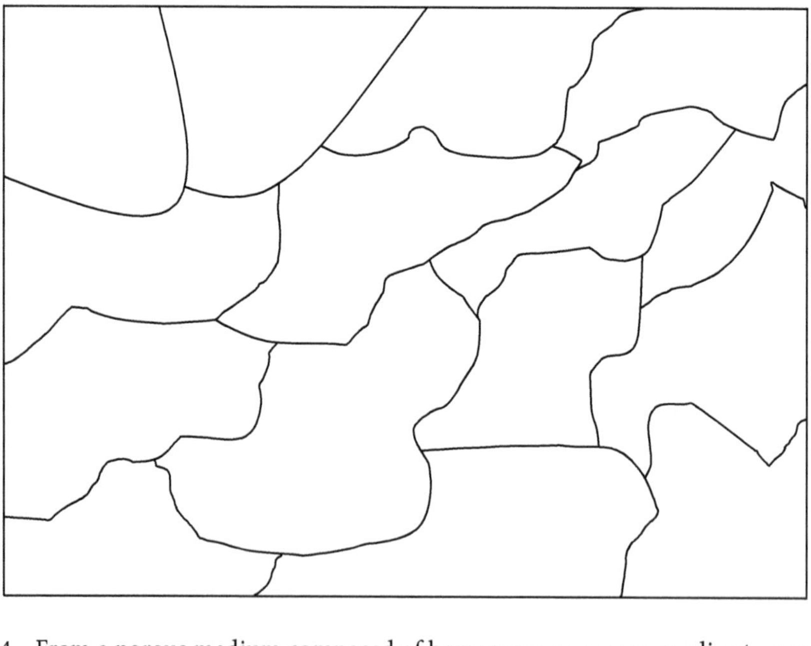

FIGURE 5.4 From a porous medium composed of homogeneous porous media at a smaller scale, the equations of motion depend on the size and structure of the homogeneous porous media. The form of the equations of motion is not Darcy's equation and permeability is not a meaningful physical parameter.

These equations describe flow at a scale large compared to the pore scale and small in comparison to structural inhomogeneities. Now assume that these homogeneous structures occur within volume elements Ω. Applying averaging to these equations up to a gigascale larger than the homogeneous structures yields

$$0=\frac{1}{\Omega}\int_{\Omega}\left(\frac{\mu}{K}\vec{q}+\vec{\nabla}p\right)d\Omega \tag{5.16}$$

and thus

$$0=\frac{\mu}{\Omega}\int_{\Omega}\frac{\vec{q}}{K}d\Omega+\vec{\nabla}\frac{1}{\Omega}\int_{\Omega}p\ d\Omega \tag{5.17}$$

Now assume there are a number of homogeneous pieces β = 1, …, n within Ω. Then

$$0=\frac{\mu}{\Omega}\sum_{\beta}\int_{\Omega_\beta}\frac{\vec{q}}{K}d\Omega+\vec{\nabla}\bar{P}_o \tag{5.18}$$

where

$$\bar{P}=\vec{\nabla}\frac{1}{\Omega}\int_{\Omega}p\ d\Omega \tag{5.19}$$

This result is slightly disturbing at first sight since Equation 5.18 is not Darcian in form. However, in the case where Ω is sufficiently large relative to the homogeneous volume elements Ω_β, then starting from the pore scale and applying volume averaging yields Darcy's equation

$$0=\frac{\mu}{K'}\bar{\vec{q}}+\vec{\nabla}\bar{P} \tag{5.20}$$

This seems to indicate the existence of a permeability parameter K′ given by

$$\frac{1}{K'}\bar{\vec{q}}=\frac{1}{\Omega}\sum_{\beta}\frac{1}{K_\beta}\int_{\Omega_\beta}\vec{q}d\Omega \tag{5.21}$$

In the limit, as Ω_β become infinitesimally small, K′ becomes a parameter, but in general K′ is not a parameter at all. Here, $\vec{q}$ may be written as $\vec{q}=\bar{\vec{q}}+\vec{q}$ and thus

$$\frac{1}{K'}\bar{\vec{q}}=\frac{1}{\Omega}\sum_{\beta}\frac{1}{K_\beta}\int_{\Omega_\beta}\left(\bar{\vec{q}}+\bar{\vec{q}}'\right)d\Omega \tag{5.22}$$

or

$$\frac{1}{K'}\bar{\vec{q}}=\frac{1}{\Omega}\bar{\vec{q}}\sum_{\beta}\frac{\Omega_\beta}{K_\beta}+\sum_{\beta}\frac{1}{K_\beta}\frac{1}{\Omega}\int_{\Omega_\beta}\bar{\vec{q}}'d\Omega \tag{5.23}$$

This relation clearly demonstrates that, in general, permeability cannot be defined as a parameter for a megascopic heterogeneous porous medium.

All naturally occurring porous media are heterogeneous at the pore scale. Laboratory samples may be constructed from glass beads of uniform size in a specific packing. This would appear to be the only example of a porous medium homogeneous at this scale. However, random heterogeneities at this scale are averaged out when going to the megascale, provided no additional structure is introduced between these scales. This is the scale at which permeability is defined. From the previous argument it is evident that if new structure is now introduced then permeability is no longer defined as a parameter at larger scales.

Often engineers and geologists refer to permeability as being scale dependent or to a reservoir having two types of permeability (say, fracture and medium). From the previous analysis it can be seen that that simply means that Darcy's equation is not the equation of motion for flow in such media and permeability is not a physical parameter. Another common error is to describe permeability as being porosity dependent, but it is possible to simply take a hexagonal packing of identical glass beads. Then choosing a different size of glass beads for each different porous medium, permeability may be varied by many orders of magnitude while keeping the porosity exactly the same. This simple experiment illustrates that these two parameters may be varied independently. However it has been observed that, for a given rock type, naturally occurring porous media have a correlation between permeability and porosity. This is the origin of numerous empirical permeability versus porosity relations.

In reality the concept of writing simple linear equations of motion and trying to solve them to describe a physical process in a complex medium such as the Earth's crust is a method completely devoid of physical content. Equations of motion for homogeneous media or media with specific heterogeneities may be used to understand physical processes that occur. In order to describe large-scale physical processes, the construction must go from the small scale to the large scale, and when this isn't possible the construction must go from the fundamental to the applied.

MULTIPHASE FLOW IN POROUS MEDIA

The equations of multiphase flow in homogeneous porous media were first constructed by de la Cruz and Spanos (1983). Previous constructions of flow in porous media in the petroleum engineering literature and in ground water hydrology literature contain numerous errors in physics; they do not have enough degrees of freedom and make fundamental errors in thermodynamics. A summary of the errors made in constructing the equations in those fields will be given here.

Here, the discussion of multiphase flow in porous media will begin by considering miscible flow in porous media. The equations of motion for the flow of miscible fluid phases has generally been modelled by the convection diffusion equation. This equation is based on the mixing of fluid phases at the molecular scale and completely misses the dynamic processes that occur at various other scales in porous media. However, the problem is much greater than the convection (or advection) diffusion equation not being the correct equation to describe the flow of miscible fluid flow in porous media. The system of equations being used is also self-contradictory and is not complete.

Degrees of freedom, which can be observed in simple thought experiments, are not accounted for because there aren't enough equations. The origin of this error is that the thermodynamics being used does not apply to porous media at the scale at which it is being applied. A detailed description of the miscible flow problem in porous media is given in Chapters 6 and 7, extending previous work by Udey and Spanos (1993), Spanos (2002) and Udey (2009).

There, it is shown that the correct equations of motion for modelling miscible flow in porous media are as follows:

1. In the case of high flow rates with negligible molecular diffusion, the equations of motion as well as a dynamic pressure equation, which says that if one phase is displacing another there is a pressure difference between them.
2. In the case of low flow rates with negligible molecular diffusion, a Fokker–Planck equation is required, which unlike the convection diffusion equation allows for a natural description of longitudinal and lateral dispersion.
3. In the case where molecular diffusion becomes dominant, two coupled Fokker–Planck equations are required. As well, component velocities and volume fractions must be replaced by mass fractions and momenta.

Cases 1 and 2 are considered in Chapter 7 and Case 3 is considered in Chapter 8.

The importance of the dynamic pressure equation mentioned above can be easily demonstrated with a simple thought experiment. Assume that one fluid is displacing another in a porous medium. Further assume that although the phases are miscible a negligible amount of molecular diffusion occurs during the course of the experiment. Since the fluids remain distinct they may be considered in terms of their volume fractions rather than their mass fractions (i.e. saturation instead of concentration). Here, if one phase is displacing another then the amount of the displacing phase within a volume element must increase in time and the amount of displaced phase must be reduced. This requires a change in saturation in time within the volume element. The average pressure of the displaced fluid must be less than that of the displacing fluid, since the pressure gradient is in the direction of flow. So if there is a saturation gradient then the saturation in the volume element must be changing in time and the average pressure in the two phases must be different. This may be expressed as

$$P_1 - P_2 = -\beta \frac{\partial S_1}{\partial t} \tag{5.24}$$

A simple experiment to illustrate the dispersional motion predicted by this theory can be performed by constructing a homogenous porous medium of about 1 darcy permeability and then displacing clear water with blue dyed water with about a metre head. During the course of the experiment little diffusion will occur; however, it can be observed that the saturation contours will initially separate in an unstable fashion and then reach a terminal velocity, at which point they will propagate as waves. The behaviour of this very simple experiment cannot be described by the convection diffusion models used by hydrologists. A theoretical description of this process is given by Udey (2009).

Including surface tension alters the values of some of the parameters in the equations of motion. It also alters the form of the dynamic pressure equation (Equation 5.24). The dynamic pressure equation now takes the following form (see Spanos 2002):

$$\frac{\partial (P_1 - P_2)}{\partial t} = -\beta_1 \frac{\partial S_1}{\partial t} - \beta_2 \frac{\partial^2 S_1}{\partial t^2} \tag{5.25}$$

Here, β_1 is proportional to surface tension (see de la Cruz et al. 1995) and thus vanishes in the miscible limit given by Equation 5.24.

The equations of motion for steady-state multiphase flow in porous media are given by the following (de la Cruz and Spanos 1983):

$$Q_{11}\vec{q}_1 - Q_{12}\vec{q}_2 = -\vec{\nabla} P_1 - \rho_1 \vec{g} \tag{5.26}$$

$$Q_{22}\vec{q}_2 - Q_{21}\vec{q}_1 = -\vec{\nabla} P_2 - \rho_2 \vec{g} \tag{5.27}$$

In petroleum engineering it is common to introduce a quantity called *relative permeability* (Richarson et al. 1952). This was a property proposed by Muscat (1946) in the 1940s in analogy with permeability.

$$\frac{\mu_1}{KK_{r1}} \vec{q}_1 = -\vec{\nabla} P_1 - \rho_1 \vec{g} \tag{5.28}$$

$$\frac{\mu_1}{KK_{r2}} \vec{q}_2 = -\vec{\nabla} P_2 - \rho_2 \vec{g} \tag{5.29}$$

and capillary pressure is imposed as a static constraint. Here, K_{r1} and K_{r2} are not physical parameters and in fact are functions of variables when constructed from Equations 5.25 through 5.27. For a quantity like viscosity to be a physical parameter, it must be able to be measured by any flow that the Navier–Stokes equation describes. Thus it may be measured by an infinite number of different experiments and one must always obtain the same result. Relative permeabilities in contrast are simply chosen to simulate flows described by Equations 5.25 through 5.27. However, they cannot simulate dispersion, for example. This can be seen by observing that Equations 5.28 and 5.29 each have only one flow velocity, meaning there is no relative motion of the fluids. If there were relative motion, then there would be a force associated with that relative motion as is evident from Equations 5.26 and 5.27. The model Equations 5.28 and 5.29 are associated with is based on molecular-scale thermostatics. Furthermore, with Equations 5.28 and 5.29 as the sole equations of motion, that is, without Equation 5.25, the equations state that saturation is uniform and no displacement is taking place (i.e. $\frac{\partial S_1}{\partial t} = 0$ and $\frac{\partial^2 S_1}{\partial t^2} = 0$). This leads to what is called the *Buckley–Leverett paradox* (if a similar sharpening of the front occurs when physical equations are being used, it is an indication that viscous fingering will be the dominant process and broadening of the front leads to dispersion). Here, since the relative permeability equations are making an unphysical statement, they yield an unphysical solution when a displacement takes place. Here, Figures 5.5 and 5.6 represent oil and water relative permeability curves measured under steady-state conditions (constant saturation and steady flow).

However, Kro and Krw can be calculated from Equations 5.25 through 5.27 and are found to depend on q_1, q_2, Q_{11}, Q_{12}, Q_{21}, Q_{22}, β_1, and β_2 (Spanos et al. 1986). Now, using Equations 5.28 and 5.29, the fractional flow of water for a water displacement process is given by Figure 5.7.

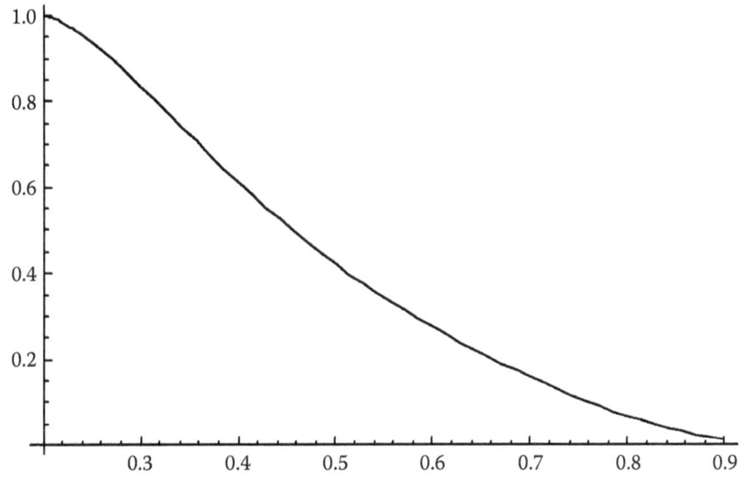

FIGURE 5.5 The relative permeability to oil, Kro, versus saturation as measured in constant saturation steady-state experiments.

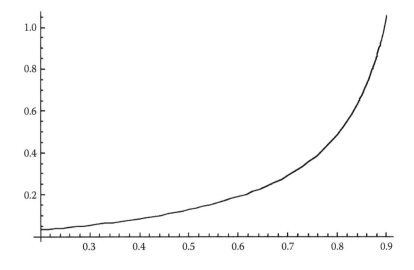

FIGURE 5.6 The relative permeability to water, Krw, versus saturation as measured in constant saturation steady-state experiments.

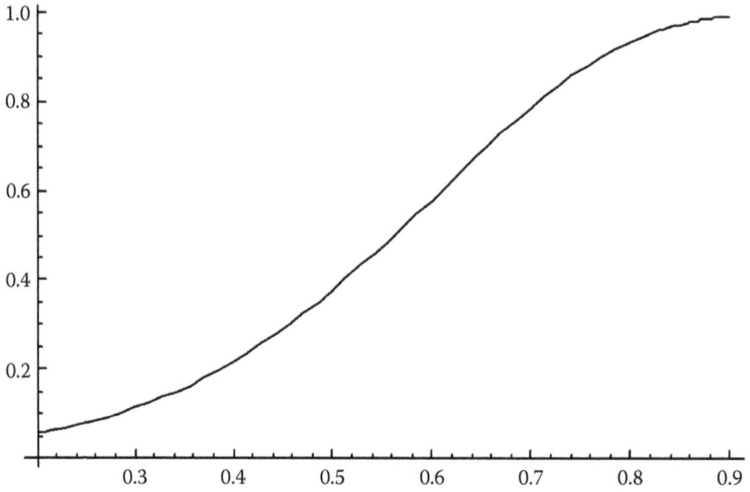

FIGURE 5.7 The fractional flow of water, fw, versus saturation as determined from the above relative permeabilities.

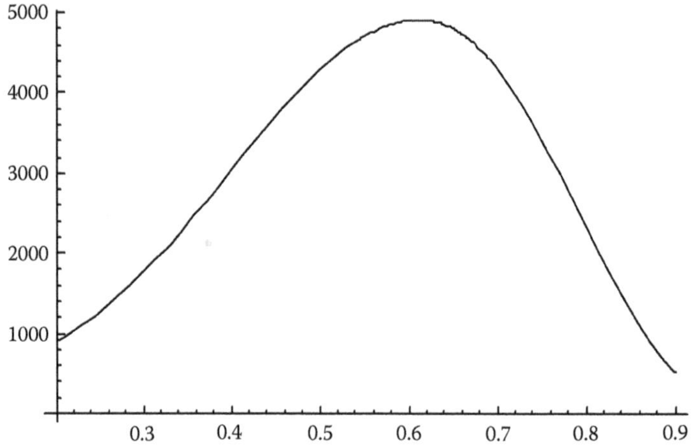

FIGURE 5.8 The position of each saturation, z(Sw), as determined from the fractional flow.

Here, the inflection point in this graph causes front sharpening during in the flow. This may be seen by plotting the position as a function of saturation in Figure 5.8.

Here, for this one dimensional model multiple values of saturation are predicted at the same position in space. When the velocity of saturation contours is plotted, it is observed that low values of saturation are moving slower than some higher values of saturation, this indicates the assumption that a one dimensional model may not be valid (as a front sharpens viscous fingers occur).

One possible origin of this problem may be observed by plotting the velocity of the saturation contours as a function of saturation, where it is observed that some higher water saturation contours move faster than lower saturation contours and thus would pass them, as can be seen in Figure 5.9.

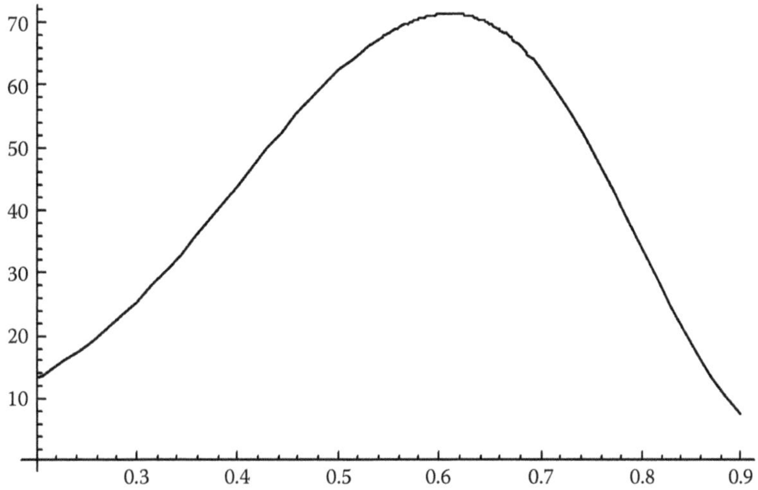

FIGURE 5.9 The velocity, v(Sw), of each saturation contour is shown. Here, higher saturation contours move faster than lower saturation contours, a result that is unphysical.

A resolution of this paradox is transparent when studied in terms of Equations 5.25 through 5.27. When this is done, it is observed that the relative permeabilities are functions of q1, q2, Q11, Q12, Q21, Q22, β1, and β2. As a result, the values of the relative permeabilities for frontal displacement are different than they are for constant saturation steady-state flow. When the new values for frontal displacement are calculated, they yield dynamic relative permeability values, given by Figures 5.10 and 5.11.

When these values are used for the relative permeabilities, the fractional flow becomes as shown in Figure 5.12.

Here, the inflection point present when the constant saturation steady-state values were used may be removed. As a result, the saturation as a function of position becomes monotonic, as can be seen from the z(Sw) graph shown in Figure 5.13.

The velocity as a function of saturation in Figure 5.14 shows a disperse front, with the amount of dispersion depending on the connate water saturation.

Now, extending this theory to include inertial terms in the fluid equations of motion, viscous dissipation within the fluids and compressibility, the equations of motion become (Spanos 2002):

$$\frac{\partial}{\partial t}(\eta_1\rho_1\mathbf{v}_1)+[\mu_1\nabla^2\mathbf{v}_1+(\xi_1+{}^1\!/_3\mu_1)\nabla(\nabla\cdot\mathbf{v}_1)]+\nabla\left[\frac{\xi_1}{\eta_1^o}\frac{\partial\eta_1}{\partial t}\right]-(Q_{11}q_1-Q_{12}q_2)=\nabla p_1 \quad (5.30)$$

$$\frac{\partial}{\partial t}(\eta_2\rho_2\mathbf{v}_2)+[\mu_2\nabla^2\mathbf{v}_2+(\xi_2+{}^1\!/_3\mu_2)\nabla(\nabla\cdot\mathbf{v}_2)]+\nabla\left[\frac{\xi_2}{\eta_2^o}\frac{\partial\eta_2}{\partial t}\right]-(Q_{22}q_2-Q_{21}q_1)=\nabla p_2 \quad (5.31)$$

$$\frac{\partial\eta_1}{\partial t}=\delta_2\nabla\cdot\mathbf{v}_2-\delta_1\nabla\cdot\mathbf{v}_1 \quad (5.32)$$

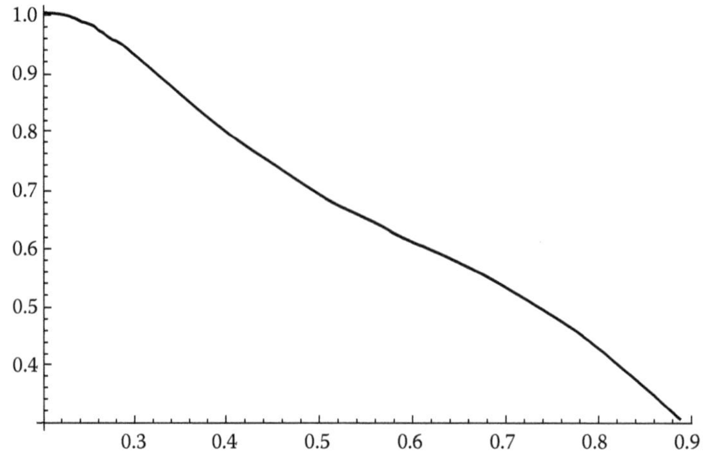

FIGURE 5.10 The relative permeability to oil, Kro, versus saturation as calculated using Equations 5.25 through 5.27 under dynamic conditions.

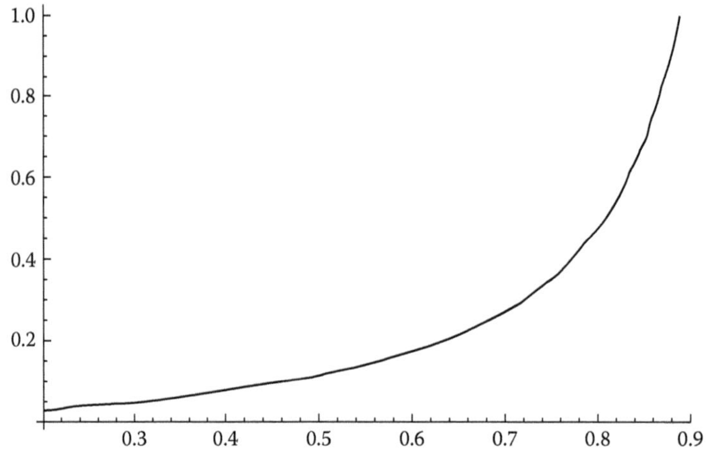

FIGURE 5.11 The relative permeability to water, Krw, versus saturation as calculated using Equations 5.25 through 5.27 under dynamic conditions.

The continuity equations state

$$\frac{1}{\rho_1^o}\frac{\partial}{\partial t}\rho_1+\frac{1}{\eta_1^o}\frac{\partial}{\partial t}\eta_1+\nabla\cdot\mathbf{v}_1=0 \tag{5.33}$$

$$\frac{1}{\rho_2^o}\frac{\partial}{\partial t}\rho_2+\frac{1}{\eta_2^o}\frac{\partial}{\partial t}\eta_2+\nabla\cdot\mathbf{v}_2=0 \tag{5.34}$$

The pressure equations state

$$\frac{1}{K_{f1}}\frac{\partial}{\partial t}p_1=-\nabla\cdot\mathbf{v}_1-\frac{1}{\eta_1^o}\frac{\partial}{\partial t}\eta_1 \tag{5.35}$$

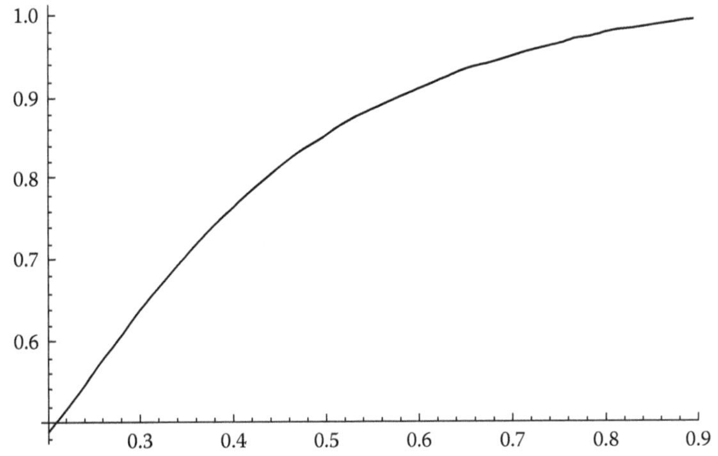

FIGURE 5.12 The fractional flow of water, fw, versus saturation as determined from the above dynamic relative permeabilities shown in figures 5.10 and 5.11.

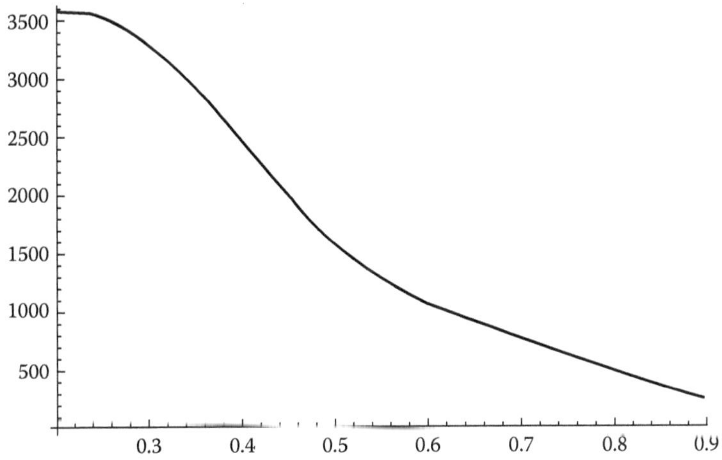

FIGURE 5.13 The position of each saturation, z(Sw), as determined from the above fractional flow shown in Figure 5.12.

$$\frac{1}{K_{f_2}}\frac{\partial}{\partial t}p_2 = -\nabla\cdot\mathbf{v}_2 - \frac{1}{\eta_2^o}\frac{\partial}{\partial t}\eta_2 \tag{5.36}$$

Now, taking the divergence of Equations 5.30 and 5.31, eliminating $\mathbf{v}_1$ and $\mathbf{v}_2$ using the pressure equations (Equations 5.35 and 5.36), the following three equations for the three unknowns η_1, p_1 and p_2 may be written

$$\rho_1\frac{\partial^2\eta_1}{\partial t^2} - (Q_{11}+Q_{12})\frac{\partial\eta_1}{\partial t} + \frac{1}{\eta_1^o}(4/3\mu_1)\nabla^2\frac{\partial\eta_1}{\partial t}$$
$$= -\frac{\eta_1\rho_1}{K_{f_1}}\frac{\partial^2 p_1}{\partial t^2} + \left(\frac{Q_{11}\eta_1}{K_{f_1}} + \frac{Q_{12}\eta_2}{K_{f_2}}\right)\frac{\partial p_1}{\partial t} - (\xi_1 + 4/3\mu_1)\frac{1}{K_{f_1}}\nabla^2\frac{\partial p_1}{\partial t} - \nabla^2 p_1 \tag{5.37}$$

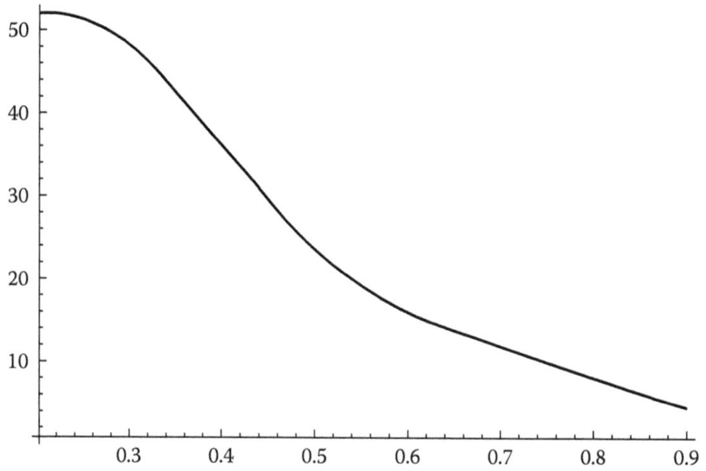

FIGURE 5.14 This graph illustrates that frontal dispersion lower water saturation contours move faster than higher water saturation contours. This process is not allowed by Equations 5.28 and 5.29 or the associated capillary pressure assumptions.

$$\rho_2 \frac{\partial^2 \eta_2}{\partial t^2} - (Q_{22} + Q_{21}) \frac{\partial \eta_2}{\partial t} + \frac{1}{\eta_2^o} (4/3\mu_2) \nabla^2 \frac{\partial \eta_2}{\partial t}$$

$$= -\frac{\eta_2 \rho_2}{K_{f_2}} \frac{\partial^2 p_2}{\partial t^2} + \left(\frac{Q_{22}\eta_2}{K_{f_2}} + \frac{Q_{21}\eta_1}{K_{f_1}} \right) \frac{\partial p_2}{\partial t} - (\xi_2 + 4/3\mu_2) \frac{1}{K_{f_2}} \nabla^2 \frac{\partial p_2}{\partial t} - \nabla^2 p_2 \tag{5.38}$$

$$\left(1 - \frac{\delta_2}{\eta_2^o} - \frac{\delta_1}{\eta_1^o} \right) \frac{\partial \eta_1}{\partial t} = \frac{\delta_1}{K_{f_1}} \frac{\partial p_1}{\partial t} - \frac{\delta_2}{K_{f_2}} \frac{\partial p_2}{\partial t} \tag{5.39}$$

Eliminating η_1 in Equations 5.37 and 5.38 using Equation 5.39 and assuming the inertial terms and bulk attenuation terms are small, two-coupled diffusion equations for pressure are obtained, yielding

$$B \frac{\partial p_2}{\partial t} = \frac{\partial p_1}{\partial t} - D_1 \nabla^2 p_1 \tag{5.40}$$

$$C \frac{\partial p_1}{\partial t} = \frac{\partial p_2}{\partial t} - D_2 \nabla^2 p_2 \tag{5.41}$$

where

$$B = \frac{(Q_{11} + Q_{12}) \frac{\delta_2}{K_{f_2}}}{\left((Q_{11} + Q_{12}) \frac{\delta_1}{K_1} + \left(1 - \frac{\delta_2}{\eta_2^o} - \frac{\delta_1}{\eta_1^o} \right) \left(\frac{Q_{11}\eta_1}{K_{f_1}} + \frac{Q_{12}\eta_2}{K_{f_2}} \right) \right)} \tag{5.42}$$

$$C = \cfrac{(Q_{22}+Q_{21})\dfrac{\delta_1}{K_{f_1}}}{\left((Q_{22}+Q_{21})\dfrac{\delta_2}{K_2} + \left(1 - \dfrac{\delta_2}{\eta_2^o} - \dfrac{\delta_1}{\eta_1^o}\right)\left(\dfrac{Q_{22}\eta_2}{K_{f_2}} + \dfrac{Q_{21}\eta_1}{K_{f_1}}\right)\right)} \tag{5.43}$$

$$D_1 = \cfrac{\left(1 - \dfrac{\delta_2}{\eta_2^o} - \dfrac{\delta_1}{\eta_1^o}\right)}{\left((Q_{11}+Q_{12})\dfrac{\delta_1}{K_1} + \left(1 - \dfrac{\delta_2}{\eta_2^o} - \dfrac{\delta_1}{\eta_1^o}\right)\left(\dfrac{Q_{11}\eta_1}{K_{f_1}} + \dfrac{Q_{12}\eta_2}{K_{f_2}}\right)\right)} \tag{5.44}$$

$$D_2 = \cfrac{\left(1 - \dfrac{\delta_2}{\eta_2^o} - \dfrac{\delta_1}{\eta_1^o}\right)}{\left((Q_{22}+Q_{21})\dfrac{\delta_2}{K_2} + \left(1 - \dfrac{\delta_2}{\eta_2^o} - \dfrac{\delta_1}{\eta_1^o}\right)\left(\dfrac{Q_{22}\eta_2}{K_{f_2}} + \dfrac{Q_{21}\eta_1}{K_{f_1}}\right)\right)} \tag{5.45}$$

From these equations the change in saturation with time across the front may be determined. Note the parameters in these equations are constrained by the generalized Onsager's relations

$$BD_2 = CD_1 \tag{5.46}$$

which yields

$$(Q_{22}+Q_{21})\frac{\delta_1}{K_{f_1}} = (Q_{11}+Q_{12})\frac{\delta_2}{K_{f_2}} \tag{5.47}$$

It is important to note that if this condition is not met, then the equations of motion will yield an unphysical result. Since experiments are not currently being performed to measure these parameters, such constraints are important for the construction of numerical studies.

DISPERSION

Figures 5.15 and 5.16 illustrate that saturation contours evolve to constant velocity waves. Also note that low water saturation contours move faster than higher water saturation contours. This causes the front to tend to spread out or diffuse. This behaviour is called *dispersion* and represents the displacing phase, tending to bypass the displaced phase at the pore scale. Here, complex interfacial phenomena are occurring between the fluids at the pore scale. For the case of immiscible displacement, the motion of fluid–fluid interfaces brings capillarity into the description of the dynamics. In general, dispersion is a very useful process since it allows the injected fluid to flow through all parts of the porous medium. As was observed in Chapter 3, dispersion is the dominant flow process when the

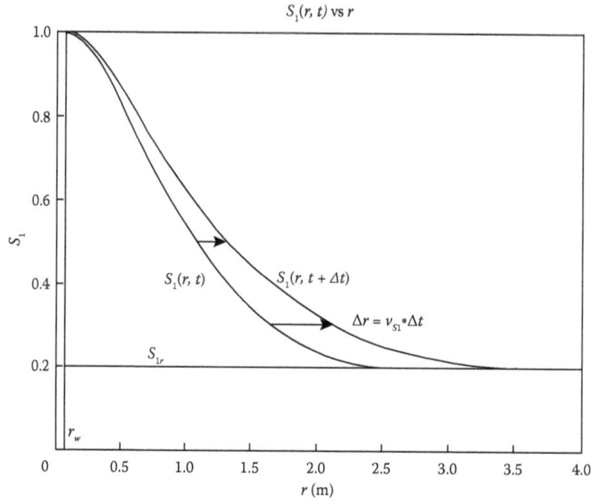

FIGURE 5.15 This graph shows a gradual change in the saturation of water at the front, which causes dispersion to dominate and a broad or diffuse front to form.

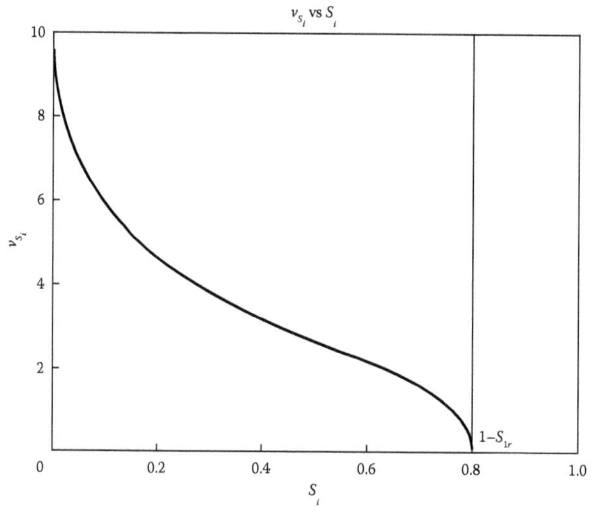

FIGURE 5.16 Note each saturation contour moves with a different constant speed, with the speed increasing as the saturation decreases.

injected fluid is pulsed into the porous medium with a rise time in the pulse that generates a porosity wave.

If there is little connate water, the oil viscosity is quite high or the surface tension is quite large, then the front will tend to be sharp due to a rapid change in the effective viscosity of the saturation contours. This causes viscous fingering to dominate at the front, as described in the section 'Viscous Fingering'.

VISCOUS FINGERING

Dispersion describes the bypassing of one fluid by another at the pore scale. This process causes the amount of oil–water contact at the pore scale to increase because new fluid interfaces are being generated at the pore scale. Thus the amount of oil displaced tends to be many times greater than if viscous fingering occurs. Basically, viscous fingering causes the water to bypass the oil along channels formed by an instability process caused by a sharp mobility change across a frontal region. If viscous fingering occurs, then the size and distance between fingers may be described by doing a stability analysis of the front. The stability analysis then describes the onset of fingering, following which the fingers grow at a constant rate. Sometimes during the onset of fingering, two fingers will start to grow too close to one another or during the growth of a finger it will start to follow two different paths. In those cases one path or finger will cease to grow; these terminated paths are called *dendrites.*

For a sharp front, the boundary condition may be written as follows (Spanos and de la Cruz 1984):

$$p_1 - p_2 = \Lambda q_n \tag{5.48}$$

where p_1, p_2 are the pressures on each side of the front, q_n is the Darcy velocity of the front and Λ is the resistance due to capillary forces and the change in viscosity across the front. Here, Equation 5.48 may be constructed by straightforward physical arguments and may also be derived by integrating across a front described by a region of multiphase flow and taking the limit as the thickness goes to 0 (or as one moves further away so the transition region appears sharper). It is also interesting to note that it is of the same form as Darcy's equation. In the stability analysis presented in Spanos and de la Cruz (1983), only the onset of fingering is presented. As a result the inertial terms do not appear in either the equations of motion or the boundary conditions and the viscous fingers grow exponentially in that analysis (Figure 5.17).

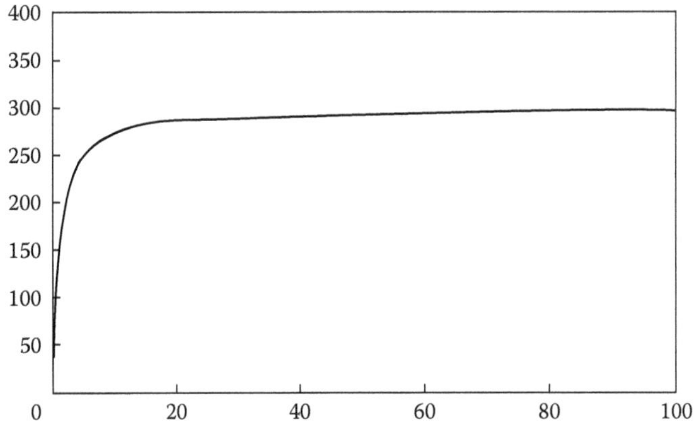

FIGURE 5.17 This graph shows the rate of growth of each wave number.

When the inertial terms are included, the equations of motion take the form of a wave equation and the fingers grow at a constant rate.

Here, the standard stability analysis without the inertial terms is presented first.

Once a sharp front forms, perturbations of the front are of the following form (see Chandrasekhar 1981):

$$\zeta = \varepsilon\xi(t)E(x,y) \tag{5.49}$$

where

$$E(x,y) \equiv \exp(i(k_x x + k_y y)) \tag{5.50}$$

and

$$\xi(t) = \exp(nt) \tag{5.51}$$

The equation of motion on each side of the front is given by Darcy's equation.

$$\frac{\mu}{K}\vec{q} = -\vec{\nabla}(\bar{p} - \rho g z) \tag{5.52}$$

and for incompressible fluids

$$\vec{\nabla}\cdot\vec{q} = 0 \tag{5.53}$$

Here, Equation 5.52 is valid for downward displacements; however, changing the sign of g describes upward displacements and setting $g = 0$ removes gravity from the problem. The unperturbed motion of the front corresponds to a constant filter velocity U of the fluid in the z direction. The front moves with the velocity

$$V = U/\eta_f \tag{5.54}$$

where η_f is the volume fraction of mobile fluid in the frontal region. The position of the front at time t is

$$z = \ell + Vt \tag{5.55}$$

where ℓ is the position of the front along the z-axis at time $t = 0$.

Once the front is perturbed, the actual filter velocity $\vec{q}_{(i)}$ is related to U through the equation

$$\vec{q}_{(i)} = U\hat{z} + \vec{q}\,'_{(i)} \tag{5.56}$$

where $\hat{z}$ is the unit vector in the z direction and $\vec{q}\,'_{(i)}$ is the perturbation velocity. The subscript i refers to the phase under consideration.

Combining Equations 5.52 and 5.56 and assuming a homogeneous porous medium yields

$$\vec{q}'_{(i)} = -\vec{\nabla}\chi_{(i)} \tag{5.57}$$

where the velocity potential $\chi_{(i)}$ determines the pressure through

$$\bar{p}_{(i)} = \frac{\mu_{(i)}}{K_{(i)}}(\chi_{(i)} - Uz) + \rho_{(i)} gz + F_{(i)}(t) \tag{5.58}$$

$F_{(i)}(t)$ being a function of time independent of the perturbation.

Assuming incompressibility

$$\begin{aligned} 0 &= \vec{\nabla}\cdot\vec{q} = \vec{\nabla}\cdot(\vec{q}'_{(i)} - \vec{U}) \\ &= \vec{\nabla}\cdot\vec{q}'_{(i)} \end{aligned} \tag{5.59}$$

one obtains the result that $\chi_{(i)}$ satisfies the Laplace equation

$$\nabla^2\chi_{(i)} = 0 \tag{5.60}$$

From the boundary condition (Equation 6.48), the following equation is obtained

$$n = \frac{1}{\eta}\frac{\left[\dfrac{\mu_{(2)}}{K_{(2)}} - \dfrac{\mu_{(1)}}{K_{(1)}}\right]U + (\rho_{(1)} - \rho_{(2)})g}{\left(\dfrac{\mu_{(2)}}{K_{(2)}} + \dfrac{\mu_{(1)}}{K_{(1)}}\right)\dfrac{1}{k} + C} \tag{5.61}$$

and note that for k much larger than k_1 where

$$k_1 = \frac{\left(\dfrac{\mu_{(2)}}{K_{(2)}} + \dfrac{\mu_{(1)}}{K_{(1)}}\right)}{C} \tag{5.62}$$

n approaches a finite constant

$$n \approx \frac{1}{\eta}\frac{\left[\dfrac{\mu_{(2)}}{K_{(2)}} - \dfrac{\mu_{(1)}}{K_{(1)}}\right]U + (\rho_{(1)} - \rho_{(2)})g}{C} \tag{5.63}$$

Thus it can be seen that the instabilities will initially grow exponentially and the dominant instabilities will have wave numbers slightly larger than k_1. However, this description only describes the onset of fingering. Fingers do not continue to grow exponentially but achieve a constant rate of growth shortly after formation. Thus the inertial terms that have been neglected in the previous analysis must be included in order to describe the growth of viscous fingers.

The equation of motion on each side of the front is now given by

$$\frac{\rho}{\eta}\frac{\partial \vec{q}}{\partial t}+\frac{\mu}{K}\vec{q}=-\vec{\nabla}(\bar{p}-\rho g z) \tag{5.64}$$

Thus the pressure is now given by

$$\bar{p}_{(i)}=\frac{\rho_{(i)}}{\eta}\frac{\partial \chi_{(i)}}{\partial t}+\frac{\mu_{(i)}}{K_{(i)}}(\chi_{(i)}-Uz)+\rho_{(i)}gz+F_{(i)}(t) \tag{5.65}$$

Equation 5.48 is the boundary condition for steady flow and allows for a description of the onset of fingering. Initially, fingers grow exponentially; however, they quickly evolve to steady-state growth. A description of this behaviour requires that inertial terms be included in both the equations of motion on each side of the front and the boundary conditions at the front.

$$p_1-p_2=Cq_n+D\frac{\partial q_n}{\partial t} \tag{5.66}$$

Using Equation 5.66 as the boundary condition and including the inertial terms in the equations of motion, the following relation is obtained

$$\begin{aligned}&\left[\left(\frac{\rho_1}{\eta}+\frac{\rho_2}{\eta}\right)+Dk\right]n^2+\eta\left[\left(\frac{\mu_{(2)}}{K_{(2)}}+\frac{\mu_{(1)}}{K_{(1)}}\right)+Ck\right]n\\&+\left[\frac{\mu_{(2)}}{K_{(2)}}-\frac{\mu_{(1)}}{K_{(1)}}\right]Uk+(\rho_{(1)}-\rho_{(2)})gk=0\end{aligned} \tag{5.67}$$

In the present case two solutions for n are obtained. If the displacement is stable, then both solutions are negative. If the displacement is unstable, then one solution is positive and the other is negative. The sum of the two yields a solution that asymptotes at 0. This means that after a finger initially grows at an exponential rate, it approaches a constant growth rate. This constant rate of growth of the fingers is also required to satisfy material balance (i.e. a constant injection rate must be associated with a constant growth rate for the fingers) (Figure 5.18).

Another interesting result observed after analysing finger growth is that the finger size and growth rate may be determined directly from the n – k relation (say choosing the 90% height of the above graph for finger width and material balance for finger growth rate), avoiding a great deal of work with Fourier analysis and Bessel functions. Figure 5.19 shows a visual representation of saturation versus distance during viscous fingering. Note how much different this curve is from the dispersion curve; these two curves are plotted together in Figure 5.20. In general one of these processes will always dominate the other. As a general rule, fingering will be the dominant process during steady-state displacements, except for anomalous situations with large connate water saturations. As will be

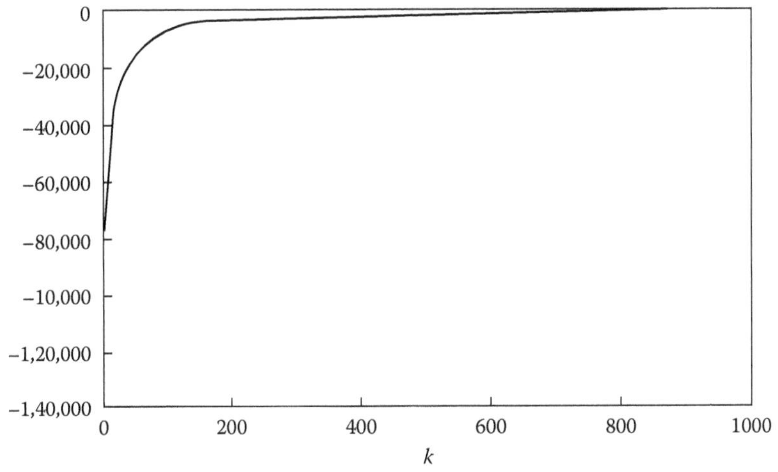

FIGURE 5.18 The second solution for the $n - k$ relation.

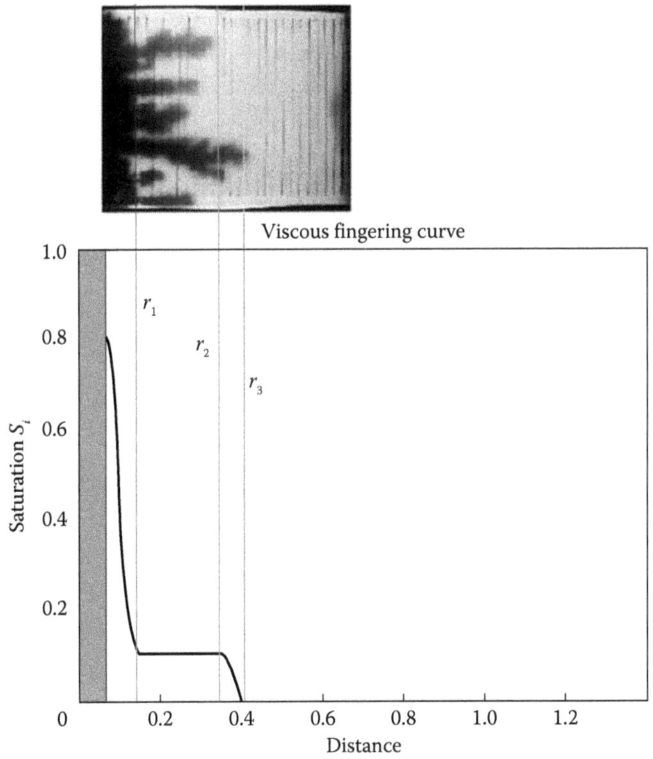

FIGURE 5.19 Typical viscous finger with a graph of the four zones.

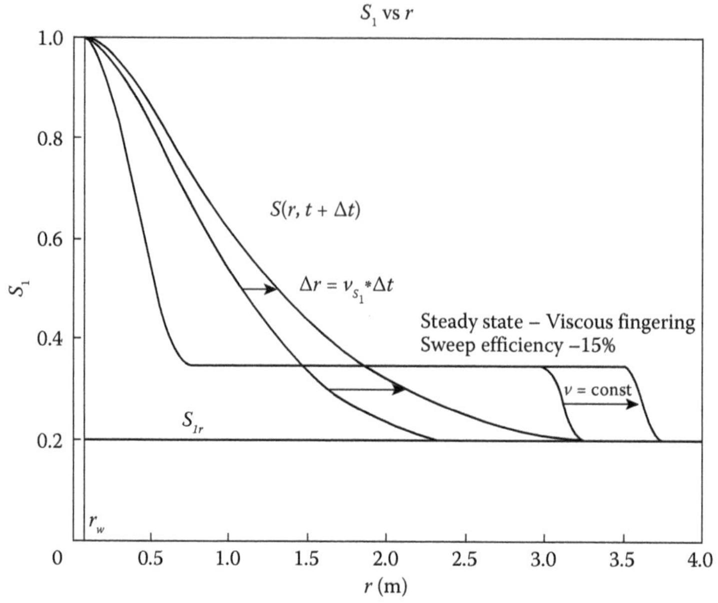

FIGURE 5.20 Shows the improvement in sweep efficiency associated with pulsed injection (dispersion) versus steady-state injection (viscous fingering).

seen in Chapter 6, dispersion will always dominate under dynamic situations associated with pressure pulsing.

For a typical fluid displacement undergoing viscous fingering, we can define four zones:

Zone 0: R 2 [RW, R0]
Zone 1: R 2 [R0, R1]
Zone 2: R 2 [R1, R2]
Zone 3: R 2 [R2, R3]

Zone 0 is the region next to the injection point r = rw where the injected fluid has completely displaced the *in situ* fluid. The saturation profile is modelled as a flat profile with Si = max(Si) = 1 – S_{wc}, where the connate water saturation is S_{wc}.

In cases where no connate water is present, dispersion is not possible. However, the sweep efficiency is still substantially enhanced by pulsing. This is observed in Figure 5.21. In Figure 5.22, the increased fluid production obtained by pulsing is shown. In the case of steady-state flow, no additional oil is produced after water breakthrough; however, in the case of pulsing, the emulsion shown in Figure 5.23 is produced after water breakthrough.

THE STABILITY OF A STEAM–WATER FRONT IN A POROUS MEDIUM

This section is an introduction to phase transitions in porous media. This is a subject that will be considered in more general terms following the discussion of non-equilibrium thermodynamics. Since this is an introduction to this topic, an attempt is made to keep this discussion as simple as possible by assuming a rigid homogeneous porous medium,

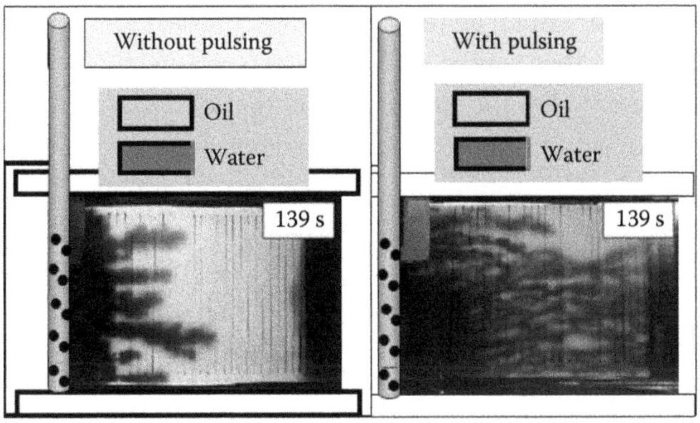

FIGURE 5.21 Experiments in the cell shown in Figure 3.14 where oil is displaced by water.

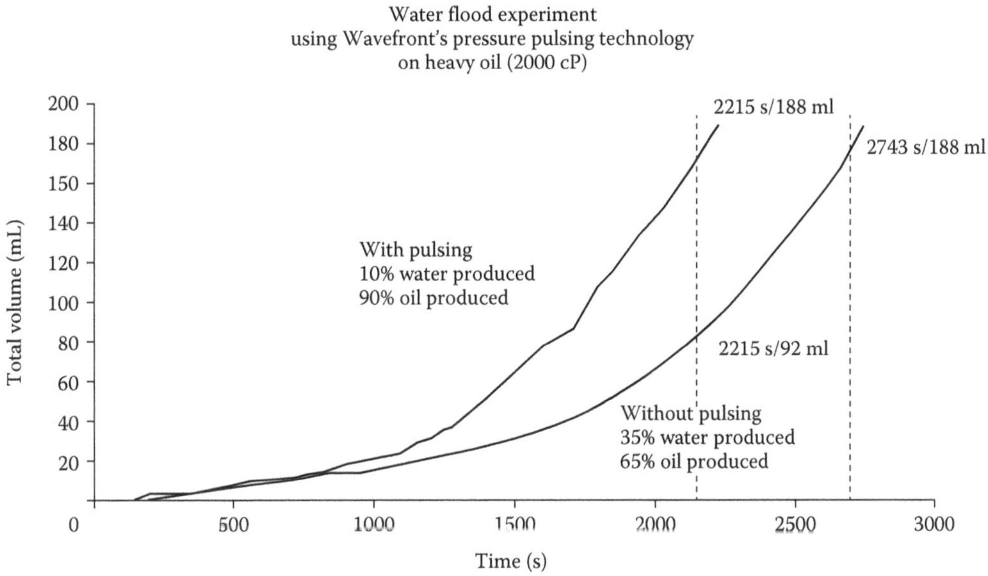

FIGURE 5.22 Shows the fluid production versus time, with and without pulsing (with the same average pressure).

FIGURE 5.23 After breakthrough in the water flood, no more oil is produced. After breakthrough when pulsing, the oil ganglia are broken up into an emulsion that passes through the pore throats and is produced with the water.

and all physical parameters (such as thermal conductivities) are treated as specified constants. Also it is assumed that compressibility effects are negligible and there is no local temperature difference between the each fluid and the solid matrix.

The equations of motion, continuity and heat transfer are given by

$$\frac{\mu_i}{K_i}\vec{q}_i = -\vec{\nabla}p_i + \rho_i\vec{g} \tag{5.68}$$

$$\vec{\nabla}\cdot\vec{q}_i = 0 \tag{5.69}$$

and (Landau and Lifshitz 1975)

$$(\rho c)_i \frac{\partial T_i}{\partial t} + \rho_i c_i \vec{q}_i \cdot \vec{\nabla}T_i - \kappa_i \nabla^2 T_i = 0 \tag{6.70}$$

where the index 1 refers to the steam zone and the index 2 refers to the water zone, so T_i refers to the temperature of the respective zone. κ_i denotes the effective thermal conductivities of the two zones and c_i the heat capacities of the two fluids. The term $(\rho c)_i$ is defined by

$$(\rho c)_i = \eta\rho_i c_i + (1-\eta)\rho_s c_s \tag{5.71}$$

and describes the heat capacity per unit zone volume. If surface tension is included at the boundary between the steam and water zones, then the frontal boundary conditions are given by

$$T_1 = T_2 \tag{5.72}$$

$$p_1 = p_2 + \chi_1 q_{1n} + \chi_2 q_{2n} + \pi(t) \tag{5.73}$$

$$\rho_1(q_{1n} - \eta v_n) = \rho_2(q_{2n} - \eta v_n) \tag{5.74}$$

Here, v_n is the normal component of the velocity of propagation of the steam–water front; χ_1 and χ_2 are required to specify Newton's third law (see Spanos and de la Cruz 1984; de la Cruz et al. 1985). Equation 5.72 is a statement of continuity of temperature across the front; Equation 5.74 expresses material balance assuming that each region is filled with only one fluid. $\pi(t)$ is a given perturbation independent pressure discontinuity and will play no role in the following analysis, so it will be set to 0.

The energy balance at the interface is expressed in the following form (Miller 1975; Eastwood 1992):

$$\rho_1 h_1 q_{1n} - \rho_1\varepsilon_1\eta v_n - \kappa_1\nabla_n T_1 = \rho_2 h_2 q_{2n} - \rho_2\varepsilon_2\eta v_n - \kappa_2\nabla_n T_2 \tag{5.75}$$

Here, h_i and ε_i are the enthalpy and the internal energy per unit mass for phase i. The approximation $h_i \approx \varepsilon_i$ will be made in the following analysis. Here, steam and water must coexist at the front. Hence a phase equilibrium relation between pressure and temperature must be satisfied there (de la Cruz et al. 1985)

$$p_1 = p_{eq}(T_1) - [\rho_1/(\rho_2 - \rho_1)][\gamma_1 q_{1n} + \gamma_2 q_{2n}] \tag{5.76}$$

The unperturbed front is assumed to be a plane surface moving through the porous medium at a constant speed, v_n. This analysis has also been done for a stationary horizontal front of water over steam in a gravitational field (de la Cruz et al. 1985; Eastwood 1992). In the present analysis, gravitational forces are negligible in comparison to mobility forces.

The unperturbed fluid velocities are

$$\vec{U}_i = U_i\vec{n} \tag{5.77}$$

where U_i are constants and $\vec{n}$ is a unit vector in the direction of flow. The equation of the front is

$$z = vt \tag{5.78}$$

The steam and water temperatures take the form

$$T_1 = T_{1\infty} + (T_f - T_{1\infty})\exp[-\psi_1(z - vt)] \tag{5.79}$$

$$T_2 = T_{2\infty} + (T_f - T_{2\infty})\exp[-\psi_2(z - vt)] \tag{5.80}$$

where

$$\psi_i = [(\rho c)_i v - \rho_i c_i U_i]/\kappa_i \tag{5.81}$$

Here, $T_1 = T_2 = T_f$ at the front, $T_1 \to T_{1\infty} = const$ as $z \to -\infty$ and $T_2 \to T_{2\infty} = const$ as $z \to \infty$. The parameter $T_{1\infty}$ controls the slope of the temperature at the front

$$dT_1/d(z - vt) = -\psi_1(T_f - T_{1\infty}) \tag{5.82}$$

which is a negative quantity, meaning that the parameter ψ_1 is negative if $T_{1\infty}$ is to be interpreted as the asymptotic value of the steam temperature.

Of the three velocities U_1, U_2 and v, suppose that one, say the steam flow rate U_1, is given. The water flow rate U_2 and the frontal velocity v are then determined using energy balance and material balance. The boundary condition (Equation 5.74) yields

$$\rho_1(U_1 - \eta v) = \rho_2(U_2 - \eta v) \tag{5.83}$$

Now using energy balance at the boundary, given by Equation 5.75, and substituting Equations 5.80 and 5.83 yields

$$\rho_1(\varepsilon_1 - \varepsilon_2)(U_1 - \eta v) = \kappa_2\psi_2(T_f - T_{2\infty}) \tag{5.84}$$

Integrating the equation of motion (Equation 5.68) yields

$$p_i = (\rho_i g\cos\theta - \mu_i U_i / K_i)(z - vt) + p_o \tag{5.85}$$

where p_o is the pressure at the front $z = vt$.

A first-order stability analysis may now be performed by perturbing the velocity, pressure and temperatures at the front while accounting for phase transitions. The perturbations of the flow rate and pressure will be denoted by u_i and p_i', respectively. Thus the actual flow rates and pressures are

$$q_i = U_i + u_i \tag{5.86}$$

and

$$p_i = p_i^o + p_i' \tag{5.87}$$

where p_i^o is given in Equation 5.85. Substituting these terms into the equations of motion and cancelling the unperturbed parts yields

$$\vec{u}_i = -\vec{\nabla}\varphi_i \tag{5.88}$$

where

$$\varphi_i = (K_i / \mu) p_i' \tag{5.89}$$

Applying the incompressibility condition results in

$$\nabla^2 \varphi_i = 0 \tag{5.90}$$

Now analysing the perturbations in terms of normal modes it is assumed that the perturbations may be decomposed into modes proportional to

$$E(x, y) \equiv \exp[i(k_x x + k_y y)] \tag{5.91}$$

Therefore Equation 5.90 may be written as

$$[(\partial^2 / \partial z^2) - k^2]\varphi_i = 0 \tag{5.92}$$

where

$$k^2 = k_x^2 + k_y^2 \tag{5.93}$$

and thus φ_i are of the form

$$\varphi_1 = f_1(t)\exp[k(z - vt)]E(x, y) \tag{5.94}$$

$$\varphi_2 = f_2(t)\exp[-k(z - vt)]E(x, y) \tag{5.95}$$

The perturbed front is given by

$$z = vt + \zeta(x, y, t) \tag{5.96}$$

where

$$\zeta(x, y, t) = \varepsilon \exp(nt) E(x, y) \tag{5.97}$$

measures the deviation from the flat surface $z = vt$. The initial amplitude, ε, of a perturbation is assumed small and the actual amplitude at time t is $\varepsilon \exp(nt)$. The index n measures the rate of growth ($n > 0$) or decay ($n < 0$). The objective is to find the value of n for each k.

From Equation 5.73

$$p_1^o + p_1' = p_2^o + p_2' + \chi_1 q_{1n} + \chi_2 q_{2n} \tag{5.98}$$

for $z = vt + \zeta$. Substituting for the zeroth order pressures as given by Equation 5.85 and for p_i' from Equation 5.89 yields

$$(\mu_1/K_1 + \chi_1 k) f_1(t) - (\mu_2/K_2 + \chi_2 k) f_2(t) +$$

$$[(\mu_2/K_2)U_2 - (\mu_1/K_1)U_1 - (\rho_2 - \rho_1) g \cos\theta] = 0 \tag{5.99}$$

Material balance as given by Equation 5.74 may be expressed as

$$\rho_1(-\partial\varphi_1/\partial z - \eta\dot{\zeta}) = \rho_2(-\partial\varphi_2/\partial z - \eta\dot{\zeta}) \tag{5.100}$$

which yields

$$\rho_1[-kf_1(t) - \varepsilon\eta n \exp(nt)] = \rho_2[-kf_2(t) - \varepsilon\eta n \exp(nt)] \tag{5.101}$$

Equations 5.99 and 5.101 together show that $f_i(t)$ are proportional to $\exp(nt)$. Hence

$$\varphi_1 = \varepsilon\bar{c}_1 \exp(nt) \exp[k(z - vt)] E(x, y) \tag{5.102}$$

$$\varphi_2 = \varepsilon\bar{c}_2 \exp(nt) \exp[k(z - vt)] E(x, y) \tag{5.103}$$

where $\bar{c}_1$ and $\bar{c}_2$ are determined from

$$(\mu_1/K_1 + \chi_1 k)\bar{c}_1 + (\mu_2/K_2 + \chi_2 k)\bar{c}_2$$

$$+ [(\mu_2/K_2)U_2 - (\mu_1/K_1)U_1 - (\rho_2 - \rho_1) g \cos\theta] = 0 \tag{5.104}$$

and

$$\rho_1 k\bar{c}_1 + \rho_2 k\bar{c}_2 + \eta n(\rho_1 - \rho_2) = 0 \tag{5.105}$$

The temperatures in the steam and water are given by

$$T_i = T_i^{(0)} + T_i' \tag{5.106}$$

where $T_1^{(0)}$, $T_2^{(0)}$ are given by Equations 5.79 and 5.80, which are the unperturbed temperature distributions in the steam and water, respectively. Here, T_1', T_2' are the perturbations in temperature associated with the velocity perturbations at the front. These perturbations in temperature are determined from the equations

$$\frac{\partial T_i'}{\partial t}+a_iU_i\frac{\partial T_i'}{\partial z}-b_i\left(\frac{\partial^2}{\partial z^2}-k^2\right)T_i'=a_i\frac{\partial \varphi_i}{\partial z}\frac{\partial T_i^{(0)}}{\partial z} \tag{5.107}$$

where

$$a_i=\rho_i c_i/(\rho c)_i \tag{5.108}$$

$$b_i=\kappa_i/(\rho c)_i \tag{5.109}$$

Now making the coordinate transformation $z'=z-vt$, Equation 5.107 may be rewritten in co-moving coordinates as

$$\frac{\partial T_1'}{\partial t}+(a_1U_1-v)\frac{\partial T_1'}{\partial z'}-b_1\frac{\partial^2 T_1'}{\partial z'^2}+b_1k^2T_1'=a_1\frac{\partial \varphi_1}{\partial z'}\frac{\partial T_1^{(0)}}{\partial z'} \tag{5.110}$$

and

$$\frac{\partial T_2'}{\partial t}+(a_2U_2-v)\frac{\partial T_2'}{\partial z'}-b_2\frac{\partial^2 T_2'}{\partial z'^2}+b_2k^2T_2'=a_2\frac{\partial \varphi_2}{\partial z'}\frac{\partial T_2^{(0)}}{\partial z'} \tag{5.111}$$

Writing the temperature perturbations in terms of normal modes yields solutions of the form

$$T_i'(x,y,z',t)=T_i'(z')\exp(nt)E(x,y) \tag{5.112}$$

Now using Equation 5.79 and 5.80, it is straightforward to write

$$T_1'=\varepsilon[\bar{\lambda}_1\exp(D_1z')+\bar{c}_1\beta_1\exp[(k-\varphi_1)z']\exp(nt)E(x,y) \tag{5.113}$$

$$T_2'=\varepsilon[\bar{\lambda}_2\exp(D_2z')+\bar{c}_2\beta_2\exp[(k-\varphi_2)z']\exp(nt)E(x,y) \tag{5.114}$$

for some constants $\bar{\lambda}_1$ and $\bar{\lambda}_2$. Here

$$D_1=-0.5\{\varphi_1-[\varphi_1^2+4k^2+4(\rho c)_1n/\kappa_1]^{0.5}\} \tag{5.115}$$

$$D_2=-0.5\{\varphi_2-[\varphi_2^2+4k^2+4(\rho c)_2n/\kappa_2]^{0.5}\} \tag{5.116}$$

$$\beta_1=\rho_1c_1\varphi_1(T_{1\infty}-T_f)k/[(\rho c)_1n+\kappa_1\varphi_1k] \tag{5.117}$$

$$\beta_2=\rho_2c_2\varphi_2(T_f-T_{2\infty})k/[(\rho c)_2n+\kappa_2\varphi_2k] \tag{5.118}$$

Continuity of temperature at the front leads to the condition

$$-\bar{\lambda}_2-\bar{c}_2\beta_2+\varphi_2(T_f-T_{2\infty})+\bar{\lambda}_1+\bar{c}_1\beta_1+\varphi_1(T_{1\infty}-T_f)=0 \tag{5.119}$$

and energy balance yields

$$[\rho_1\varepsilon_1 k+\kappa_1\beta_1(k-\varphi_1)]\overline{c_1}+[\rho_2\varepsilon_2 k+\kappa_2\beta_2(k+\varphi_2)]\overline{c_2}+\kappa_1 D_1\lambda_1-\kappa_2 D_2\lambda_2$$
$$+\eta n(\rho_1\varepsilon_1-\rho_2\varepsilon_2)-\kappa_1\varphi_1^2(T_{1\infty}-T_f)-\kappa_2\varphi_2^2(T_f-T_{2\infty})=0 \quad (5.120)$$

Now note that four equations (Equations 5.104, 5.105, 5.119 and 5.120) have been constructed that describe velocity and temperature perturbations at the front. In addition to these equations, the Clapeyron equation

$$p_1=p_{eq}(T_1)-[\rho_1/(\rho_2-\rho_1)][\gamma_1 q_{1n}+\gamma_2 q_{2n}] \quad (5.121)$$

is required to describe the phase transitions that occur as a result of these perturbations. This equation may be written to first order as

$$p_1^0+p_1'=p_{eq}(T_f)+\left(\frac{\partial p_{eq}}{\partial T}\right)T'-[\rho_1/(\rho_2-\rho_1)][\gamma_1 q_{1n}+\gamma_2 q_{2n}] \quad (5.122)$$

where p_{eq} and $\frac{\partial p_{eq}}{\partial T}$ are evaluated at T_f. Thus

$$p_0+[\rho_1 g\cos\theta-(\mu_1/K_1)U_1+(\mu_1/K_1)\overline{c_1}]\zeta=p_{eq}(T_f)+\left(\frac{\partial p_{eq}}{\partial T}\right)[\overline{\lambda}_1+\beta_1\overline{c_1}]\zeta$$
$$-[\rho_1/(\rho_2-\rho_1)][k\gamma_1\overline{c_1}+k\gamma_2\overline{c_2}]\zeta-[\rho_1/(\rho_2-\rho_1)][\gamma_1 U_1+\gamma_2 U_2] \quad (5.123)$$
$$+\left(\frac{\partial p_{eq}}{\partial T}\right)\varphi_1(T_f-T_{1\infty})$$

Here the zeroth order terms yield

$$p_0=p_{eq}(T_f)-[\rho_1/(\rho_2-\rho_1)][\gamma_1 U_1+\gamma_2 U_2] \quad (5.124)$$

Thus the perturbation terms yield the equation

$$\{(\mu_1/K_1)-\left(\frac{\partial p_{eq}}{\partial T}\right)\beta_1-[\rho_1/(\rho_2-\rho_1)]k\gamma_1\}\overline{c_1}+[\rho_1/(\rho_2-\rho_1)]k\gamma_2\overline{c_2}$$
$$-\left(\frac{\partial p_{eq}}{\partial T}\right)\overline{\lambda}_1+[\rho_1 g\cos\theta-(\mu_1/K_1)U_1]+\left(\frac{\partial p_{eq}}{\partial T}\right)\varphi_1(T_f-T_{1\infty}) \quad (5.125)$$

Equations 5.104, 5.105, 5.119, 5.120 and 5.125 therefore yield five equations of the form

$$a_{i1}\overline{c_1}+a_{i2}\overline{c_2}+a_{i3}\overline{\lambda}_1+a_{i4}\overline{\lambda}_2+a_{i5}=0 \quad (5.126)$$

$i = 1$ to 5. The condition for non-trivial solutions to this system of equations is the vanishing of the 5×5 determinant

$$\det\left|a_{ij}\right| = 0 \tag{5.127}$$

which yields a relationship for n as a function of k.

Here

$$a_{11} = \mu_1/K_1 + \chi_1 k \tag{5.128}$$

$$a_{12} = \mu_2/K_2 + \chi_2 k \tag{5.129}$$

$$a_{13} = 0 \tag{5.130}$$

$$a_{14} = 0 \tag{5.131}$$

$$a_{15} = (\mu_2/K_2)U_2 - (\mu_1/K_1)U_1 - (\rho_2 - \rho_1)g\cos\theta \tag{5.132}$$

$$a_{21} = \rho_1 k \tag{5.133}$$

$$a_{22} = \rho_2 k \tag{5.134}$$

$$a_{23} = 0 \tag{5.135}$$

$$a_{24} = 0 \tag{5.136}$$

$$a_{25} = \eta n(\rho_1 - \rho_2) \tag{5.137}$$

$$a_{31} = \beta_1 \tag{5.138}$$

$$a_{32} = \beta_2 \tag{5.139}$$

$$a_{33} = 1 \tag{5.140}$$

$$a_{34} = -1 \tag{5.141}$$

$$a_{35} = \varphi_1(T_{1\infty} - T_f) + \varphi_2(T_f - T_{2\infty}) \tag{5.142}$$

$$a_{41} = \rho_1\varepsilon_1 k + \kappa_1\beta_1(k - \varphi_1) \tag{5.143}$$

$$a_{42} = \rho_2\varepsilon_2 k + \kappa_2\beta_2(k + \varphi_2) \tag{5.144}$$

$$a_{43} = \kappa_1 D_1 \tag{5.145}$$

$$a_{44} = -\kappa_2 D_2 \tag{5.146}$$

$$a_{45} = \eta n(\rho_1\varepsilon_1 - \rho_2\varepsilon_2) - \kappa_1\varphi_1^2(T_{1\infty} - T_f) - \kappa_2\varphi_2^2(T_f - T_{2\infty}) \tag{5.147}$$

$$a_{51} = (\mu_1/K_1) - \left(\frac{\partial p_{eq}}{\partial T}\right)\beta_1 - [\rho_1/(\rho_2 - \rho_1)]k\gamma_1 \tag{5.148}$$

$$a_{52} = [\rho_1/(\rho_2 - \rho_1)]k\gamma_2 \tag{5.149}$$

$$a_{53} = -\left(\frac{\partial p_{eq}}{\partial T}\right) \tag{5.150}$$

$$a_{54} = 0 \tag{5.151}$$

$$a_{55} = [\rho_1 g\cos\theta - (\mu_1/K_1)U_1] + \left(\frac{\partial p_{eq}}{\partial T}\right)\varphi_1(T_f - T_{1\infty}) \tag{5.152}$$

Now note that this solution assumes a displacement process is occurring. In such cases as geothermal reservoirs, water overlying steam may occur. This case has been considered by Schubert and Strauss (1980), de la Cruz et al (1985) and Eastwood (1991), where rather than specifying the temperatures at infinity the temperature gradients at the front must be specified. Here

$$\kappa_1\left(\frac{dT_1}{dz}\right)_0 = \kappa_2\left(\frac{dT_2}{dz}\right)_0 \tag{5.153}$$

In order to use the results presented here, in this limiting case, the following limiting procedure must be used. The slopes of the temperature, at the front, in the steam and water regions are given by $-(T_f - T_{1\infty})\varphi_1$ and $-(T_f - T_{2\infty})\varphi_2$, respectively. Since φ_1 and φ_2 are proportional to v, the two temperature factors must diverge as v^{-1} when $v \to 0$; otherwise the temperature would become uniform throughout. So a new matrix b_{ij} is obtained from the matrix a_{ij} by rewriting $T_f - T_{1\infty}$ and $T_f - T_{2\infty}$ as *const*(1)/*v* and *const*(2)/*v*, respectively, and then taking the limit $v \to 0$. The two constants *const*(1)/*v* and *const*(2)/*v* are related to the slopes of the condensation front according to Equation 5.128 and only one may be specified. The resulting matrix elements for b_{ij} that differ from a_{ij} are

$$b_{31} = (\kappa_2/\kappa_1)[\rho_1 c_1/(\rho c)_1]\left(\frac{\partial T_2}{\partial z}\right)_0\left(\frac{k}{n}\right) \tag{5.154}$$

$$b_{32} = [\rho_2 c_2/(\rho c)_2]\left(\frac{\partial T_2}{\partial z}\right)_0\left(\frac{k}{n}\right) \tag{5.155}$$

$$a_{35} = [(\kappa_2/\kappa_1) - 1]\left(\frac{\partial T_2}{\partial z}\right)_0 \tag{5.156}$$

$$b_{41} = \rho_1 \varepsilon_1 k + \kappa_2 [\rho_1 c_1 / (\rho c)_1] \left(\frac{\partial T_2}{\partial z} \right)_0 \left(\frac{k^2}{n} \right) \tag{5.157}$$

$$b_{42} = \rho_2 \varepsilon_2 k + \kappa_2 [\rho_2 c_2 / (\rho c)_2] \left(\frac{\partial T_2}{\partial z} \right)_0 \left(\frac{k^2}{n} \right) \tag{5.158}$$

$$b_{43} = \kappa_1 [k^2 + (\rho c)_1 n / \kappa_1]^{0.5} \tag{5.159}$$

$$b_{44} = \kappa_2 [k^2 + (\rho c)_2 n / \kappa_2]^{0.5} \tag{5.160}$$

$$b_{45} = \eta n (\rho_1 \varepsilon_1 - \rho_2 \varepsilon_2) \tag{5.161}$$

$$b_{51} = (\mu_1 / K_1) - \left(\frac{\partial p_{eq}}{\partial T} \right) (\kappa_2 / \kappa_1) [\rho_1 c_1 / (\rho c)_1] \left(\frac{\partial T_2}{\partial z} \right)_0 \left(\frac{k}{n} \right) - [\rho_1 / (\rho_2 - \rho_1)] k \gamma_1 \tag{5.162}$$

$$b_{55} = \rho_1 g \cos\theta + \left(\frac{\partial p_{eq}}{\partial T} \right) (\kappa_2 / \kappa_1) \left(\frac{\partial T_2}{\partial z} \right)_0 \tag{5.163}$$

Using this theory, it is straightforward to determine the behaviour of a steam–water front in a porous medium. Here, the mechanical stability of the front is determined not only by the mobility ratio, as was done previously for immiscible displacement, but now phase transitions must also be accounted for. Numerical representations of the general solution equation (Equation 5.127) make it possible, through the use of computer graphics, to observe the effect of changing various physical parameters. Figures 5.24 and 5.25 show the effect of increasing the steam velocity on the (n/k) versus k relationship when $T_{1\infty}$ - T_f. For the values of the parameters used, the (n/k) versus k relationship shows that the various modes become less stable as the velocity is increased. The heat capacity of the porous medium is now increased and the same increase in steam velocity on the stability of the condensation front is illustrated. In this case, Figures 5.26 and 5.27 illustrate that the front becomes more stable as the steam velocity increases. This occurs because the loss in heat to the medium causes the phase transition to dominate. Figures 5.28 and 5.29 illustrate the time evolution of an isolated perturbation whose normal modes are described by the (n/k) versus k relationship given in Figure 5.24. Similarly Figures 5.29 and 5.30 illustrate the time evolution of the identical initial perturbation when the steam velocity is increased. The time evolution of the same perturbation is illustrated in Figure 5.31 when the heat capacity is increased and the steam velocity is the same as in Figure 5.29. When the steam velocity is increased to the same as in Figure 5.31, the front becomes stable; Figure 5.32 illustrates how the initial perturbation is damping out.

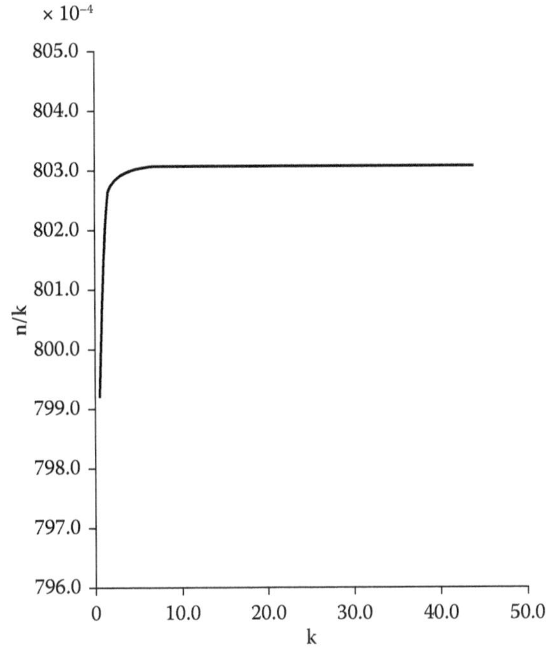

FIGURE 5.24 (n/k) versus k relationship. $U_1 = 1.5 \times 10^{-5}$ m/s; $c_m = 260$ J/kg°C; $c_1 = 1960$ J/kg°C; $c_2 = 4200$ J/kg°C; $\varepsilon_1 = 155 \times 10^5$ J/kg; $\varepsilon_2 = 4.45 \times 10^4$ J/kg; $\eta = 0.02$; $K_1 = K_2 = 10^{-11}$ m^2; $\kappa_1 = \kappa_2 = 5W$/m°C; $\mu_1 = 1.6 \times 10^{-5}$ N·s/m^2; $\mu_2 = 8.1 \times 10^{-4}$ N·s/m^2; $\rho_1 = 7.9$ kg/m^3; $\rho_2 = 996$ kg/m^3; $\rho_m = 5 \times 10^3$ kg/m^3; g cos $\theta = 0.1$ m/s^2; $dP_{eq}/dT = 3.1 \times 10^4$ N/m^2°C; $T_{1f} = 100$°C; $T_2 = 30$°C; $T_s = 101.3$°C; note that T_s is the steam temperature 1 m from the front.

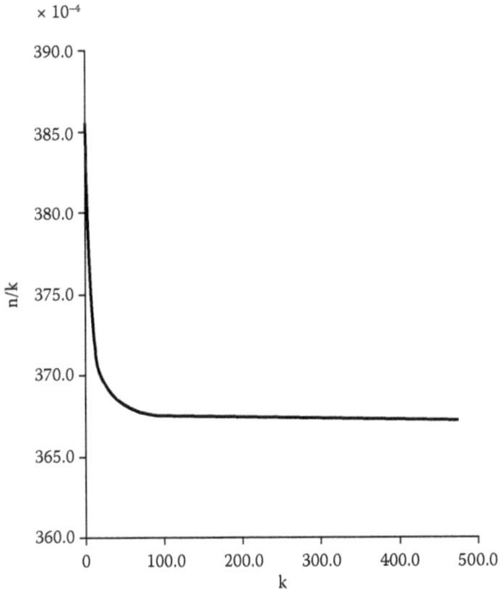

FIGURE 5.25 (n/k) versus k relationship. $U_1 = 1.5 \times 10^{-4}$ m/s; all other parameters are the same as Figure 5.21.

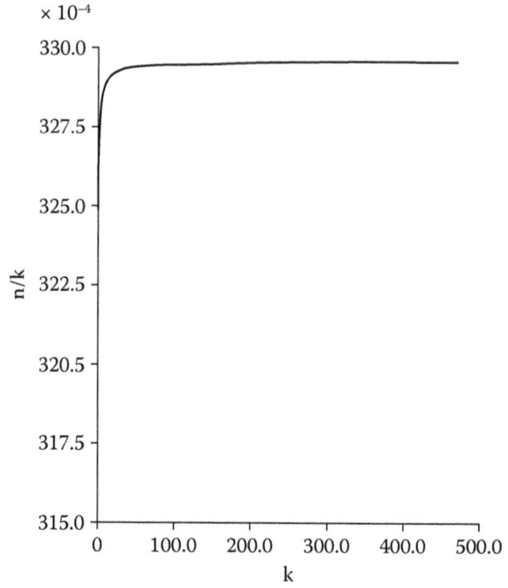

FIGURE 5.26 (n/k) versus k relationship for $c_m = 860$ J/kg°C; all other parameters are the same as Figure 5.21.

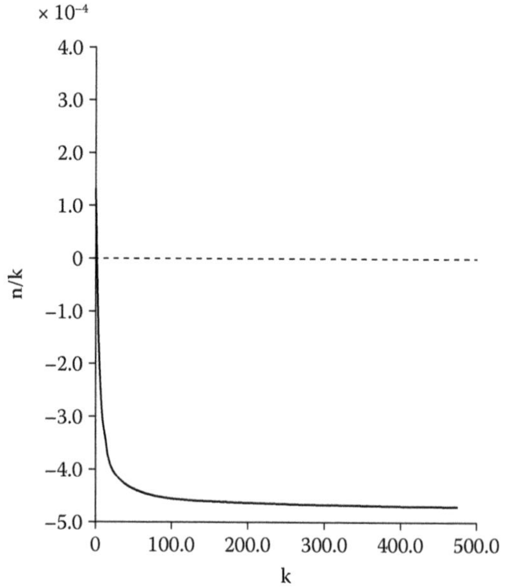

FIGURE 5.27 (n/k) versus k relationship for $U^1 = 1.5 \times 10^{-4}$ m/s; all other parameters are the same as Figure 5.23.

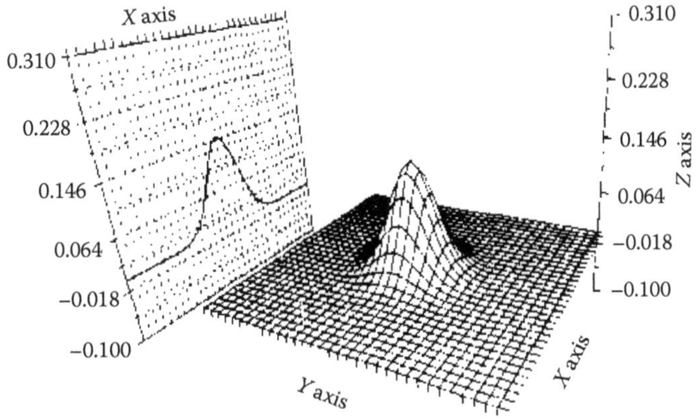

FIGURE 5.28 Steam–water front (time $t = 0$ s). A bell perturbation of the condensation front.

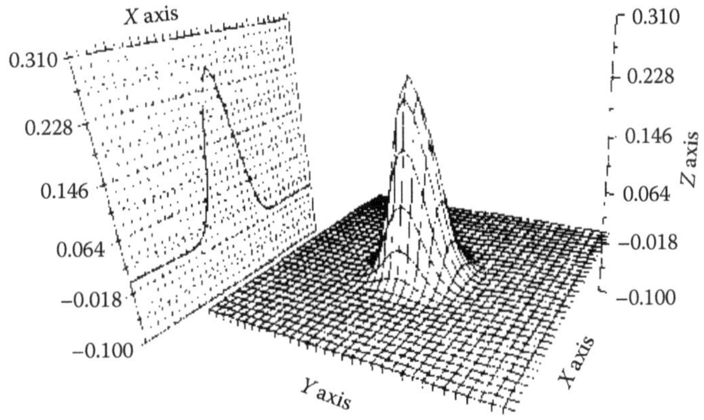

FIGURE 5.29 Steam–water front. The shape of the perturbation given in Figure 6.24 after 900 s when the (n/k) versus k relation given in Figure 5.24 is used to describe the time evolution of the various modes.

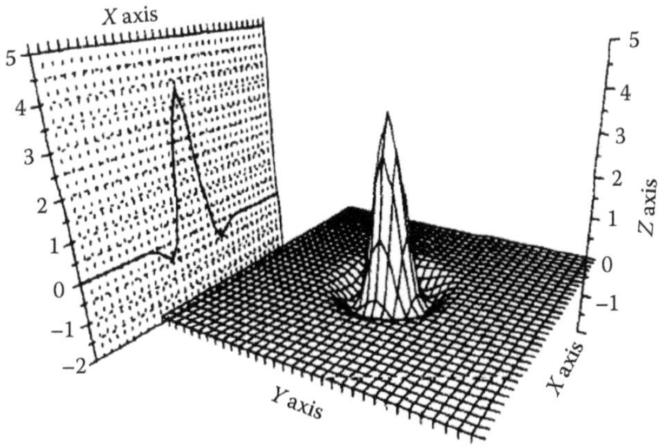

FIGURE 5.30 Steam–water front. The shape of the initial perturbation given in Figure 5.28 after 900 s when the (n/k) versus k relation given in Figure 5.25 is used to describe the time evolution of the various modes.

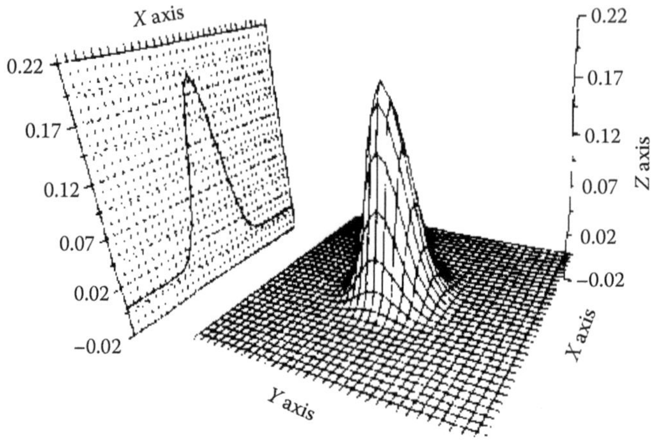

FIGURE 5.31 Steam–water front. The shape of the initial perturbation given in Figure 5.28 after 900 s when the (n/k) versus k relation given in Figure 5.26 is used to describe the time evolution of the various modes.

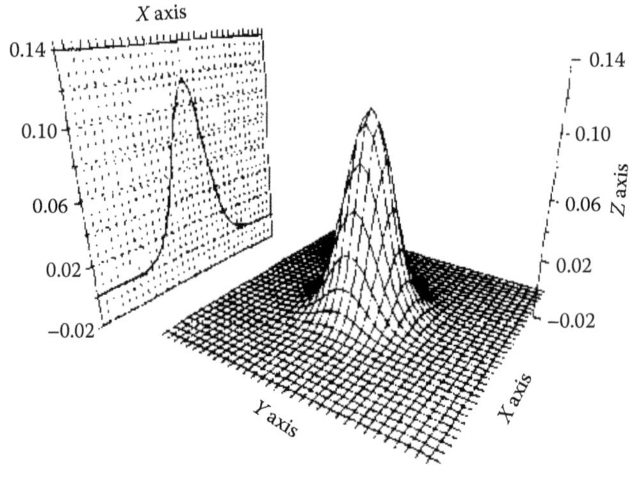

FIGURE 5.32 Steam–water front. The shape of the initial perturbation given in Figure 5.28 after 900 s when the (n/k) versus k relation given in Figure 5.27 is used to describe the time evolution of the various modes.

SUMMARY

Equations describing the flow of two immiscible fluid phases were constructed. It was shown that if the displacing fluid has a lower viscosity than the displaced fluid, then one of two instability processes will dominate at the front separating these two fluids. In the case of steady-state flow, viscous fingering will generally dominate. This occurs because the front becomes very sharp at the positions where fingering is initiated. As the channels form, their size may be calculated using a stability analysis. Once formed, the channelling then proceeds at a constant rate resulting from the volume of fluid being injected into the medium. Individual fingers may follow different directions as they grow, with one of the channels continuing to grow and the other stopping and forming a dendrite. In cases where a high concentration of water (for example) exists in a medium containing oil (for example). The water may bypass the oil by flowing through the exiting water and moving the oil through dispersional flow. In cases where the water is pulsed into the medium (at the rates described), dispersion will always dominate if connate water is present, causing increased sweep efficiency. If connate water is not present, the water will still attempt to disperse through the oil in the form of fine fingers and also increase oil production. In the case of dispersional flow, it is observed that the saturation contours propagate as waves, with lower saturation water waves moving faster than higher saturation waves. In the case of steam displacing water in a porous medium, it is observed that phase transitions from steam to water may stabilize the front. This may be seen in two ways. If the displacement rate is increased or the heat capacity of the medium is increased, the front will stabilize. Another observation is that if a region of water is placed over steam, gravitational instability will cause the water to channel in the steam. However, if the permeability is reduced, then the front will stabilize as the rate of growth of the instabilities is reduced until phase transition overcomes the gravitational instability.

REFERENCES

Anderson, T.B., and Jackson, R., 1967, Fluid mechanical description of fluidized beds equations of motion, *Ind. Eng. Chem. Fundam.*, **6**, 527–539.

Barenblatt, G.I., Entov, V.M., and Ryzhik, V.M., 1990, *Theory of Fluid Flows Through Natural Rocks, Theory and Applications of Transport in Porous Media*, Kluwer Academic, Dordrecht.

Bear, J., and Bachmat, Y., 1990, *Introduction to Modeling of Transport Phenomena in Porous Media, Applications of Transport in Porous Media*, Kluwer Academic, Dordrecht.

Bentsen, R.G., 1994, Effect of hydrodynamic forces on the pressure-difference equation, *Trans. Porous Media*, **17**, 121–135.

Chandrasekhar, S., 1981, *Hydrodynamic and Hydromagnetic Stability*, Dover, New York.

de Gennes, P.G., 1983, Theory of slow biphasic flows in porous media, *Physico-Chem. Hydrodyn.*, **4**, 175–185.

de la Cruz, V., and Spanos, T.J.T., 1983, Mobilization of Oil Ganglia, *AIChE J.*, **29**(7), 854–858.

de la Cruz, V., Sahay, P.N., and Spanos, T.J.T., 1993, Thermodynamics of Porous Media, *Proceedings of the Royal Society of London (rapid communication)*, **443**(1917), 247–255.

de la Cruz, V., Spanos, T.J.T., and Sharma, R.C., 1985, The stability of a steam-water front in a porous medium, *Can. J. Chem. Eng.*, **63**, 735–746.

de la Cruz, V., Spanos, T.J.T., and Yang, D.S., 1995, Macroscopic capillary pressure, *Trans. Porous Media*, **19**, 67–77.

Dullien, F.A.L., 1992, *Porous Media Fluid Transport and Pore Structure*, Academic Press, San Diego, CA.

Eastwood, J.E., 1991, Thermomechanics of porous media, PhD dissertation, University of Alberta.

Eastwood, J.E., 1992, Thermomechanics of Porous Media, Ph.D. Dissertation in Physics, University of Alberta.

Eastwood, J.E., and Spanos, T.J.T., 1991, Steady-state countercurrent flow in one dimension, *Trans. Porous Media*, **6**, 173–182.

Hickey, C.J., 1994, Mechanics of porous media, PhD dissertation, University of Alberta.

Hickey, C.J., Spanos, T.J.T., and de la Cruz, V., 1995, Deformation parameters of permeable media, *Geophys. J. Int.*, **121**, 359–370.

Krueger, D.A., 1982a, Stability of piston-like displacements of water by steam and nitrogen in porous media. *Soc. Pet. Eng. J.*, **22,** 625–634.

Krueger, D.A., 1982b, *Stability of Steam Plus Nitrogen Displacements in a Porous Medium*, Contract 74–6746 Sandia Laboratories, Albuquerque, NM.

Landau, L.D., and Lifshitz E.M., 1975, *Fluid Mechanics*, Pergamon, Toronto.

Lenormand, R., Zarcone, C., and Sarr, A., 1983, Mechanisms of the displacement of one fluid by another in a network of capillary ducts, *J. Fluid Mech.*, **135**, 337–353.

Miller, C.A., 1975, Stability of moving surfaces in fluid systems with heat and mass transport, III. Stability of displacement fronts in porous media, *AICHE J.*, **21**, 474.

Muskat, M., 1946. *The Flow of Homogeneous Fluids Through Porous Media*, J.W. Edwards, Ann Arbor, MI.

Newman, S.P., 1977, Theoretical derivation of Darcy's Law, *Acta. Mech.*, **25**, 153–170.

Richardson, J.G., Kerver, J.K., Hafford, J.A., and Osoba, J.S., 1952, Laboratory determination of relative permeability, *Trans AIME*, **195**, 187.

Scheidegger, A.E., 1974, *The Physics of Flow Through Porous Media*, University of Toronto Press.

Schubert, G., and Strauss, J.M., 1980, Gravitational stability of water over steam in a vapour dominated geothermal system, *J. Geophys. Res.*, **85**, 6505–6512.

Slattery, J.C., 1967, Flow of viscoelastic fluids through porous media, *AIChE J.*, **13**, 1066–1071.

Slattery, J.C., 1969. Single phase flow through porous media, *J. Am. Inst. Chem. Eng.*, **15**, 866–872.

Spanos, T.J.T., 2002, *The Thermophysics of Porous Media*, Monographs and Surveys in Pure and Applied Mathematics, 126, Chapman and Hall, Boca Raton, FL.

Spanos, T.J.T., and de la Cruz, V., 1984, Some stability problems during immiscible displacement in a porous medium, *AOSTRA J. Res.*, **V1**, 63–80.

Spanos, T.J.T., de la Cruz, V., Hube, J., and Sharma, R.C., 1986, An analysis of Buckley–Leverett theory, *J. Can. Pet. Tech.*, 71–75.

Udey, N., 2009, Dispersion waves of two fluids in a porous medium, *Trans. Porous Media*, **79**, 107–115.

Udey, S.N., and Spanos, T.J.T., 1993, The equations of miscible flow with negligible molecular diffusion, *Trans. Porous Media*, **10**, 1–41.

Whitaker, S., 1967, Diffusion and dispersion in porous media, *AIChE* J., 13, 420–427.

Whitaker, S., 1969, Advances in the theory of fluid motion in porous media, *Ind. Eng. Chem.*, **61**(12), 14–28.

CHAPTER 6

Porosity–Pressure Waves and Dispersion

OBJECTIVE OF THIS CHAPTER

The primary objective of this chapter is to present very simple descriptions of porosity–pressure waves and saturation waves. This includes an introductory description of the coupling between these waves. However, this work exposes a weakness in the theory presented in the earlier chapters. Although this theory based on volume averaging is on firm physical grounds when describing two phases mixed at the macroscale (pore scale), it is observed that it does not generalize to fluids mixed at multiple scales or more than two phases. So this discussion will proceed with the observation that saturation waves are strongly coupled to porosity waves and how the limitation mentioned above is overcome will be presented in subsequent chapters.

The first descriptions of porosity–pressure waves were developed in 1993. This observation resulted from the construction of the non-linear field theory discussed in the previous chapters, which describes multiphase fluid flow and seismic wave propagation in porous media. Following the development of this theory it was noticed that the new dynamic parameter porosity predicted additional processes; the first process to be studied was porosity–pressure diffusion in porous media (Geilikman et al. 1993), which described many commonly observed phenomena such as the diffusional propagation of microseismic activity following the loading of reservoirs (Talwani and Acree 1985; Talwani 2000). Immediately following this construction, it was noticed that including the inertial terms in the equations yielded an additional wave process propagating close to the incompressible limit of fluid motions (about 100 m/s for water in a silica matrix). This wave moved fluid through the porous medium by coupling the elastic motions of the elastic solid to the (incompressible) flow of the fluid, similar to the way blood flows through veins. Furthermore, this wave may extract mechanical energy from the stress field of the matrix. If the energy extracted is as large as the loss to viscous dissipation, then a soliton is predicted. If the mechanical energy extracted exceeds the loss due to viscous dissipation at some positions, then rock bursts or earthquakes can be generated. A similar saturation–pressure wave propagates in multifluid flow processes in porous media.

These observations attracted the interest of researchers at the University of Waterloo in 1997 (B. Davidson and M. Dusseault), who saw practical applications for these results. This started an experimental program that quantified these predictions experimentally. One of the first observations was that for unconsolidated media, stress was required and this stress had to be applied by rigid material such as steel. Substances such as rubber or neoprene would absorb the energy of the wave and it would not propagate. Some examples of these experiments are given in Chapter 3. At this point a new company was registered (PE-TECH, now publically traded as Wavefront Technology Solutions); a number of patents were submitted and subsequently accepted (Davidson et al. 2001, 2002a, 2002b, 2005, 2011) then turned over to the company and applied to environmental remediation (Grey 2001), enhanced oil recovery (Groenenboom et al. 2003) and workovers (here fines, larger hydrocarbon molecules, etc., are trapped around a well and lowers the permeability—pulsing then re-establishes the original permeability) (Dusseault et al. 2001). At present the Company has 40 patents and 11 others under examination. Another important experimental observation was the coupling between porosity waves and dispersion. These coupled processes are observed to suppress viscous fingering, enhance dispersion and thus enhance sweep efficiency both in the lab and in reservoirs. In fact, even in cases of highly viscous fluids (e.g. heavy oil and creosol), pulsing of a surfactant was observed to get up to 100% recovery.

In this chapter, very simplistic approximations of this theory are investigated under the assumption that saturation waves are strongly coupled to porosity–pressure waves. This assumption is based on experimental observations that clearly illustrate that saturation waves are completely dominant over viscous fingering in the presence of porosity–pressure waves. Thus this chapter explores a very simple approximation that can be applied to practical applications of this theory subject to specific assumptions. As well, all injection processes studied have an adverse mobility ratio. As is observed in the lab, steady-state injection will be assumed to yield viscous fingering and pulsed injection processes are observed to yield dispersion. One of the most striking examples of this assumption is the following simple experiment in a visual cell. Take a highly viscous fluid of about 10,000 cp (say, a heavy oil, creosote or a nano-iron in corn oil emulsion). The Plexiglas cell is wet-packed using a 30 mesh Ottawa silica sand in water and a vibrator to about 0.30 porosity. The cell is then filled to about halfway, with the highly viscous fluid displacing the water. If the highly viscous fluid is now displaced by water at a constant pressure, then the water channels through the highly viscous fluid, leaving the slug of highly viscous fluid in place. If the highly viscous fluid is now displaced by water with the pulser shown in Chapter 3, then the slug of fluid spreads across the other half of the cell due to the dispersional flow of the water (see Figures 6.1 through 6.3). Now consider similar experiments where the displacement of heavy oil fills the entire porous medium. In the case of displacement by water with about a 1 meter head, 8% to 10% of the heavy oil is displaced. In the case of pulsed water injection with the same average pressure, about two to three times as much heavy oil is displaced.

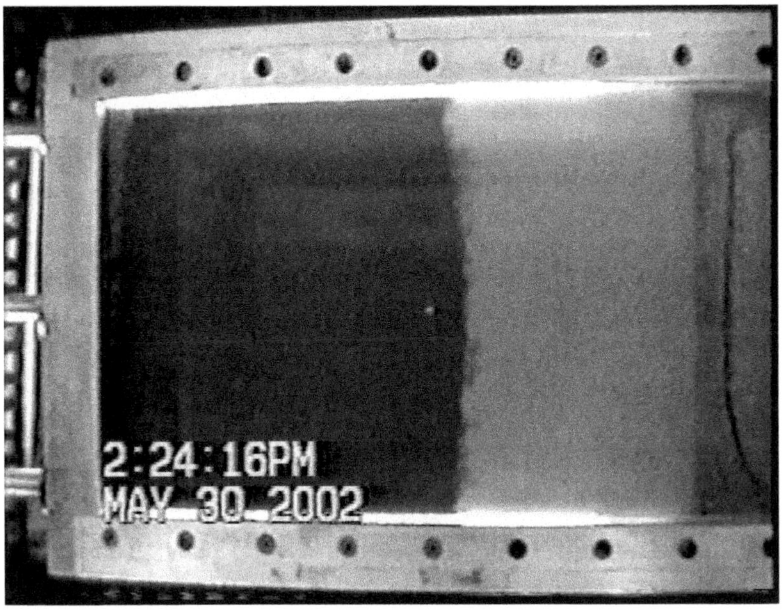

FIGURE 6.1 A 10,000 cp nano-iron in corn oil emulsion prior to water displacement.

FIGURE 6.2 Displacement of nano-iron emulsion by pulsing water. Note that without pulsing the displacing water just channels through the emulsion.

FIGURE 6.3 Pulsing the injected water causes the nano-iron emulsion to spread over the entire porous medium.

Next, consider a similar experiment where creosote is injected into the entire porous medium; then 1/2 pore volume of surfactant is injected by pulsing and left to soak for 17 hours. An additional 4.5 pore volumes of surfactant is then pulsed through the column at 8 psi peak pressure and 2.7 psi average pressure for a total of 5 pore volumes of surfactant throughput. This is followed by 4 pore volumes of water pulsing (8 psi peak, 1.8 psi average). In these experiments 100% of the creosote is recovered from the test sample. When a similar procedure was followed without pulsing, 48% recovery was obtained.

The purpose of these descriptions is to encourage readers to do experiments. There are a couple of warnings, though: do not have any rubber in contact with the porous medium (Tygon tubing connecting to the cell is fine) and do not use a standard syringe pump to pulse. The use of solenoids, magnets, etc., can yield a sudden increase in pressure, which may be followed by a slow reduction in pressure of about 10 × as long. These have been the most effective pulsers both in the lab and the field.

POROSITY–PRESSURE WAVES

Equations 3.97 and 3.98 may be rewritten as a wave equation for porosity coupled to pressure changes and a wave equation for pressure coupled to porosity changes

$$\frac{\partial^2 \eta}{\partial t^2} + C_\eta \frac{\partial \eta}{\partial t} + C_\gamma \nabla^2 \eta = C_p \frac{\partial p_f}{\partial t} - C_b \nabla^2 \frac{\partial p_f}{\partial t} + C_\chi \nabla^2 p_f \tag{6.1}$$

$$\frac{\partial^2 p_f}{\partial t^2} + C_p \frac{\partial p_f}{\partial t} - C_b \nabla^2 \frac{\partial p_f}{\partial t} + C_\chi \nabla^2 p_f = C_\eta \frac{\partial \eta}{\partial t} + C_\gamma \nabla^2 \eta \tag{6.2}$$

These waves are so strongly coupled they propagate at almost the same speed. So the simple approximations discussed at the end of Chapter 3 may yield a reasonable calculation for the wave speed.

Solutions for these equations have been obtained using the Wavefront Technology Solutions porosity wave analyser, which solves these coupled porosity-pressure equations. The first observation presented here is that the solution depends on the medium permeability and the shape of the pulse at the source. An optimum pulsing source is obtained when the pressure increase (rise time) is determined from the permeability, the elastic properties of the matrix, the properties of the viscous fluid and is about one tenth of the time over which the pressure is decreased. This pulse shape is obtained due to the fact that a porosity–pressure wave may be created by either a pressure increase or a pressure decrease. Here the pressure decrease is done at a rate slow enough not to create a negative porosity–pressure wave. The shape of the pulsing source is shown in Figure 6.4.

After determining the shape of this source, it is interesting to note that it is identical in shape to the pulse generated by a heart. This shape is the same for a hummingbird's heart, which beats 100 times per second; a blue whale's heart, which beats three to five times per minute; and all hearts in between (Figure 6.5).

For laboratory experiments and environmental remediation projects where the permeability is around a darcy, the injected fluid has properties close to water and the solid is assumed to be silica, the rise time is about a tenth of a second (Wang et al. 1998; Davidson et al. 1999). For oil field applications at a depth of 4 km below the Earth's surface, the rise time is usually under a hundredth of a second. Pulse times right up to that of an acoustic wave may be obtained by the explosive release of the injected fluid. Figure 6.6 shows the rise time versus permeability for waves propagating at 80 m/sec and 350 m/sec (Spanos et al. 1999, Samaroo et al. 1999, Dusseault et al. 1998, 1999, 2000a, 2000b, 2000c, 2002a, 2002b); the optimum pulse generates a wave moving at about 100 m/sec. Here, the fluid is water and the matrix is silica.

Figure 6.7 shows the wave speed versus period for four different permeabilities. Here, the period required to create a porosity wave becomes much shorter as the permeability is reduced.

Figure 6.8 shows how the wave speed for the first P wave, the second P wave and the porosity–pressure wave change as a function of frequency. For experiments in the

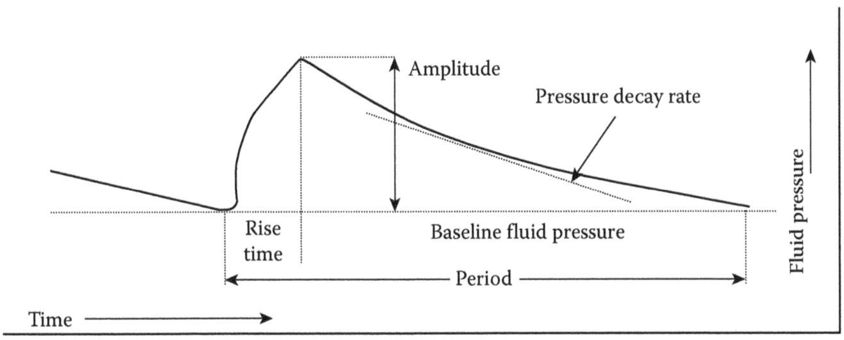

FIGURE 6.4 The shape of the pulse source for creating the coupled porosity–pressure waves.

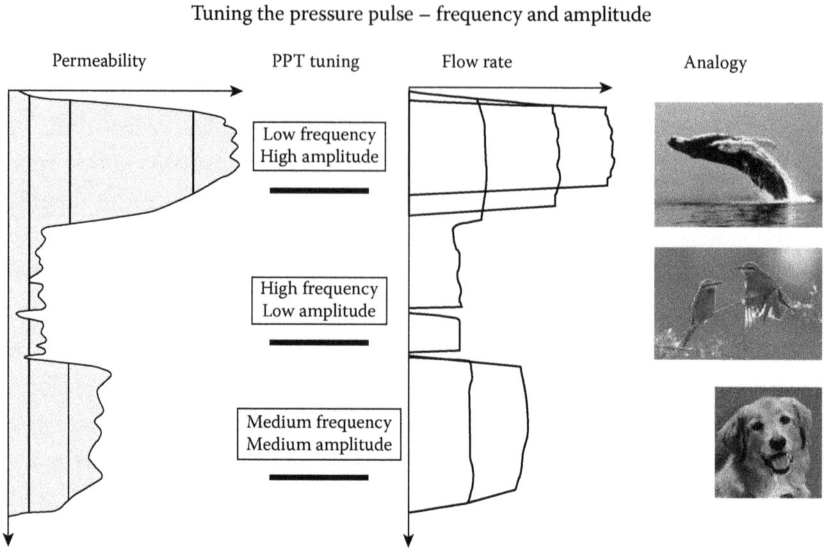

FIGURE 6.5 Analogy between the pulsing source for the porosity wave at various permeabilities and the beating of hearts for various sizes of animals.

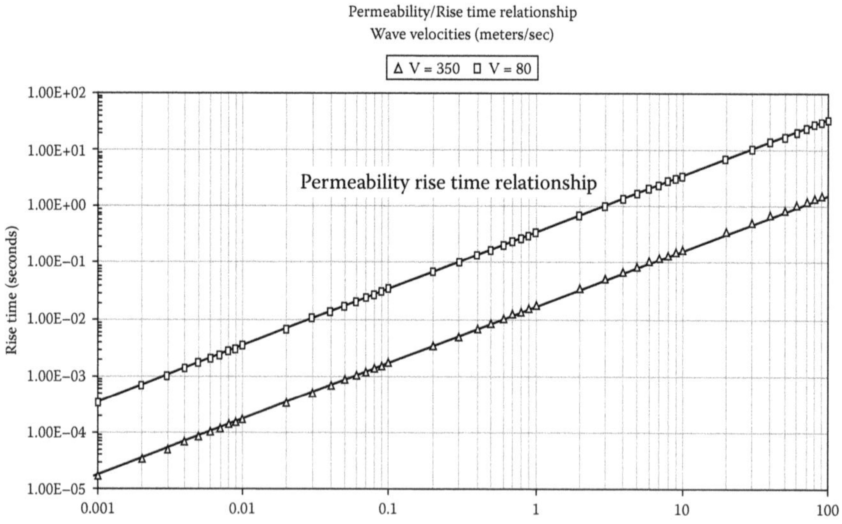

FIGURE 6.6 The rise time is plotted for a source that generates porosity–pressure waves propagating at 80 m/sec and 350 m/sec as the permeability is increased. Note that for 1 darcy permeability the optimum rise time is around a tenth of a second. For 100 millidarcys, it is around a hundredth of a second.

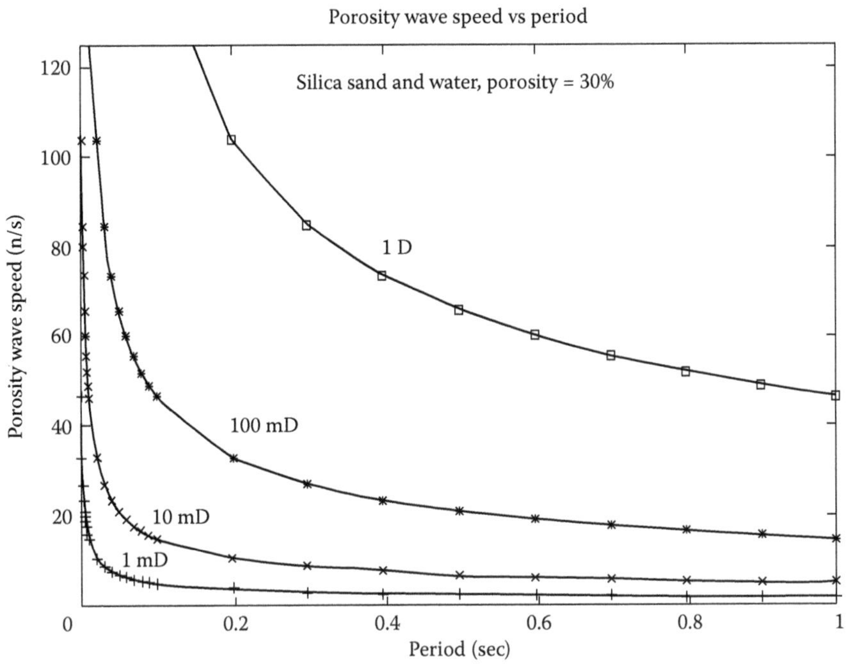

FIGURE 6.7 A plot of the porosity wave speed versus period for four different permeabilities.

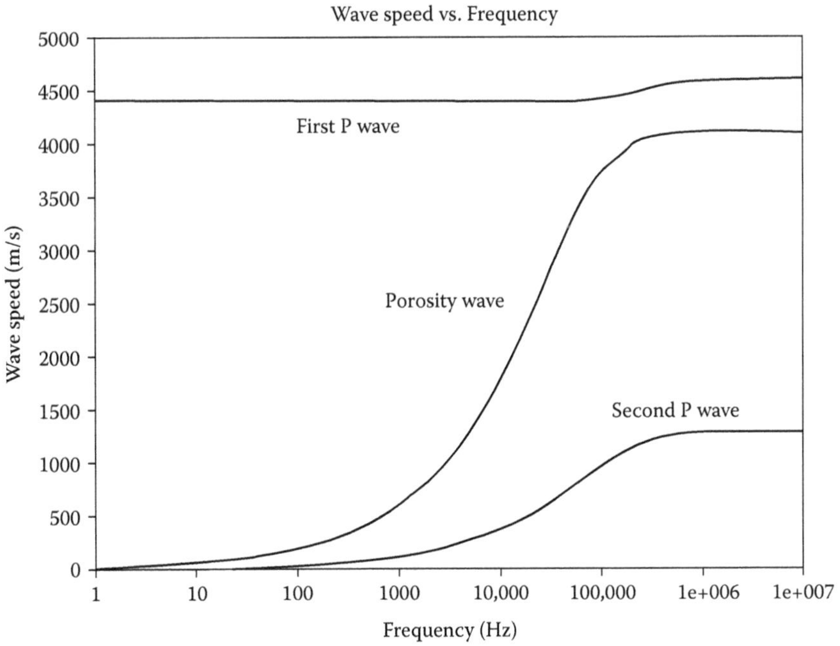

FIGURE 6.8 Plots of the wave speed as predicted by the equations of motion for the first P wave, the second P wave and the porosity wave for frequencies from 1 hertz up to megahertz.

megahertz range performed by Bouzidi (2003) a wave was observed moving at the speed of the porosity wave predicted here but was not identified.

PULSED DISPERSION

It has been predicted theoretically and observed in the laboratory that when a less viscous fluid, like water, is pulsed into a more viscous fluid, like oil, that the pulsed injected fluid disperses through the *in situ* fluid. Here, a description of this process is derived and an algorithm is developed to compute the frontal advancement due to dispersion. A much more detailed theoretical description of dispersion and the coupling between porosity waves and dispersion is presented in Chapter 10. Here, a more practical description of laboratory observations and commercial applications is described.

The system of equations considered here is isothermal, so the fluid densities depend only on pressure. Then

$$\frac{1}{\rho_1}\frac{\partial \rho_1}{\partial t}=\frac{c_{f1}}{K_1}\frac{\partial P_1}{\partial t} \qquad \frac{1}{\rho_2}\frac{\partial \rho_2}{\partial t}=\frac{c_{f2}}{K_2}\frac{\partial P_2}{\partial t} \tag{6.3}$$

here the compressibility factors c_{f1} and c_{f2} are utilized as bookkeeping parameters to indicate the compressible or incompressible case. They have the values

$$c_{f1}=c_{f2}=\begin{cases} 1 & \text{compressible case} \\ 0 & \text{incompressible case} \end{cases} \tag{6.4}$$

Substituting these pressure equations into the fluid divergence equations yields

$$\nabla\cdot\vec{v}_1=-\frac{1}{\eta}\frac{\partial \eta}{\partial t}-\frac{1}{S_1}\frac{\partial S_1}{\partial t}-\frac{c_{f1}}{K_1}\frac{\partial P_1}{\partial t} \tag{6.5}$$

$$\nabla\cdot\vec{v}_2=-\frac{1}{\eta}\frac{\partial \eta}{\partial t}-\frac{1}{S_2}\frac{\partial S_2}{\partial t}-\frac{c_{f2}}{K_2}\frac{\partial P_2}{\partial t} \tag{6.6}$$

Then in the incompressible limit the fluid divergences become

$$\nabla\cdot\vec{v}_1=-\frac{1}{\eta}\frac{\partial \eta}{\partial t}-\frac{1}{S_1}\frac{\partial S_1}{\partial t} \tag{6.7}$$

$$\nabla\cdot\vec{v}_2=-\frac{1}{\eta}\frac{\partial \eta}{\partial t}-\frac{1}{S_2}\frac{\partial S_2}{\partial t} \tag{6.8}$$

Since the fluids are being considered incompressible, it is useful to use the volumetric flow in expressing some of the results. The volumetric flow for each fluid is

$$\vec{q}_1=\eta_1\vec{v}_1 \tag{6.9}$$

$$\vec{q}_2=\eta_2\vec{v}_2 \tag{6.10}$$

and the total volumetric flow is a given by

$$\vec{q} = \vec{q}_1 + \vec{q}_2 \tag{6.11}$$

The volumetric filter velocity is given by

$$\vec{v}_q = \frac{1}{\eta}\vec{q} = S_1\vec{v}_1 + S_2\vec{v}_2 \tag{6.12}$$

Since the divergences of the volumetric flows are given by

$$\nabla \cdot \vec{q}_1 = -S_1\frac{\partial \eta}{\partial t} - \eta\frac{\partial S_1}{\partial t} \tag{6.13}$$

$$\nabla \cdot \vec{q}_2 = -S_2\frac{\partial \eta}{\partial t} + \eta\frac{\partial S_1}{\partial t} \tag{6.14}$$

then

$$\nabla \cdot \vec{q} = -\frac{\partial \eta}{\partial t} \tag{6.15}$$

and

$$\nabla \cdot \vec{v}_q = -\frac{1}{\eta}\frac{\partial \eta}{\partial t} \tag{6.16}$$

In the absence of gravity, the megascopic equations of motion for the solid and each fluid are as follows (Hickey 1994; Spanos 2002) (the current equations differ in form by simple algebraic transformations):

$$\begin{aligned}\eta_s^o\rho_s^o\frac{\partial}{\partial t}v_s &= \eta_s^o K_s\nabla(\nabla\cdot u_s) + \mu_{Mss}[\nabla^2 u_s + \frac{1}{3}\nabla(\nabla\cdot u_s)] - K_s\nabla\eta \\ &+ Q_{s1}(v_1 - v_s) + Q_{s2}(v_2 - v_s) + \rho_{s1}\frac{\partial}{\partial t}(v_1 - v_s) + \rho_{s2}\frac{\partial}{\partial t}(v_2 - v_s)\end{aligned} \tag{6.17}$$

$$\begin{aligned}\eta_1^o\rho_1^o\frac{\partial}{\partial t}\vec{v}_1 &= -\eta_1^o\nabla P_1 + \mu_{M11}[\nabla^2\vec{v}_1 + \frac{1}{3}\nabla(\nabla\cdot\vec{v}_1)] + \eta_1^o\xi_1\nabla(\nabla\cdot\vec{v}_1) + \xi_1\nabla\frac{\partial}{\partial t}\eta_1 \\ &+ \mu_{M1s}[\nabla^2\vec{v}_s + \frac{1}{3}\nabla(\nabla\cdot\vec{v}_s)] + \mu_{M12}[\nabla^2\vec{v}_2 + \frac{1}{3}\nabla(\nabla\cdot\vec{v}_2)] \\ &- Q_{11}(v_1 - v_s) + Q_{12}(v_2 - v_s) - \rho_{11}\frac{\partial}{\partial t}(v_1 - v_s) + \rho_{12}\frac{\partial}{\partial t}(v_2 - v_s)\end{aligned} \tag{6.18}$$

$$\begin{aligned}\eta_2^o\rho_2^o\frac{\partial}{\partial t}\vec{v}_2 &= -\eta_2^o\nabla P_2 + \mu_{M22}[\nabla^2\vec{v}_2 + \frac{1}{3}\nabla(\nabla\cdot\vec{v}_2)] + \eta_2^o\xi_2\nabla(\nabla\cdot\vec{v}_2) + \xi_2\nabla\frac{\partial}{\partial t}\eta_2 \\ &+ \mu_{M2s}[\nabla^2\vec{v}_s + \frac{1}{3}\nabla(\nabla\cdot\vec{v}_s)] + \mu_{M21}[\nabla^2\vec{v}_1 + \frac{1}{3}\nabla(\nabla\cdot\vec{v}_1)] \\ &+ Q_{21}(v_1 - v_s) - Q_{22}(v_2 - v_s) + \rho_{21}\frac{\partial}{\partial t}(v_1 - v_s) - \rho_{22}\frac{\partial}{\partial t}(v_2 - v_s)\end{aligned} \tag{6.19}$$

In these equations, the megascopic shear modulus μ_{Mss} and the megascopic viscosities μ_{Mij} are given by

$$\begin{bmatrix} \mu_{Mss} & & \\ \mu_{M1s} & \mu_{M11} & \mu_{M12} \\ \mu_{M2s} & \mu_{M21} & \mu_{M22} \end{bmatrix} = \begin{bmatrix} \eta_s\mu_s(1-(\lambda_{s1}+\lambda_{s2}) & & \\ \eta_s\mu_1\lambda_{s1} & \eta_1\mu_1(1-\lambda_{12}) & \eta_2\mu_1\lambda_{21} \\ \eta_s\mu_2\lambda_{s2} & \eta_1\mu_2\lambda_{12} & \eta_2\mu_2(1-\lambda_{21}) \end{bmatrix} \tag{6.20}$$

where the λ_{ij} are shear and viscosity strength parameters. The λ_{ij} parameters, the flow resistance coefficients Q_{ij} and the induced mass coefficients ρ_{ij} are defined by the surface averages that arise in volume averaging and are presumed to be single-valued functions of saturation S_1. In the following analysis, the induced mass effects are considered negligible and will be ignored.

The porosity equation (de la Cruz and Spanos 1989; Hickey 1994) is

$$\frac{\partial\eta}{\partial t} = \delta_s\nabla\cdot\vec{v}_s - \delta_1\nabla\cdot\vec{v}_1 - \delta_2\nabla\cdot\vec{v}_2 \tag{6.21}$$

and it describes the relative volumetric changes of the solid and two fluids during dilatation. The choice of parameters δ_s, δ_1 and δ_2 selects the specific thermodynamic process of dilatation (e.g. static compression involves only compressions of the phases, and the δ's may be determined from an infinite number of different compression experiments—see Hickey 1994; seismic P waves involve compression and local fluid flow and the δ's may be calculated from the P wave velocities; porosity waves involve a net flux of fluid in the direction of the wave coupled to incompressible elastic deformations of the matrix and the δ's may be calculated from the porosity wave velocities).

For steady-state flow, the megascopic capillary pressure $P_c(S_1)$ is utilized to account for surface tension effects. In such situations the megascopic pressure difference $P_{21} = P_2 - P_1$ is normally taken to be the megascopic capillary pressure, that is, $P_{21} = P_c$. However, for dynamic processes like wave motion and fluid flow the value of P_{21} can differ substantially from $P_c = P$. The dynamic value of P_{21} is now related to temporal changes in saturation given by Equation 5.25, which for the current analysis will be approximated by the following (de la Cruz et al. 1995):

$$\frac{\partial P_{21}}{\partial t} = \frac{\partial(P_2 - P_1)}{\partial t} = -\beta_1\frac{\partial S_1}{\partial t} \tag{6.22}$$

where β_1 is assumed to be a single-valued function of saturation. Here β_1 is taken to be

$$\beta_1 = \beta_c + \beta_h(S_1) \tag{6.23}$$

where

$$\beta_c = -\frac{dP_c(S_1)}{dS_1} \tag{6.24}$$

and β_h is an empirically determined function of saturation that accounts for dynamic pressure hysteresis.

The porosity and saturation waves are dilatational waves since they are describing changes in volume fractions with time. Taking the divergence of the equations of motion for the solid and each fluid yields

$$\eta_s^o \rho_s^o \frac{\partial}{\partial t} \nabla \cdot v_s = (\eta_s^o K_s + \frac{4}{3} \mu_{Mss}) \nabla^2 \nabla \cdot u_s - K_s \nabla^2 \eta$$
$$+ Q_{s1} \nabla \cdot v_1 + Q_{s2} \nabla \cdot v_2 - (Q_{s1} + Q_{s2}) \nabla \cdot v_s) \tag{6.25}$$

$$\eta_1^o \rho_1^o \frac{\partial}{\partial t} \nabla \cdot \vec{v}_1 = -\eta_1^o \nabla^2 P_1 + (\frac{4}{3} \mu_{M11} + \eta_1^o \xi_1) \nabla^2 \nabla \cdot \vec{v}_1 + \xi_1 \nabla^2 \frac{\partial}{\partial t} \eta_1$$
$$+ \frac{4}{3} \mu_{M1s} \nabla^2 \nabla \cdot \vec{v}_s + \frac{4}{3} \mu_{M12} \nabla^2 \nabla \cdot \vec{v}_2 - Q_{11} \nabla \cdot v_1 + Q_{12} \nabla \cdot v_2 + (Q_{11} - Q_{12}) \nabla \cdot v_s \tag{6.26}$$

$$\eta_2^o \rho_2^o \frac{\partial}{\partial t} \nabla \cdot \vec{v}_2 = -\eta_2^o \nabla^2 P_2 + \frac{4}{3} \mu_{M22} \nabla^2 \nabla \cdot \vec{v}_2 + \eta_2^o \xi_2 \nabla^2 (\nabla \cdot \vec{v}_2) + \xi_2 \nabla^2 \frac{\partial}{\partial t} \eta_2$$
$$- \frac{4}{3} \mu_{M2s} \nabla^2 \nabla \cdot \vec{v}_s + \frac{4}{3} \mu_{M21} \nabla^2 \nabla \cdot \vec{v}_1 + Q_{21} \nabla \cdot v_1 - Q_{22} \nabla \cdot v_2 + (Q_{22} - Q_{21}) \nabla \cdot v_s \tag{6.27}$$

Then, substituting the incompressible fluid divergences into these intermediate equations yields

$$\eta_s^o \rho_s^o \frac{\partial}{\partial t} \nabla \cdot v_s = (\eta_s^o K_s + \frac{4}{3} \eta_s^o \mu_{Mss}) \nabla^2 (\nabla \cdot u_s) - K_s \nabla^2 \eta$$
$$-(Q_{s1} + Q_{s2}) \frac{1}{\eta} \frac{\partial \eta}{\partial t} - \left(\frac{Q_{s1}}{S_1} + \frac{Q_{s2}}{(1 - S_1)} \right) \frac{\partial S_1}{\partial t} - (Q_{s1} + Q_{s2}) \nabla \cdot v_s \tag{6.28}$$

$$S_1 \rho_1^o \frac{\partial^2 \eta}{\partial t^2} + \eta \rho_1^o \frac{\partial^2 S_1}{\partial t^2} = \eta S_1 \nabla^2 P_1 + \left[\frac{4}{3} (\mu_{M11} + \mu_{M12}) + \eta_1^o \xi_1 \right] \nabla^2 \frac{1}{\eta} \frac{\partial \eta}{\partial t}$$
$$+ \frac{4}{3} (\mu_{M11} + \mu_{M12}) \nabla^2 \frac{1}{S_1} \frac{\partial S_1}{\partial t} - \frac{4}{3} \mu_{M1s} \nabla^2 \nabla \cdot \vec{v}_s - (Q_{11} - Q_{12}) \frac{1}{\eta} \frac{\partial \eta}{\partial t} \tag{6.29}$$
$$- \left(\frac{Q_{11}}{S_1} + \frac{Q_{12}}{(1 - S_1)} \right) \frac{\partial S_1}{\partial t} - (Q_{11} - Q_{12}) \nabla \cdot v_s$$

$$(1 - S_1) \rho_2^o \frac{\partial^2 \eta}{\partial t^2} + \eta \rho_2^o \frac{\partial^2 S_2}{\partial t^2} = \eta_2^o \nabla^2 P_2 + \left[\frac{4}{3} (\mu_{M22} + \mu_{M21}] + \eta (1 - S_1) \xi_2 \right] \nabla^2 \frac{1}{\eta} \frac{\partial \eta}{\partial t}$$
$$- \frac{4}{3} \left[\frac{\mu_{M22}}{(1 - S_1)} - \frac{\mu_{M21}}{S_1} \right] \nabla^2 \partial S + \frac{4}{3} \mu_{M2s} \nabla^2 \nabla \cdot \vec{v}_s + (Q_{21} - Q_{22}) \frac{1}{\eta} \frac{\partial \eta}{\partial t} \tag{6.30}$$
$$+ \left[\frac{Q_{21}}{S_1} + \frac{Q_{22}}{(1 - S_1)} \right] \frac{\partial S_1}{\partial t} - (Q_{22} - Q_{21}) \nabla \cdot v_s$$

These equations are the dilatational wave equations. The can be rewritten in a more compact form by defining some coefficients. For the solid equation, define the coefficients

$$A_{s\eta} = \frac{1}{2}\frac{Q_{s1} + Q_{s2}}{\eta_s} \tag{6.31}$$

$$K_{Mss} = K_s + \frac{4}{3}\frac{\mu_{Mss}}{\eta_s} \tag{6.32}$$

$$K_{s\eta} = K_s \frac{\eta}{\eta_s} \tag{6.33}$$

$$A_{sS_1} = \frac{1}{2}\frac{1}{\eta_s}\left(\frac{Q_{s1}}{S_1} + \frac{Q_{s2}}{(1-S_1)}\right) \tag{6.34}$$

Then the solid equation becomes

$$\begin{aligned} &\rho_s^o \frac{\partial^2}{\partial t^2}(\nabla \cdot \mathrm{u}_s) + 2A_{s\eta}\frac{\partial}{\partial t}(\nabla \cdot \mathrm{u}_s) - K_{Mss}\nabla^2(\nabla \cdot \mathrm{u}_s) \\ &+ 2A_{s\eta}\frac{1}{\eta}\frac{\partial \eta}{\partial t} + K_{s\eta}\nabla^2\eta + 2A_{sS_1}\frac{\partial S_1}{\partial t} = 0 \end{aligned} \tag{6.35}$$

For Fluid 1 define

$$A_{1\eta} = \frac{1}{2}(Q_{11} + Q_{12}) \tag{6.36}$$

$$B_{1s} = \frac{2}{3}\frac{\mu_{M1s}}{\eta} \tag{6.37}$$

$$B_{1\eta} = \frac{2}{3}\frac{\mu_{M11} + \mu_{M12}}{\eta} \tag{6.38}$$

$$A_{1S_1} = \frac{1}{2}\left(\frac{Q_{11}}{S_1} + \frac{Q_{12}}{(1-S_1)}\right) \tag{6.39}$$

$$B_{1S_1} = \frac{1}{2}\frac{1}{\eta}\left(\frac{\mu_{M11}}{S_1} + \frac{\mu_{M12}}{(1-S_1)}\right) \tag{6.40}$$

Then the simplified form of the equation for Fluid 1 may be written as

$$\begin{aligned} &2A_{1\eta}\frac{\partial}{\partial t}(\nabla \cdot \mathrm{u}_s) + 2B_{1s}\nabla^2\frac{\partial}{\partial t}(\nabla \cdot \mathrm{u}_s) + S_1\rho_1^o\frac{1}{\eta}\frac{\partial^2 \eta}{\partial t^2} + 2A_{1\eta}\frac{1}{\eta}\frac{\partial \eta}{\partial t} \\ &-2B_{1\eta}\frac{1}{\eta}\nabla^2\frac{\partial \eta}{\partial t} + \rho_1\frac{\partial^2 S_1}{\partial t^2} + 2A_{1S_1}\frac{\partial S_1}{\partial t} - 2B_{1S_1}\nabla^2\frac{\partial^2 S_1}{\partial t^2} - S_1\nabla^2 P_1 = 0 \end{aligned} \tag{6.41}$$

For Fluid 2 define

$$A_{2\eta} = \frac{1}{2}(Q_{22} + Q_{21}) \tag{6.42}$$

$$B_{2s} = \frac{2}{3}\frac{\mu_{M2s}}{\eta} \tag{6.43}$$

$$B_{2\eta} = \frac{2}{3}\frac{\mu_{M22} + \mu_{M21}}{\eta} \tag{6.44}$$

$$A_{2S_1} = \frac{1}{2}\left(\frac{Q_{21}}{S_1} + \frac{Q_{22}}{(1-S_1)}\right) \tag{6.45}$$

$$B_{2S_1} = \frac{1}{2}\frac{1}{\eta}\left(\frac{\mu_{M21}}{S_1} + \frac{\mu_{M22}}{(1-S_1)}\right) \tag{6.46}$$

Then the simplified form of the equation for Fluid 2 may be written as

$$\begin{aligned} &2A_{2\eta}\frac{\partial}{\partial t}(\nabla\cdot \mathrm{u}_s) + 2B_{2s}\nabla^2\frac{\partial}{\partial t}(\nabla\cdot \mathrm{u}_s) + (1-S_1)\rho_2\frac{1}{\eta}\frac{\partial^2\eta}{\partial t^2} + 2A_{2\eta}\frac{1}{\eta}\frac{\partial\eta}{\partial t} \\ &-2B_{2\eta}\frac{1}{\eta}\nabla^2\frac{\partial\eta}{\partial t} - \rho_2\frac{\partial^2 S_1}{\partial t^2} - 2A_{2S_1}\frac{\partial S_1}{\partial t} + 2B_{2S_1}\nabla^2\frac{\partial^2 S_1}{\partial t^2} - (1-S_1)\nabla^2 P_2 = 0 \end{aligned} \tag{6.47}$$

In these equations the terms of the form A_{ij} are attenuation densities with dimensionality [attenuation (s^{-1}) × density (kg/m^3)]. Terms of the form B_{ij} are bulk attenuation densities with dimensionality [attenuation density × area (m^2)]. The coefficients of the form K_{ij} are bulk moduli with units of pressure (Pa).

The porosity dilatational wave equation is obtained by substituting the fluid divergences into the porosity equation (Equation 6.21), which yields

$$\frac{\partial}{\partial t}\nabla\cdot \mathrm{u}_s - \frac{\alpha_\eta}{\eta}\frac{\partial\eta}{\partial t} + \alpha_{S_1}\frac{\partial S_1}{\partial t} = 0 \tag{6.48}$$

where

$$\alpha_\eta = \frac{\eta - \delta_1 - \delta_2}{\delta_s} \tag{6.49}$$

$$\alpha_{f1} = \frac{\delta_1}{\delta_s} \tag{6.50}$$

$$\alpha_{f1} = \frac{\delta_2}{\delta_s} \tag{6.51}$$

$$\alpha_{S_1} = \frac{\alpha_{f1}}{S_1} - \frac{\alpha_{f2}}{(1-S_1)} \tag{6.52}$$

The saturation equation (Equation 6.22) is now rewritten by expanding terms and putting them all on the left-hand side

$$\beta_1 \frac{\partial S_1}{\partial t} + \frac{\partial P_2}{\partial t} - \frac{\partial P_1}{\partial t} = 0 \tag{6.53}$$

Here, five dilatational wave equations have been obtained. These equations govern the wave behaviour of five variables, namely the solids divergence of displacement $\nabla \cdot \mathbf{u}_s$, the porosity η, the saturation S_1, and the two fluid pressures P_1 and P_2. The analysis of these dilatational wave equations will be facilitated by casting them into a wave operator format.

To establish the wave operator formalism, define a differential wave operator basis W_b by letting

$$W_b = \left[\frac{\partial^2}{\partial t^2}, \frac{\partial}{\partial t}, \nabla^2 \frac{\partial}{\partial t}, \nabla^2 \right] \tag{6.54}$$

Let $\psi = \psi(\vec{x}, t)$ be a wave variable associated with a wave operator W_ψ, defined formally by

$$\begin{aligned} W_\psi &= [w_o, w_1, w_2, w_3] \cdot W_b \\ &= w_o \frac{\partial^2}{\partial t^2} + w_1 \frac{\partial}{\partial t} + w_2 \nabla^2 \frac{\partial}{\partial t} + w_3 \nabla^2 \end{aligned} \tag{6.55}$$

Then the application of the wave operator W_ψ to the variable ψ is

$$\begin{aligned} W_\psi \cdot \psi &= [w_o, w_1, w_2, w_3] \cdot W_b \cdot \psi \\ &= w_o \frac{\partial^2}{\partial t^2} \psi + w_1 \frac{\partial}{\partial t} \psi + w_2 \nabla^2 \frac{\partial}{\partial t} \psi + w_3 \nabla^2 \psi \end{aligned} \tag{6.56}$$

For a practical example, suppose the variable ψ has associated with it a density ρ, an attenuation density A, a bulk attenuation density B and a bulk modulus K_b. Then the damped wave equation operator W_d for ψ is

$$W_d = [\rho, 2\mathrm{A}, -2\mathrm{B}, -\mathrm{K}_b] \cdot W_b \tag{6.57}$$

and the damped wave equation is then

$$W_d \cdot \psi = 0 \tag{6.58}$$

or

$$\rho \frac{\partial^2}{\partial t^2} \psi + 2\mathrm{A} \frac{\partial}{\partial t} \psi - 2\mathrm{B} \nabla^2 \frac{\partial}{\partial t} \psi - \mathrm{K}_b \nabla^2 \psi = 0 \tag{6.59}$$

The dilatational wave equations for the solid and two fluids can now be rewritten using this wave operator formalism. The solid's dilatational wave equation becomes

$$W_{ss} \nabla \cdot \vec{u}_s + W_{s\eta} \frac{\eta}{\eta_o} + 2\mathrm{A}_{sS_1} \frac{\partial S_1}{\partial t} = 0 \tag{6.60}$$

where the wave operators W_{ss} and $W_{s\eta}$ are defined by

$$W_{ss} = [\rho_s, 2A_{s\eta}, 0, -K_{M_{ss}}]\cdot W_b \tag{6.61}$$

$$W_{s\eta} = [0, 2A_{s\eta}, 0, -K_{s\eta}]\cdot W_b \tag{6.62}$$

The dilatational wave equation for Fluid 1 becomes

$$W_{1s}\nabla\cdot\vec{u}_s + W_{1\eta}\frac{\eta}{\eta_o} + W_{1S_1}S_1 - S_1\nabla^2 P_1 = 0 \tag{6.63}$$

where

$$W_{1s} = [0, 2A_{1\eta}, 2B_{1s}, 0]\cdot W_b \tag{6.64}$$

$$W_{1\eta} = [S_1\rho_1, 2A_{1\eta}, -2B_{1\eta}, 0]\cdot W_b \tag{6.65}$$

$$W_{1S_1} = [\rho_1, 2A_{1S_1}, -2B_{1S_1}, 0]\cdot W_b \tag{6.66}$$

The dilatational wave equation for Fluid 2 is

$$W_{2s}\nabla\cdot\vec{u}_s + W_{2\eta}\frac{\eta}{\eta_o} - W_{2S_1}S_1 - S_2\nabla^2 P_2 = 0 \tag{6.67}$$

where

$$W_{2s} = [0, 2A_{2\eta}, -2B_{2s}, 0]\cdot W_b \tag{6.68}$$

$$W_{2\eta} = [S_2\rho_2, 2A_{2\eta}, -2B_{2\eta}, 0]\cdot W_b \tag{6.69}$$

$$W_{2S_1} = [\rho_2, 2A_{2S_1}, -2B_{2S_1}, 0]\cdot W_b \tag{6.70}$$

The three rewritten dilatational wave equations, the porosity dilatational wave equation and the saturation equation may now be assembled into a wave operator matrix equation. The result is

$$\begin{bmatrix} W_{ss} & W_{s\eta} & 2A_{2S_1}\dfrac{\partial}{\partial t} & 0 & 0 \\ W_{1s} & W_{1\eta} & W_{1S_1} & -S_1\nabla^2 & 0 \\ W_{2s} & W_{2\eta} & W_{2S_1} & 0 & -S_2\nabla^2 \\ \dfrac{\partial}{\partial t} & -\alpha_\eta\dfrac{\partial}{\partial t} & \alpha_{S_1}\dfrac{\partial}{\partial t} & 0 & 0 \\ 0 & 0 & \beta_1\dfrac{\partial}{\partial t} & -\dfrac{\partial}{\partial t} & \dfrac{\partial}{\partial t} \end{bmatrix} \begin{bmatrix} \nabla\cdot u_s \\ \dfrac{\eta}{\eta_o} \\ S_1 \\ P_1 \\ P_2 \end{bmatrix} = \begin{bmatrix} 0 \\ 0 \\ 0 \\ 0 \\ 0 \end{bmatrix} \tag{6.71}$$

The wave solutions of this matrix wave equation are intended to describe coupled porosity and saturation waves. However, this analysis will expose a flaw in the theory. That will allow for a new insight into the non-equilibrium thermodynamics and an understanding of the true thermodynamic variables, which will be discussed in Chapters 8, 9 and 10. A detailed solution of this equation is presented in Udey (2012) and yields a prediction that is inconsistent with experimental observations: Those solutions can only be obtained when the waves are decoupled, the problem being that this component and volume averaging construction can only be used for two coupled phases. The three components can couple four different ways to yield in-phase and out-of-phase momentums. Thus this construction is not complete when considering three components. Another way of observing this result is to attempt to construct Onsager's relations. In doing this construction, it is observed that the present system of equations cannot be used to construct Onsager's relations because it is not complete. Experimental observations indicate that the porosity and saturation waves are so strongly coupled that the passing of porosity waves causes dispersion to completely dominate viscous fingering. So the rest of this chapter proceeds with that assumption.

A standard technique to analyse waves governed by a wave equation is to substitute a trial solution, typically a homogeneous plane wave, into the wave equation to obtain a dispersion relation. An analysis of the dispersion relation reveals valuable information about the wave, such as attenuation and speed of propagation.

Recall the example of a damped wave equation for the variable ψ, that is,

$$W_d \cdot \psi = [\rho, 2A, -2B, -K_b] \cdot W_b \cdot \psi = 0 \tag{6.72}$$

Let the trial solution of this damped wave equation for ψ consist of an unperturbed value ψ^o plus a one-dimensional wave perturbation. This may be written as

$$\psi(\vec{x},t) = \psi^o + \delta\psi \quad \delta\psi = A_\psi^o e^{nx+wt} = 0 \tag{6.73}$$

where the perturbation $\delta\psi$ represents a homogeneous plane wave of amplitude A_ψ^o moving in the x direction with complex wave number n and complex frequency w.

Applying the wave operator basis W_b to the trial solution yields

$$\begin{aligned} W_d &= \left[\frac{\partial^2}{\partial t^2}, \frac{\partial}{\partial t}, \nabla^2 \frac{\partial}{\partial t}, \nabla^2\right] \cdot \delta\psi \\ &= w^2 + 2aw - 2bwn^2 - v_o^2 n^2 \\ &= \rho_b(n,w) \cdot \delta\psi \end{aligned} \tag{6.74}$$

where the dispersion polynomial $p_d(n, \omega)$ corresponding to the wave operator W_d is

$$\begin{aligned} \rho_d(n,w) &= [\rho, 2A, -2B, -K_b] \cdot \rho_b(n,w) \\ &= [\rho, 2A, -2B, -K_b] \cdot [w^2, w, wn^2, n^2] \\ &= \rho w^2 + 2Aw - 2Bwn^2 - K_b n^2 \end{aligned} \tag{6.75}$$

So substituting the trial solution into the damped wave equation $W_d \cdot \psi = 0$ yields

$$\rho(n,w)A_{\psi}^{o}e^{nx+wt} = 0 \tag{6.76}$$

A non-trivial solution $A_{\psi}^{o} \neq 0$ now requires

$$\rho(n,w) = 0 \tag{6.77}$$

and this produces the dispersion relation for the trial solution, namely

$$\rho w^2 + 2\mathrm{A}w - 2\mathrm{B}wn^2 - \mathrm{K}_b n^2 = 0 \tag{6.78}$$

Dividing by ρ yields

$$w^2 + 2aw - 2\mathrm{b}wn^2 - \mathrm{v}_o^2 n^2 = 0 \tag{6.79}$$

where the attenuation a, bulk attenuation b and undamped wave speed v_o^2 are defined by

$$a = \frac{A}{\rho} \tag{6.80}$$

$$b = \frac{B}{\rho} \tag{6.81}$$

$$\mathrm{v}_o^2 = \frac{\mathrm{K}_b}{\rho} \tag{6.82}$$

Choosing ω as the independent variable, the solution for n^2 is

$$n^2 = \frac{w^2 + 2aw}{\mathrm{v}_o^2 + 2bw} \tag{6.83}$$

The frequency domain for this trial solution is selected by letting the complex wave frequency be $w = -i\omega$, where ω is the angular frequency of the wave. The equation for n becomes

$$n^2 = -\frac{\omega^2 + 2a\omega i}{\mathrm{v}_o^2 - 2b\omega i} \tag{6.84}$$

Solutions for n are of the form

$$n = -\kappa + ik \tag{6.85}$$

where κ and k are the special attenuation and real wave number, respectively. Then the trial solution is

$$\psi(\vec{x},t)=\psi^{o}+A_{\psi}^{o}e^{-\kappa x}e^{i(kx-\omega t)} \tag{6.86}$$

which is a damped homogeneous plane wave 'travelling to the right in the x direction' with the wave speed v given by

$$\mathrm{v}=\frac{\omega}{k} \tag{6.87}$$

If b ≠ 0, then the solution for n yields a wave speed v whose asymptotic behaviour is

$$\lim_{\omega\to\infty} v=2\sqrt{|b|}\sqrt{\omega} \tag{6.88}$$

That is, the maximum wave speed is frequency dependent. Such behaviour is unphysical, since all wave speeds are bounded above by the unattenuated wave speed v_o. Now setting b = 0, the solution to

$$n^{2}=-\frac{\omega^{2}+2a\omega i}{\mathrm{v}_o^{2}} \tag{6.89}$$

is

$$n=-\kappa+ik \tag{6.90}$$

$$\kappa=\frac{1}{\mathrm{v}_o\sqrt{2}}\sqrt{\sqrt{\omega^{4}+4a^{2}\omega^{2}}-\omega^{2}} \tag{6.91}$$

$$k=\frac{1}{\mathrm{v}_o\sqrt{2}}\sqrt{\sqrt{\omega^{4}+4a^{2}\omega^{2}}+\omega^{2}} \tag{6.92}$$

and the wave speed is

$$\mathrm{v}=\frac{\omega}{k}=\mathrm{v}_o\frac{\sqrt{2}\omega}{\sqrt{\sqrt{\omega^{4}+4a^{2}\omega^{2}}+\omega^{2}}} \tag{6.93}$$

Here v(ω) < v_o and v(ω) approaches v_o as ω → ∞. Also, when the attenuation is set to 0, v(ω) = v_o. Now, assuming $S_1 = 1$, a damped wave equation for porosity is obtained

$$\rho_\eta\frac{\partial^{2}\eta}{\partial t^{2}}+2A_\eta\frac{\partial\eta}{\partial t}-K_\eta\nabla^{2}\eta=0 \tag{6.94}$$

where the porosity's density, attenuation density and bulk modulus are

$$\rho_\eta=\alpha_\eta\rho_s \tag{6.95}$$

$$A_\eta=\frac{1}{2}\frac{(\alpha_\eta+1)\eta^{2}}{\eta_s}\frac{\mu_f}{K} \tag{6.96}$$

$$K_\eta = K_s(\alpha_\eta - \frac{\eta}{\eta_s}) + \frac{4}{3}\alpha_\eta \frac{\mu_{Mss}}{\eta_s} \tag{6.97}$$

In wave operator format, this equation is

$$W_\eta \cdot \eta = 0 \tag{6.98}$$

where the porosity wave operator is

$$W_\eta = [\rho_\eta, 2A_\eta, 0, -K_\eta \nabla^2 \eta] \tag{6.99}$$

The solution of this wave equation is a porosity wave.

The fluid's response as described by Equation 3.159 under the current assumptions is

$$\nabla^2 P_f = \frac{1}{\eta_o}\left(\rho_f \frac{\partial^2 \eta}{\partial t^2} + 2A_{f\eta}\frac{\partial \eta}{\partial t} - B_{f\eta}\nabla^2 \frac{\partial \eta}{\partial t}\right) \tag{6.100}$$

where the attenuation density $A_{f\eta}$ and bulk attenuation density $B_{f\eta}$ are given by

$$A_{f\eta} = \eta(\alpha_\eta + 1)\frac{\mu_f}{K} \tag{6.101}$$

$$B_{f\eta} = \frac{2}{3}(\mu_f - \alpha_\eta \mu_{Mfs}) \tag{6.102}$$

The porosity wave's attenuation a_η and undamped wave speed v_o are given by

$$a_\eta = \frac{A_\eta}{\rho_\eta} \tag{6.103}$$

$$v_{o\eta}^2 = \frac{K_\eta}{\rho_\eta} \tag{6.104}$$

In the diffusion limit of these equations, the inertial and bulk attenuation density terms are considered negligible and therefore

$$\frac{\partial \eta}{\partial t} = D_\eta \nabla^2 \eta \tag{6.105}$$

$$\nabla^2 P_f = \frac{2A_{f\eta}}{\eta}\frac{\partial \eta}{\partial t} \tag{6.106}$$

where the porosity diffusion coefficient is

$$D_\eta = \frac{K_\eta}{2A_\eta} = \frac{v_o^2}{2a_\eta} = \frac{\eta_s}{\eta^2}\frac{K_s(\alpha_\eta - \frac{\eta}{\eta_s}) + \frac{4}{3}\alpha_\eta \frac{\mu_{Mss}}{\eta_s}}{(\alpha_\eta + 1)}\frac{K}{\mu_f} \tag{6.107}$$

This is identical to the porosity diffusion results originally obtained by Geilikman et al. (1993).

In general, porosity and pressure are separate but coupled diffusion processes (Spanos 2002).

Now consider a trial solution of the form

$$\eta = \eta_o + A_\eta^o e^{nx-wt} \tag{6.108}$$

Substitution of this trial solution into the porosity wave equation $W_\eta \cdot \eta = 0$ produces the dispersion relation

$$\rho_\eta w^2 + 2A_\eta w - K_\eta n^2 = 0 \tag{6.109}$$

or, dividing by ρ_n

$$\frac{\partial^2 \eta}{\partial t^2} + 2a_\eta \frac{\partial \eta}{\partial t} - v_{o\eta}^2 \nabla^2 \eta = 0 \tag{6.110}$$

For a given $w = i\omega$, the solution of the dispersion relation is

$$n = -\kappa + ik \tag{6.111}$$

$$\kappa = \frac{1}{v_{o\eta}\sqrt{2}} \sqrt{\sqrt{\omega^4 + 4a_\eta^2\omega^2} - \omega^2} \tag{6.112}$$

$$k = \frac{1}{v_{o\eta}\sqrt{2}} \sqrt{\sqrt{\omega^4 + 4a_\eta^2\omega^2} + \omega^2} \tag{6.113}$$

$$v = \frac{\omega}{k} = v_{o\eta} \frac{\sqrt{2}\omega}{\sqrt{\sqrt{\omega^4 + 4a_\eta^2\omega^2} + \omega^2}} \tag{6.114}$$

This illustrates the dependence of the phase velocity of the porosity–pressure wave on frequency, permeability and viscosity.

The limiting case of pressure diffusion may be obtained by first taking the solid to be perfectly rigid, that is, η is constant, and neglecting viscous dissipation and induced mass effects. Then the fluid's equation of motion is then governed by the competition between the pressure gradient and the Darcy force,

$$\eta\rho_f \frac{\partial v_f}{\partial t} = -\eta\nabla P_f - \eta^2 \frac{\mu_f}{K} v_f \tag{6.115}$$

The divergence of this equation yields the intermediate wave equation

$$\eta\rho_f \frac{\partial(\nabla \cdot v_f)}{\partial t} = -\eta\nabla^2 P_f - \eta^2 \frac{\mu_f}{K} \nabla \cdot v_f \tag{6.116}$$

For a single fluid, the fluid divergence in the compressional case is

$$\nabla \cdot v_f = -\frac{1}{K_f}\frac{\partial P_f}{\partial t} \tag{6.117}$$

When this result is substituted into the intermediate wave equation, a damped wave equation for the pressure is obtained

$$\rho_f \frac{\partial^2 P_f}{\partial t^2} + \eta \frac{\mu_f}{K}\frac{\partial P_f}{\partial t} - K_f \nabla^2 P_f = 0 \tag{6.118}$$

The diffusion limit of this equation is the pressure diffusion equation

$$\frac{\partial P_f}{\partial t} = D_{P_f} \nabla^2 P_f \tag{6.119}$$

where the pressure diffusion coefficient D_{P_f} is

$$D_{P_f} = \frac{K_f K}{\eta \mu_f} \tag{6.120}$$

The pressure can diffuse either by this pressure diffusion process or by porosity diffusion (where an increase in pressure changes the porosity). The porosity diffusion process will be the preferred process if the porosity diffusion coefficient D_η, given by Equation 6.107, is larger than the pressure diffusion coefficient D_{P_f}, that is, when

$$\frac{D_\eta}{D_{P_f}} > 1 \tag{6.121}$$

The expression for this ratio is

$$\frac{D_\eta}{D_{P_f}} = \frac{\eta_s}{\eta} \frac{K_s(\alpha_\eta - \frac{\eta}{\eta_s}) + \frac{4}{3}\alpha_\eta \frac{\mu_{Mss}}{\eta_s}}{(\alpha_\eta + 1)K_f} \tag{6.122}$$

In cases where the right-hand side of this equation is greater than 1, the pressure will relax much more quickly than expected by pressure diffusion.

An analysis of saturation was presented by Udey 2009, 2012. That analysis will be summarized here.

The solid will be taken to be rigid, so η is a constant and its derivatives are 0. Porosity waves cannot exist in this case. The fluids are taken to be incompressible. The only forces considered are pressure gradients, the forces between the fluids and the forces between the fluids and the solid. All other forces and the induced mass effects are considered negligible. The equations of motion for the two fluids are given by

$$\eta_1^o \rho_1^o \frac{\partial}{\partial t}\vec{v}_1 = -\eta_1^o \nabla P_1 - Q_{11}(v_1 - v_s) + Q_{12}(v_2 - v_s) \tag{6.123}$$

$$\eta_2^o \rho_2^o \frac{\partial}{\partial t}\vec{v}_2 = -\eta_2^o \nabla P_2 + Q_{21}(v_1 - v_s) - Q_{22}(v_2 - v_s) \tag{6.124}$$

Taking the divergence of these equations yields

$$\eta_1^o \rho_1^o \frac{\partial}{\partial t} \nabla \cdot \vec{v}_1 = -\eta_1^o \nabla^2 P_1 - Q_{11} \nabla \cdot \vec{v}_1 + Q_{12} \nabla \cdot \vec{v}_2 \tag{6.125}$$

$$\eta_2^o \rho_2^o \frac{\partial}{\partial t} \nabla \cdot \vec{v}_2 = -\eta_2^o \nabla^2 P_2 + Q_{21} \nabla \cdot \vec{v}_1 - Q_{22} \nabla \cdot \vec{v}_2 \tag{6.126}$$

The equations for the fluid divergences in the present case simplify to

$$\nabla \cdot \vec{v}_1 = -\frac{1}{S_1} \frac{\partial S_1}{\partial t} \tag{6.127}$$

$$\nabla \cdot \vec{v}_2 = \frac{1}{S_2} \frac{\partial S_1}{\partial t} \tag{6.128}$$

Substituting these divergences into the previous equations yields the dilatational wave equations for the two fluids

$$\rho_1^o \frac{\partial^2 S_1}{\partial t^2} + 2A_{1S_1} \frac{\partial S_1}{\partial t} - S_1 \nabla^2 P_1 = 0 \tag{6.129}$$

$$-\rho_2^o \frac{\partial^2 S_1}{\partial t^2} - 2A_{2S_1} \frac{\partial S_1}{\partial t} - S_2 \nabla^2 P_2 = 0 \tag{6.130}$$

where the attenuation densities A_{1S_1} and A_{2S_1} are defined by Equations 6.39 and 6.45, respectively. These two equations and the saturation equation

$$\beta_1 \frac{\partial S_1}{\partial t} + \frac{\partial P_2}{\partial t} - \frac{\partial P_1}{\partial t} = 0 \tag{6.131}$$

constitute three wave equations in the three unknowns S_1, P_1 and P_2. These equations may be assembled into a wave operator matrix equation, namely

$$\begin{bmatrix} \rho_1^o \frac{\partial^2}{\partial t^2} + 2A_{1S_1} \frac{\partial}{\partial t} & -S_1 \nabla^2 & 0 \\ -\left(\rho_2^o \frac{\partial^2}{\partial t^2} + 2A_{2S_1} \frac{\partial}{\partial t} \right) & 0 & -S_2 \nabla^2 \\ \beta_1 \frac{\partial}{\partial t} & -\frac{\partial}{\partial t} & \frac{\partial}{\partial t} \end{bmatrix} \begin{bmatrix} S_1 \\ P_1 \\ P_2 \end{bmatrix} = \begin{bmatrix} 0 \\ 0 \\ 0 \end{bmatrix} \tag{6.132}$$

Solutions of this matrix wave equation are saturation waves; they are dilatational dispersion waves in the porous medium. Also, note this wave matrix equation is a submatrix of the more general matrix wave equation (Equation 6.71) with the bulk attenuation densities being absent.

Now consider a trial solution

$$\begin{bmatrix} S_1 & P_1 & P_2 \end{bmatrix} = \begin{bmatrix} S_1^o & P_1^o & P_2^o \end{bmatrix} + \begin{bmatrix} A_{S_1}^o & A_{P_1}^o & A_{P_2}^o \end{bmatrix} e^{nx+wt} \tag{6.133}$$

whose form was previously discussed in the context of porosity–pressure waves. Substitution of this trial solution into the matrix wave equations yields a dispersion matrix equation given by

$$\begin{bmatrix} \rho_1^o w^2 + 2A_{1S_1} w & -S_1 n^2 & 0 \\ -\left(\rho_2^o w^2 + 2A_{2S_1} w\right) & 0 & -S_2 n^2 \\ \beta_1 w & -w & w \end{bmatrix} \begin{bmatrix} A_{S_1}^o \\ A_{P_1}^o \\ A_{P_2}^o \end{bmatrix} = \begin{bmatrix} 0 \\ 0 \\ 0 \end{bmatrix} \tag{6.134}$$

However, non-trivial solutions exist only if the determinant of the matrix on the left-hand side of this equation is 0, that is,

$$\Delta(n,w) = \begin{Vmatrix} \rho_1^o w^2 + 2A_{1S_1} w & -S_1 n^2 & 0 \\ -\left(\rho_2^o w^2 + 2A_{2S_1} w\right) & 0 & -S_2 n^2 \\ \beta_1 w & -w & w \end{Vmatrix} = 0 \tag{6.135}$$

Evaluation of the determinant yields the dispersion relation

$$wn^2(\rho_{S_1} w^2 + 2A_{S_1} w - K_{S_1} n^2) = 0 \tag{6.136}$$

where the density, attenuation density and bulk modulus of the saturation wave are

$$\rho_{S_1} = S_1 \rho_2 + S_2 \rho_1 \tag{6.137}$$

$$A_{S_1} = S_1 A_{2S_1} + S_2 A_{1S_1} \tag{6.138}$$

$$K_{S_1} = S_1 S_2 \beta_1 \tag{6.139}$$

The attenuation density may be evaluated using Equations 6.39 and 6.45

$$A_{S_1} = S_1 \frac{1}{2}\left(\frac{Q_{21}}{S_1} + \frac{Q_{22}}{S_2}\right) + S_2 \frac{1}{2}\left(\frac{Q_{11}}{S_1} + \frac{Q_{12}}{S_2}\right) \tag{6.140}$$

Discarding the trivial solutions $w = 0$ and $n = 0$, the dispersion relation reduces to

$$\rho_{S_1} w^2 + 2A_{S_1} w - K_{S_1} n^2 = 0 \tag{6.141}$$

Dividing this equation by the density ρ_{S_1} yields

$$w^2 + 2a_{S_1} w - v_{oS_1}^2 n^2 = 0 \tag{6.142}$$

where the attenuation and unattenuated wave speed are

$$a_{S_1} = \frac{A_{S_1}}{\rho_{S_1}} \tag{6.143}$$

$$v_{oS_1}^2 = \frac{K_{S_1}}{\rho_{S_1}} \tag{6.144}$$

As shown previously for a given $w = -i\omega$, the solution of this dispersion relation is

$$n = -\kappa + ik \tag{6.145}$$

$$\kappa = \frac{1}{v_{oS_1}\sqrt{2}} \sqrt{\sqrt{\omega^4 + 4a_{S_1}^2 \omega^2} - \omega^2} \tag{6.146}$$

$$k = \frac{1}{v_{oS_1}\sqrt{2}} \sqrt{\sqrt{\omega^4 + 4a_{S_1}^2 \omega^2} + \omega^2} \tag{6.147}$$

$$v = \frac{\omega}{k} = v_{oS_1} \frac{\sqrt{2}\omega}{\sqrt{\sqrt{\omega^4 + 4a_{S_1}^2 \omega^2} + \omega^2}} \tag{6.148}$$

which illustrates the frequency, viscosity, saturation and permeability dependence of the wave velocity.

For the purposes of the discussion in this chapter, it is assumed that the saturation waves are strongly coupled to the porosity waves. This assumption is based on observations of the coupling between these two waves. The actual theoretical description of the coupling between these waves is given in Chapter 10. Wave solutions of this matrix wave equation (Equation 6.71) are only possible in the limit where the waves are decoupled. That is because the theory that has been constructed using volume averaging, the component velocities and volume fractions cannot be used to describe three interacting phases. That becomes apparent when attempts are made to construct Onsager's relations. When that construction is attempted using the current system of equations, it is observed that the system of equations is not complete. An illustration of the conceptual leap that must be made in describing two versus three components is easily observed in the context of the seismic waves. For two components there are two component velocities and there are two momentums associated with the first and second P waves, the (almost) in-phase motion (momentum) and the (almost) out-of-phase motion (momentum). However, with three components there are four momentums that must be considered. Thus there are four waves that must be described. That theory is constructed in the following chapters. The calculations that show that the matrix wave equation (Equation 6.71) only yields solutions when the waves are decoupled are presented in Udey (2012).

NUMERICAL ILLUSTRATIONS

For the purposes of a numerical illustration, consider a porous medium consisting of a Berea sandstone, water and a light oil. The physical properties of these materials are presented in Table 6.1, along with the model parameters required for the numerical illustration. In addition to these parameters, the saturation dependence of the total fluid viscosity μ_f, the relative Darcy resistances Q_{ij} and the megascopic capillary pressure P_c must also be specified. These functions will be presented as needed.

To illustrate the porosity wave, the total fluid viscosity is chosen to be a parallel combination of the individual fluids weighted by the saturation

$$\frac{1}{\mu_f} = \frac{S_1}{\mu_1} + \frac{S_2}{\mu_2} \tag{6.149}$$

This function, along with the parameters in Table 6.1, completely specifies the porosity wave attenuation given by

$$a_\eta = \frac{1}{2}\frac{\eta\mu_f}{\rho_\eta K} \tag{6.150}$$

TABLE 6.1 Material and Model Parameters

Quantity	Value	Units	Comment
ρ_s	2650.0	kg/m^3	
K_s	3.3×10^{10}	Pa	
μ_s	2.3×10^{10}	Pa	
ρ_1	1000	kg/m^3	
K_1	2.9×10^9	Pa	
μ_1	0.001	$Pa \cdot s$	$1cp$
ξ_1	0.002	$Pa \cdot s$	
ρ_2	900	kg/m^3	
K_2	2.9×10^9	Pa	
μ_2	0.002	$Pa \cdot s$	
ξ_2	0.0042	$Pa \cdot s$	
η	0.03		
S_{1r}	0.2		
S_{2r}	0.05		
K	9.87×10^{-13}	m^2	$1Darcy$
α_η	0.85		
λ_s	0.9		
P_e	6894.8	Pa	$1psi$
δ_e	0.01		

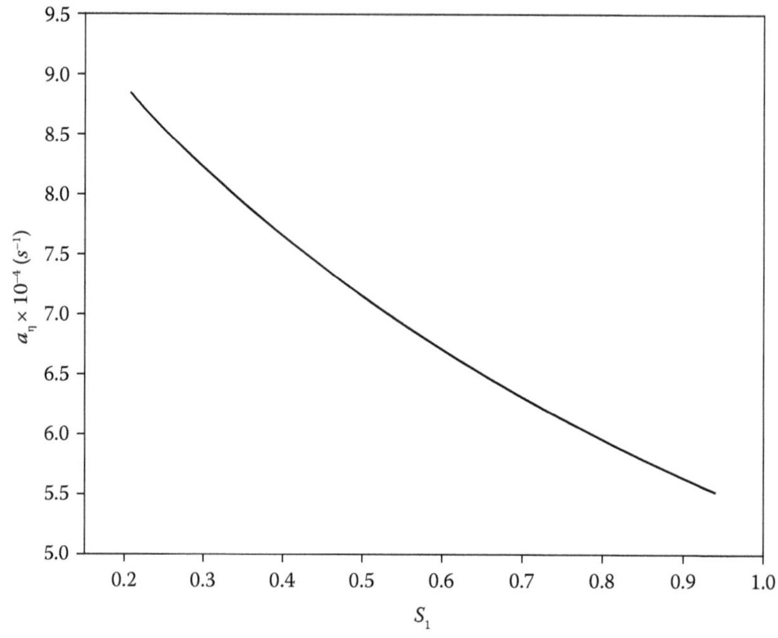

FIGURE 6.9 Porosity wave attenuation a_η versus saturation S_1.

A graph of this function is shown in Figure 6.9.

The undamped porosity wave speed $v_{o\eta}$ depends on the value of $\mu_{M_{ss}}$, which in turn depends on the values of λ_{s1} and λ_{s2}. Let

$$\lambda_{S1} = S_1\lambda_S \quad \lambda_{S2} = S_2\lambda_S \tag{6.151}$$

then

$$\begin{aligned} \mu_{M_{ss}} &= \eta_s\mu_s(1-\lambda_{S1}-\lambda_{S2}) \\ &= \eta_s\mu_s(1-\lambda_S) \end{aligned} \tag{6.152}$$

By specifying λ_s as a constant independent of saturation, then the undamped wave speed is independent of saturation. Applying the parameter values given in Table 6.1 yields

$$v_{o\eta} = 2707.64\,\text{m/s} \tag{6.153}$$

Now that a_η and $v_{o\eta}$ have been specified, using Equations 6.112 to 6.114, the spatial attenuation κ, the wave number k and the wave speed v of the trial solutions may be determined as functions of the saturation S_1 and angular frequency ω. Note the wave speed $v_{o\eta}$ is the speed the wave approaches in the megahertz range. Attempts to create a porosity wave in this frequency range results in the majority of the energy going into acoustic waves. The transition from a wave speed of hundreds of m/s to this wave speed occurs in the kilohertz range. The most effective frequencies for propagating a fluid using a porosity wave are given by Figure 6.6.

Now consider the comparison of porosity diffusion to the pressure diffusion process. Applying the parameters in Table 6.1 to the diffusion coefficient ratio, Equation 6.122 yields

$$\frac{D_\eta}{D_{P_f}} = 9.47 \tag{6.154}$$

In this case, the porosity diffusion process is clearly the preferred mechanism for the relaxation of pressure to an ambient level.

A numerical illustration of the saturation wave is given by specifying the saturation dependence of the attenuation a_{S_1} and the unattenuated wave speed v_{oS_1}. Then the wave's spatial attenuation κ, wave number k and wave speed v can be obtained from the dispersion relation solution given by Equations 6.146 through 6.148. Graphs of these functions may then be presented.

For this example, the following relations from Udey (2009) are used:

$$Q_{11} = \eta\frac{\mu_1}{K}R_{11} \tag{6.155}$$

$$Q_{12} = \eta\frac{\mu_1}{K}R_{12} \tag{6.156}$$

$$Q_{21} = \eta\frac{\mu_2}{K}R_{21} \tag{6.157}$$

$$Q_{22} = \eta\frac{\mu_2}{K}R_{22} \tag{6.158}$$

where R_{ij} are relative Darcy resistance functions, which yields

$$A_{S_1} = \frac{1}{2}\eta\frac{1}{K}\left(S_1\mu_2\left(\frac{R_{21}}{S_1}+\frac{R_{22}}{S_2}\right)+S_2\mu_1\left(\frac{R_{11}}{S_1}+\frac{R_{12}}{S_2}\right)\right) \tag{6.159}$$

Now defining the effective viscosity of the combined fluids μ_e as

$$\mu_e = S_1\mu_2\left(\frac{R_{21}}{S_1}+\frac{R_{22}}{S_2}\right)+S_2\mu_1\left(\frac{R_{11}}{S_1}+\frac{R_{12}}{S_2}\right) \tag{6.160}$$

the expression for a_{S_1} becomes

$$a_{S_1} = \frac{A_{S_1}}{\rho_{S_1}} = \frac{1}{2}\frac{\eta}{S_1\rho_2+S_2\rho_1}\frac{\mu_e}{K} \tag{6.161}$$

So a_{S_1} depends on the relative Darcy resistance functions R_{ij}. To obtain these functions, consider the equations of motion for the two fluids in the case of steady-state flow with the viscous terms considered negligible. In this case the equations of motion become

$$0 = -\eta_1 \nabla P_1 - \eta^2 \frac{\mu_1}{K} R_{11} \vec{v}_1 + \eta^2 \frac{\mu_1}{K} R_{12} \vec{v}_2 \tag{6.162}$$

$$0 = -\eta_2 \nabla P_2 + \eta^2 \frac{\mu_2}{K} R_{21} \vec{v}_1 - \eta^2 \frac{\mu_2}{K} R_{22} \vec{v}_2 \tag{6.163}$$

Solving this system of equations for the volumetric flow rates $\vec{q}_1 = \eta_1 \vec{v}_1$ and $\vec{q}_2 = \eta_2 \vec{v}_2$ yields

$$\vec{q}_1 = -\frac{K}{\mu_1} S_1^2 J_{11} \nabla P_1 - \frac{K}{\mu_2} S_1 S_2 J_{12} \nabla P_2 \tag{6.164}$$

$$\vec{q}_2 = -\frac{K}{\mu_1} S_1 S_2 J_{21} \nabla P_1 - \frac{K}{\mu_2} S_2^2 J_{22} \nabla P_2 \tag{6.165}$$

where the relative Darcy conductivities J_{ij} are defined by

$$\begin{bmatrix} J_{11} & J_{12} \\ J_{21} & J_{22} \end{bmatrix} \begin{bmatrix} R_{11} & -R_{12} \\ -R_{21} & R_{22} \end{bmatrix} = \begin{bmatrix} 1 & 0 \\ 0 & 1 \end{bmatrix} \tag{6.166}$$

So if the J_{ij} are specified, then the functions R_{ij} can be computed using Equation 6.166. The solution for the volumetric flow rates $\vec{q}_1$ and $\vec{q}_2$ may be rewritten as follows (where $P_{21} = P_2 - P_1$):

$$\vec{q}_1 = -\frac{K}{\mu_1} L_{11} \nabla P_1 - \frac{K}{\mu_1} L_{1c} \nabla P_{21} \tag{6.167}$$

$$\vec{q}_2 = -\frac{K}{\mu_2} L_{22} \nabla P_2 - \frac{K}{\mu_2} L_{2c} \nabla P_{21} \tag{6.168}$$

where the relative mobilities L_{ij} are given by

$$\begin{bmatrix} L_{11} & L_{1c} \\ L_{22} & L_{2c} \end{bmatrix} = \begin{bmatrix} S_1^2 J_{11} + \frac{\mu_1}{\mu_2} S_1 S_2 J_{12} & \frac{\mu_1}{\mu_2} S_1 S_2 J_{12} \\ \frac{\mu_2}{\mu_1} S_1 S_2 J_{21} + S_2^2 J_{22} & \frac{\mu_2}{\mu_1} S_1 S_2 J_{21} \end{bmatrix} \tag{6.169}$$

As discussed by Udey (2009), the J_{ij} functions may be specified to produce the functions L_{11} and L_{22} that are similar to the standard relative permeability curves K_{r1} and K_{r2}, that is,

$$K_{r1} \approx L_{11} \; K_{r2} \approx L_{22} \tag{6.170}$$

Furthermore, the cross terms L_{1c} and L_{2c} should be negligible at low and high saturations; however, they should be significant at intermediate saturations (van Grenabeek and Rothman 1996).

Now, defining the effective saturation S_e as

$$S_e = \frac{S_1 - S_{1r}}{1 - S_{2r} - S_{1r}} \tag{6.171}$$

the J_{ij} functions were chosen to be (Udey, 2009). These functions were chosen to have the correct values at the end points.

$$J_{11} = \frac{S_e}{(1 - S_{2r})^2} \tag{6.172}$$

$$J_{12} = S_e^2(1 - S_e) \tag{6.173}$$

$$J_{21} = S_e(1 - S_e)^2 \tag{6.174}$$

$$J_{22} = \frac{(1 - S_e)}{(1 - S_{1r})^2} \tag{6.175}$$

A graph of these functions is shown in Figure 6.10. The relative mobility curves L_{ij} may now be determined using Equation 6.169. A graph of the relative mobility curves is shown in Figure 6.11. Note that the curves for L_{11} and L_{22} are similar to standard relative permeability curves (Bear, 1972; Spanos et al., 1988). With the relative Darcy resistances R_{ij} specified by the relative Darcy conductivities J_{ij} via Equation 6.166, the effective viscosity of the fluids μ_e may be computed as a function of saturation. In turn, if the flow is dispersional, this specifies the saturation wave attenuation a_{S_1}.

The unattenuated wave speed may be calculated from the expression

$$v_{oS_1}^2 = \frac{K_{S_1}}{\rho_{S_1}} = \frac{S_1 S_2 \beta_1}{S_1\rho_2 + S_2\rho_1} \tag{6.176}$$

here β_1 is given by

$$\beta_1 = \beta_c + \beta_h(S_1) \tag{6.177}$$

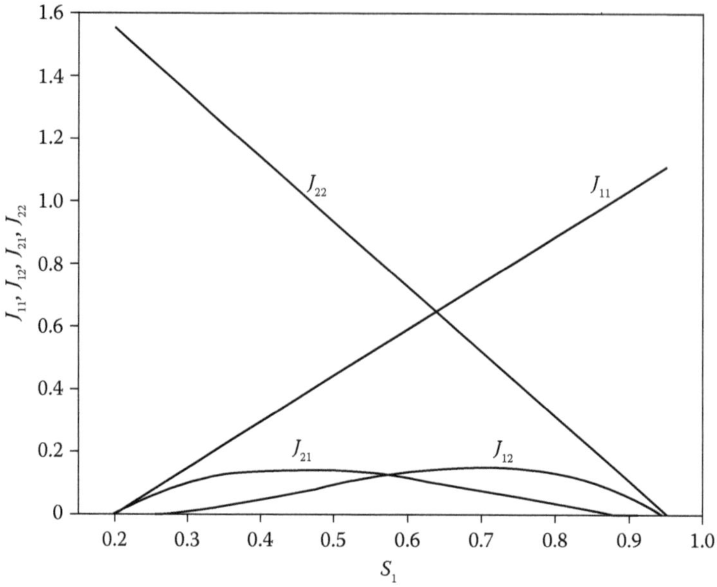

FIGURE 6.10 Plots of the functions J_{ij}. (Adapted from MAD Viera, PN Sahay, M Coronado, AO Tapia, *Applications in Geosciences*, CRC Press, July 24, 2012.)

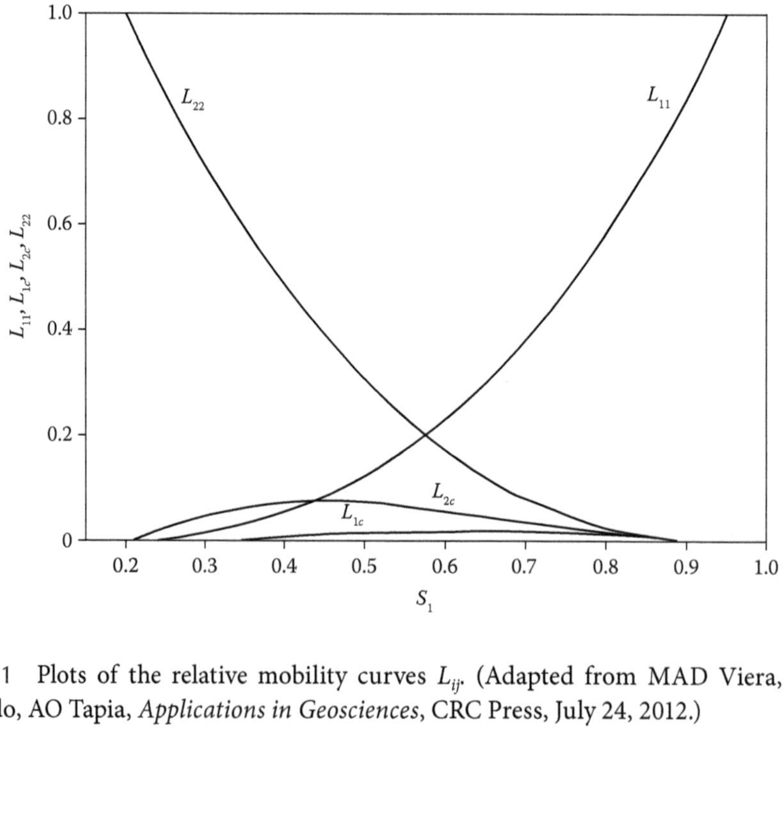

FIGURE 6.11 Plots of the relative mobility curves L_{ij}. (Adapted from MAD Viera, PN Sahay, M Coronado, AO Tapia, *Applications in Geosciences*, CRC Press, July 24, 2012.)

The megascopic capillary pressure function P_c will be taken to be the Brooks–Cory capillary function with parameter $\lambda = 2$ (Brooks and Corey 1964), namely

$$P_c = P_e\left(\frac{S_e + \delta_e}{1 + \delta_e}\right)^{-\frac{1}{2}} \tag{6.178}$$

Consequently, the function β_c is

$$\beta_c = -\frac{dP_c}{dS_1} = \frac{P_e}{2(1 - S_{2r} - S_{1r})}\left(\frac{S_e + \delta_e}{1 + \delta_e}\right)^{-\frac{3}{2}} \tag{6.179}$$

For this numerical illustration, capillary hysteresis effects will be considered negligible, so $\beta_h(S_1) = 0$. Therefore $\beta_1 = \beta_c$ and there is now enough information to compute v_{oS_1}. A graph of the unattenuated wave speed for dispersion is shown in Figure 6.12.

At this point, it is important to note that this is a two-phase flow model without pulsing. These values of permeabilities (Equations 6.172, 6.173, 6.174, 6.175) do not actually yield a physical dispersion curve. Here, upon substituting into Equation 6.148, dispersional flow is not observed. This means that such a displacement under steady-state conditions will result in viscous fingering as is observed experimentally. So in order to model dispersion, it must be asked how pulsing can be assumed to change the permeability parameters to

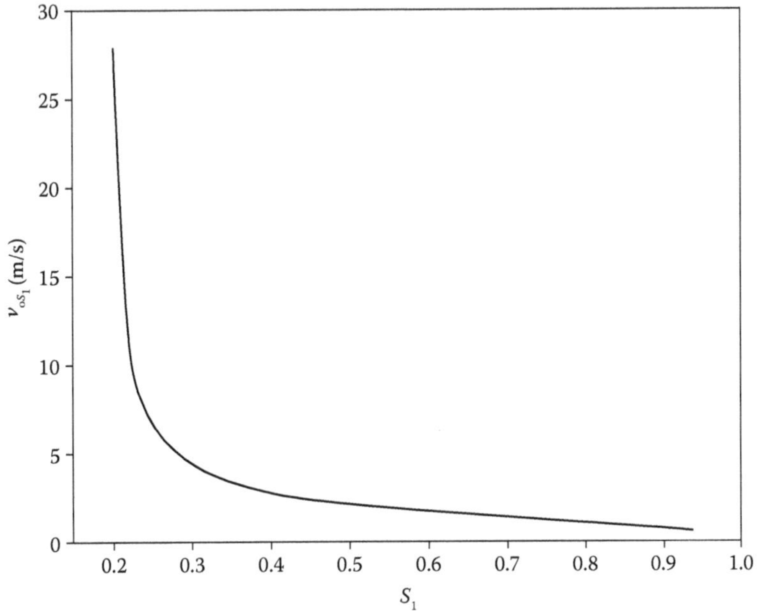

FIGURE 6.12 A plot of unattenuated wave speed versus saturation when dispersion occurs. (Adapted from MAD Viera, PN Sahay, M Coronado, AO Tapia, *Applications in Geosciences*, CRC Press, July 24, 2012.)

simulate dispersion. Here, if the slope of J_{11} is increased at low water saturations and the slope of J_{22} is reduced by a sufficient amount, dispersion is obtained (the velocity of saturation waves increase monotonically as saturation is decreased). Also, the changes in the slope of J_{22} change the shape of the dispersion curve. Since these parameters depend on the reservoir properties, it allows for reservoir properties to be crudely introduced into this very simple model. For an actual physical solution, all of the parameters and variables described in Chapter 10 must be included.

A MODEL OF FRONTAL DISPERSION FROM A WELL

The next objective is to use this theory to describe a very practical example of pulsed flow into a well and to describe the evolution of the radius of influence of the injected fluid. Here, various coordinate transformations are presented to express the description of the injected fluid in terms of saturation, radius, volume, dispersion and time. Note that in this analysis the surface pump supplies a constant volume of injected fluid Q per unit time. That is simply the reality of how oil field pumps work; it has nothing to do with the physical theory. The fluid is then pulsed in at the perforations, which results in the same average injection rate $\bar{Q}$ over a sufficiently long period of time.

Consider a layer of height h and porosity η within a reservoir that is penetrated by a well of radius r_w. Let the centre of the well be the origin of a cylindrical geometry whose coordinates are the radius r, the angle φ and the height z. Angular symmetry is assumed.

A fluid is injected into the layer, which displaces the *in situ* fluid. When the displacing fluid is less viscous than the *in situ* fluid and is pulsed into the reservoir, a transition zone is generated between the two fluids. Denoting the saturation of the displacing fluid by S_1, the front is described as a distribution of the saturation of injected fluid $S_1(r, t)$ dependent on radius r and time t. A typical front and its evolution is shown in Figure 6.13.

Since the saturation is a monotonic function of the radius within the front, the radius r may be obtained for a given saturation S_1. Therefore the front may be described as a radial function $r(S_1, t)$ that depends on saturation and time. This alternative point of view is shown in Figure 6.14.

Typically, the injected fluid is water or a chemical mixed with water. Assume there exists a connate water distribution corresponding to a constant irreducible saturation S_{1r}. The figures presented assume $S_{1r} = 0.2$ and $r_w = 2.5$ inch for illustrative purposes. The connate water and the injected fluid are treated as a single fluid with saturation S_1. Denoting the saturation of the *in situ* fluid as S_2 yields

$$S_1 + S_2 = 1 \tag{6.180}$$

In principal, PowerWave^{TM} pulsing can reduce the saturation of the *in situ* fluid to 0 (as has been illustrated in numerous laboratory experiments), so the irreducible saturation of Fluid 2 is set to $S_{2r} = 0$. When injection has not begun, $S_1 = S_{1r}$ and so the maximum value of S_2 is $1 - S_{1r}$.

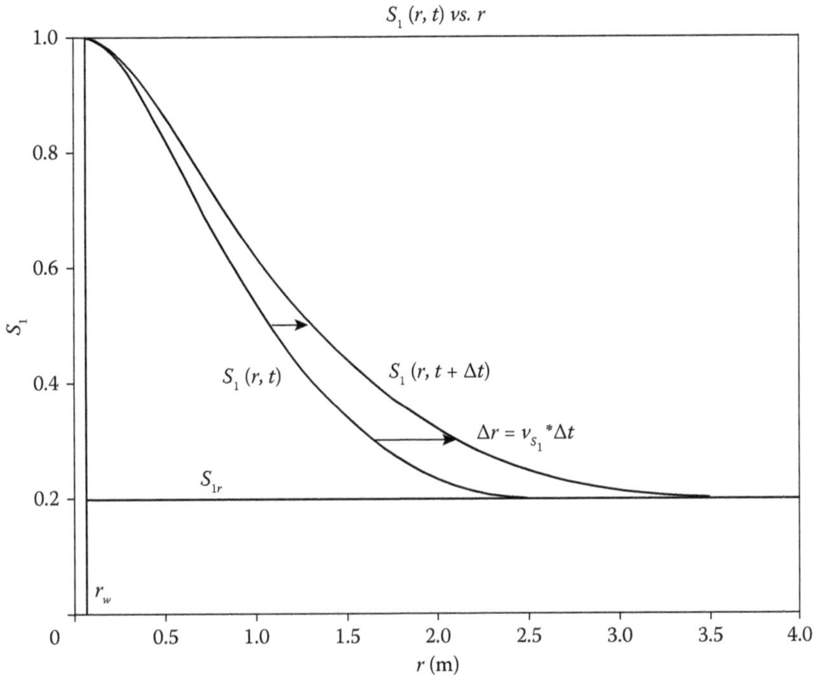

FIGURE 6.13 This graph shows a gradual change in the saturation of water at the front, which causes dispersion to dominate and a broad or diffuse front to form.

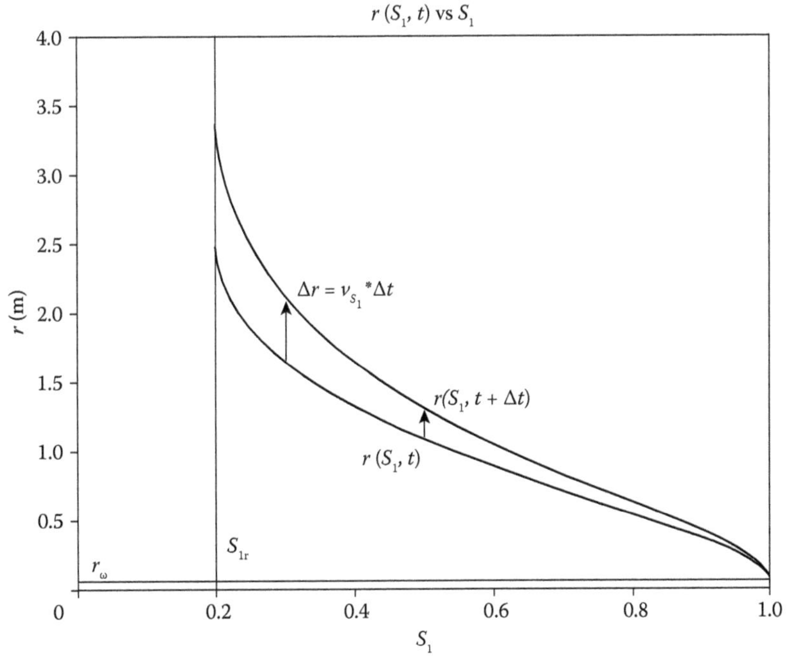

FIGURE 6.14 The front as a radial function of saturation and time.

The part of Fluid 1 that was injected may be considered distinct from the connate water, so the saturation S_i of the injected fluid may be defined as

$$S_i = S_1 - S_{1r} \tag{6.181}$$

Consequently, the front may be described in terms of S_i instead of S_1. For Figure 6.13 this shifts the graph down until the line denoting the connate water saturation coincides with the y-axis.

An element of total volume within the reservoir layer is

$$dV = r dr d\varphi dz \tag{6.182}$$

and consequently, the total volume contained within radius r is

$$V(r) = \int_{r=r_w,\varphi=0,z=0}^{r=r,\varphi=2\pi,z=h} r\,dr\,d\varphi\,dz = \pi(r^2 - r_w^2)h = \pi r^2 h - V_w \tag{6.183}$$

where $V_w = \pi r_w^2 h$ is the volume of the well within the layer.

An element of total fluid volume within the layer is

$$dV_f = \eta dV \tag{6.184}$$

By taking the porosity η to be a constant, the fluid volume contained within radius r is

$$V_f(r) = \int \eta dV = \eta V(r) = \eta\pi(r^2 - r_w^2)h \tag{6.185}$$

Then the radius may be expressed as

$$r^2 = \frac{V_f}{\eta\pi h} + r_w^2 \tag{6.186}$$

or

$$r = \sqrt{\frac{V_f}{\eta\pi h} + r_w^2} \tag{6.187}$$

Thus any function $f(r)$ that depends on r may be recast as a function $f(V_f)$ of fluid volume V_f. So, now change the description of the front in Figure 6.14 from a function $r(S_1, t)$ to a fluid volume function $V_f(S_i, t)$ of the injected fluid saturation S_i and time t. The result is shown in Figure 6.15, with h = 3 meters and η = 0.3 for illustrative purposes.

An element of volume dV_f contains an amount of injected fluid dV_i at time t given by

$$dV_i = S_i(r, t) dV_f \tag{6.188}$$

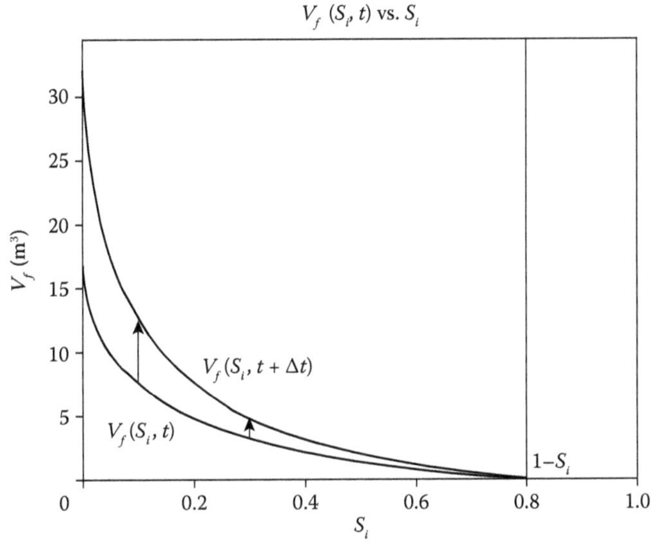

FIGURE 6.15 The front as fluid volume versus saturation.

It has been noted that the radius r can be expressed in terms of a corresponding volume V_f and consequently the element of injected volume may be rewritten as

$$dV_i = S_i(V_f, t)dV_f \tag{6.189}$$

The volume of injected fluid at time t contained within a volume V_f that corresponds to radius r is

$$V_i(t) = \int_{V_f=0}^{V_f} S_i(V_f, t)dV_f \tag{6.190}$$

Let $r_p(t)$ be the radius of penetration of the front at time t, where $S_i(r_p, t) = 0$ or alternatively $r_p(t) = r(S_i = 0, t)$. This is the distance that the front has travelled into the layer since the beginning of the injection. Let $V_f^p(t)$ be the total fluid volume enclosed within the radius $r_p(t)$. Then the total volume of injected fluid within the front at time t is

$$V_i(t) = V_i[V_f^p(t)] = \int_{V_f=0}^{V_f^p(t)} S_i(V_f, t)dV_f \tag{6.191}$$

and subsequently, the volume rate of injection $Q(t)$ is

$$Q(t) = \frac{dV_i(t)}{dt} \tag{6.192}$$

Utilizing integration by parts the injected volume may be reinterpreted as

$$\begin{aligned} V_i(t) &= \int_{V_f=0}^{V_f} S_i(V_f,t)dV_f \\ &= S_i(V_f,t)\times V_f\Big]_{S_i=1-S_{1r},V_f=0}^{S_i=0,V_f=V_f^o} - \int_{S_i=1-S_{1r}}^{S_i=0} V_f(S_i,t)dS_i \\ &= 0\times V_f^o - (1-S_{1r})\times 0 + \int_{S_i=0}^{S_i=1-S_{1r}} V_f(S_i,t)dS_i \\ &= \int_{S_i=0}^{S_i=1-S_{1r}} V_f(S_i,t)dS_i \end{aligned} \tag{6.193}$$

This is the area under the curve shown in Figure 6.15.

The volumetric injection rate now becomes

$$\begin{aligned} Q(t) &= \frac{dV_i(t)}{dt} \\ &= \int_{S_i=0}^{S_i=1-S_{1r}} \frac{dV_f(S_i,t)}{dt} dS_i \\ &= \int_{S_i=0}^{S_i=1-S_{1r}} Q(S_i,t)dS_i \end{aligned} \tag{6.194}$$

where the dispersion function $Q(S_i, t)$ has been defined by

$$Q(S_i,t) = \frac{dV_f(S_i,t)}{dt} \tag{6.195}$$

From Equation 6.185

$$\begin{aligned} \frac{dV_f(S_i,t)}{dt} &= \frac{d[\eta\pi[r^2(S_i,t)-r_w^2]h]}{dt} \\ &= \eta\pi h\frac{d[r^2(S_i,t)-r_w^2]}{dt} = \eta\pi h\frac{dr^2(S_i,t)}{dt} \end{aligned} \tag{6.196}$$

Therefore

$$Q(S_i,t) = \eta\pi h\frac{dr^2(S_i,t)}{dt} \tag{6.197}$$

or

$$\frac{dr^2(S_i,t)}{dt} = \frac{Q(S_i,t)}{\eta\pi h} \tag{6.198}$$

Integrating this last result from $t = t_o$ to t yields

$$r^2(S_i,t) = r^2(S_i,t_o) + \int_{t=t_o}^{t} \frac{Q(S_i,t)}{\eta\pi h} dt \tag{6.199}$$

The essence of solving the dispersion problem is specifying the dispersion function $Q(S_i, t)$. This function describes how a volume of injected fluid is distributed across the range of the injection saturation. In an experiment, the evolution of the front may be observed and the function $r^2(S_i, t)$ may be obtained. The dispersion function may then be obtained empirically from Equation 6.197.

This theory may be simplified even more by assuming that in many cases the saturation dependence of the dispersion function is independent of time and independent of the fluid injection rate. This hypothesis may be formulated as

$$Q(S_i,t) = Q(t) f_V(S_i) \tag{6.200}$$

where $f_V(S_i)$ is the dispersion curve. It is the fraction of an injected fluid volume V that is distributed into the saturation\ S_i. A typical dispersion curve is shown in Figure 6.16.

From Equation 6.194

$$\begin{aligned} Q(t) &= \int_{S_i=0}^{S_i=1-S_{1r}} Q(S_i,t) dS_i \\ &= \int_{S_i=0}^{S_i=1-S_{1r}} Q(t) f_V(S_i) dS_i \\ &= Q(t) \int_{S_i=0}^{S_i=1-S_{1r}} f_V(S_i) dS_i \end{aligned} \tag{6.201}$$

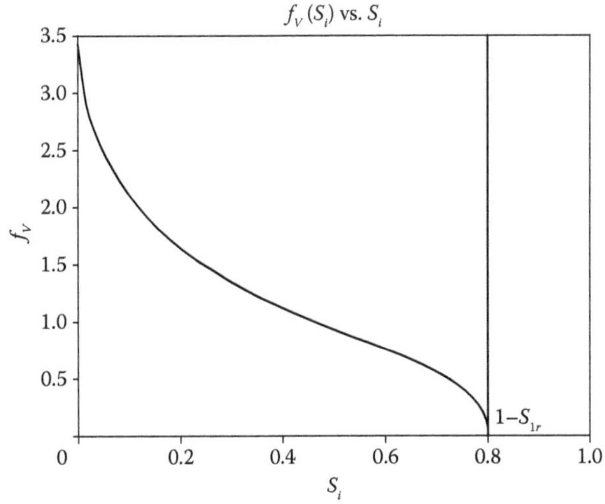

FIGURE 6.16 The dispersion curve $f_V(S_i)$.

Therefore the dispersion curve is subject to the constraint

$$\int_{S_i=0}^{S_i=1-S_{1r}} f_V(S_i)dS_i = 1 \tag{6.202}$$

From Equation 6.199, the evolution of the front is given by

$$\begin{aligned} r^2(S_1,t) &= r^2(S_1,t_o) + \int_{t=t_o}^{t} \frac{Q(S_i,t)}{\eta\pi h} dt \\ &= r^2(S_1,t_o) + \int_{t=t_o}^{t} \frac{Q(t) f_V(S_i)}{\eta\pi h} dt \\ &= r^2(S_1,t_o) + \frac{f_V(S_i)}{\eta\pi h} \int_{t=t_o}^{t} Q(t)dt \\ &= r^2(S_1,t_o) + \frac{f_V(S_i)}{\eta\pi} \frac{V_i(t)}{h} \end{aligned} \tag{6.203}$$

This convenient result expresses the radial distribution of the front in terms of the treatment volume $\frac{V_i(t)}{h}$ injected up to time t.

Since the area of the well is $A_w = 2\pi r_w h$

$$\begin{aligned} \frac{Q(S_i,t)}{\eta\pi h} &= 2r_w \frac{Q(t) f_V(S_i)}{\eta 2 r_w \pi h} = 2r_w \frac{Q(t) f_V(S_i)}{\eta A_w} \\ &= 2r_w v_{iw}(t) f_V(S_i) = 2r_w v_{S_i}(t) \end{aligned} \tag{6.204}$$

where the injected fluid's speed at the well is

$$v_{iw}(t) = \frac{Q(t)}{\eta A_w} = \frac{Q(t)}{\eta 2 r_w \pi h} \tag{6.205}$$

and the isosaturation speed is given by

$$v_{S_i}(t) = v_{iw}(t) f_V(S_i) \tag{6.206}$$

From Equation 6.199, the evolution of the front is given by

$$\begin{aligned} r^2(S_i,t) &= r^2(S_i,t_o) + \int_{t=t_o}^{t} \frac{Q(S_i,t)}{\eta\pi h} dt \\ &= r^2(S_i,t_o) + \int_{t=t_o}^{t} 2r_w v_{iw}(t) f_V(S_i) dt \\ &= r^2(S_i,t_o) + 2r_w \Delta r(t,t_o) f_V(S_i) \end{aligned} \tag{6.207}$$

where the cumulative travel distance is defined by

$$\Delta r(t,t_o)=\int_{t=t_o}^{t} v_{iw}(t)dt \tag{6.208}$$

The definition for v_{iw} (Equation 6.205) yields

$$\Delta r(t,t_o)=\int_{t=t_o}^{t}\frac{Q(t)}{\eta A_w}dt=\int_{t=t_o}^{t}\frac{Q(t)}{\eta 2r_w\pi h}dt=\frac{V(t)}{\eta 2r_w\pi h} \tag{6.209}$$

During PowerWave™ pulsing, the fluid is supplied to the tool at a constant rate (that is how oilfield pumps work). In the case of PowerWave™ pulsing, the tool operates with some fixed configuration parameters, which results in a constant pulse rate. Under these circumstances the volumetric injection rate is represented by an average value, namely

$$Q(t)=\bar{Q} \tag{6.210}$$

and the total volume injected up to time t is

$$V_i(t)=\bar{Q}(t-t_o) \tag{6.211}$$

In this situation the dispersion function is given by

$$Q(S_i,t)=\bar{Q}(t)f_V(S_i) \tag{6.212}$$

The frontal advancement is then given by

$$r^2(S_i,t)=r^2(S_i,t_o)+\frac{f_V(S_i)}{\eta\pi}\frac{\bar{Q}(t)}{h} \tag{6.213}$$

The injected fluid's speed will be constant in time

$$v_{iw}(t)=\frac{\bar{Q}(t)}{\eta A_w}=\frac{\bar{Q}(t)}{\eta 2r_w\pi h} \tag{6.214}$$

as is the isosaturation speed

$$v_{S_i}(t)=v_{iw}(t)f_V(S_i) \tag{6.215}$$

and the cumulative travel distance will be

$$\Delta r(t,t_o)=v_{iw}\times(t-t_o) \tag{6.216}$$

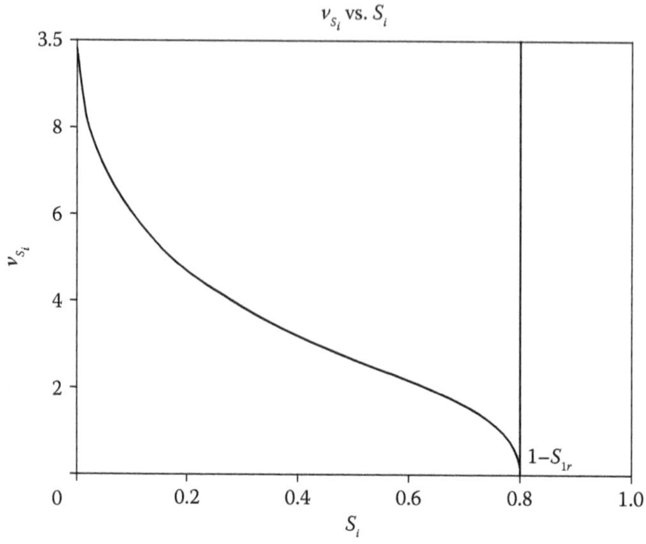

FIGURE 6.17 The isosaturation speed v_{si}.

The evolution of the front becomes

$$\begin{aligned} r^2(S_i,t) &= r^2(S_i,t_o)+2r_w[v_{iw}(t)\times(t-t_o)]f_V(S_i) \\ &= r^2(S_1,t_o)+2r_w v_{S_i}(t)\times(t-t_o) \end{aligned} \tag{6.217}$$

The dispersion problem is now specified by defining the dispersion curve $f_V(S_i)$.

An alternative approach to specifying the dispersion problem is to define the isoconcentration speed. Since the area under the isoconcentration curve is

$$\int v_{S_i} dS_i = \int v_{iw} f_V(S_i) dS_i = v_{iw} \int f_V(S_i) dS_i = v_{iw} \tag{6.218}$$

The dispersion curve is then specified from the isoconcentration curve by performing the above integration and then computing

$$f_V(S_i) = \frac{v_{S_i}}{v_{iw}} \tag{6.219}$$

Figure 6.17 plots the isosaturation speed.

RADIUS OF INFLUENCE

The radius of influence r_1 is defined here as the radius around the well that encompasses 90% of the volume of the injected fluid. This definition provides a reasonable estimate of the volume around the well that is affected by an injection treatment volume during stimulation or chemical remediation.

From Equation 6.203

$$f_V(S_i)V_i(t) = \eta\pi h[r(S_i,t) - r^2(S_1,t_o)] \tag{6.220}$$

an element of injected fluid volume at radius $r(S_i,t)$ is

$$dV_i(S_i) = \eta\pi h[r(S_i,t) - r^2(S_1,t_o)]dS_i = f_V(S_i)V_i(t)dS_i \tag{6.221}$$

Let S_i^o be the saturation that corresponds with the radius of influence and define $f_V^o = f_V(S_i^o)$. The integral of the fluid volume from the radius of penetration to the radius of influence corresponds to the saturation range $S_i \in [0, S_i^o]$. This constitutes 10% of the injected volume. The saturation interval $S_i \in [S_i^o, (1 - S_{1r})]$ corresponds with the 90% of the injected volume encompassed by the radius of influence. Performing the integral as shown in Figure 6.18 yields

$$\int_{S_i=0}^{S_i^o} dV_i(S_i) = \int_{S_i=0}^{S_i^o} V_i(t) f_V(S_i) dS_i = 0.1\ V_i(t) \tag{6.222}$$

Therefore S_i^o is determined from the constraint

$$\int_{S_i=0}^{S_i^o} f_V(S_i) dS_i = 0.1 \tag{6.223}$$

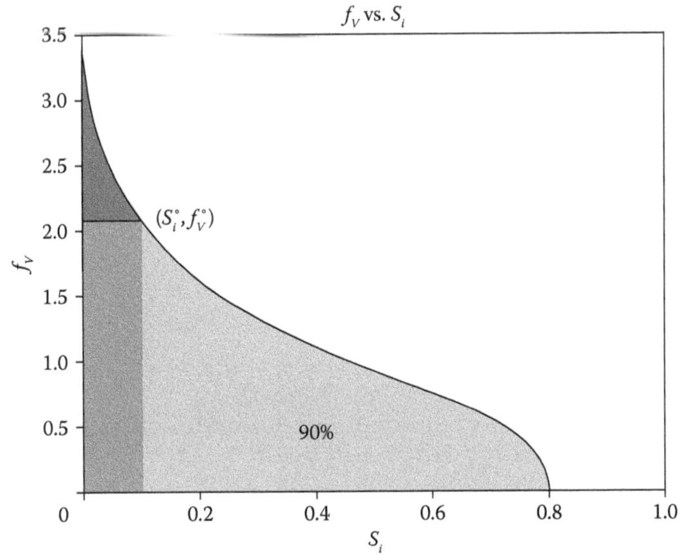

FIGURE 6.18 Integration of the fractional volume f_V (S_i).

In practice, the function $f_V(S_i)$ is represented by a set of points (S_i, f_V). Unfortunately, the steep slopes of the function at the end points makes it impractical to use the data points to define a natural cardinal spline for f_V in the variable S_i. However, a natural cardinal spline may be constructed for $S_i(f_V)$ from the points (f_V, S_i). Since the spline is piecewise cubic, numerical integration of f_V is straightforward.

Integration by parts yields

$$
\begin{aligned}
0.1 &= \int_{S_i=0}^{S_i^o} f_V(S_i)\,dS_i \\
&= S_i f_V \,\Big|_{S_i=0,\, f_V=f_V^{\max}}^{S_i^o,\, f_V^o} - \int_{f_V^{\max}}^{f_V^o} S_i\, df_V(S_i) \\
&= S_i^o f_V^o + \int_{f_V^o}^{f_V^{\max}} S_i\, df_V(S_i)
\end{aligned}
\tag{6.224}
$$

This integration is shown in Figure 6.19, where the areas of integration correspond to the areas of integration shown in Figure 6.18.

This allows for a numerical calculation of the radius of influence as a function of time for the pulsed injection process.

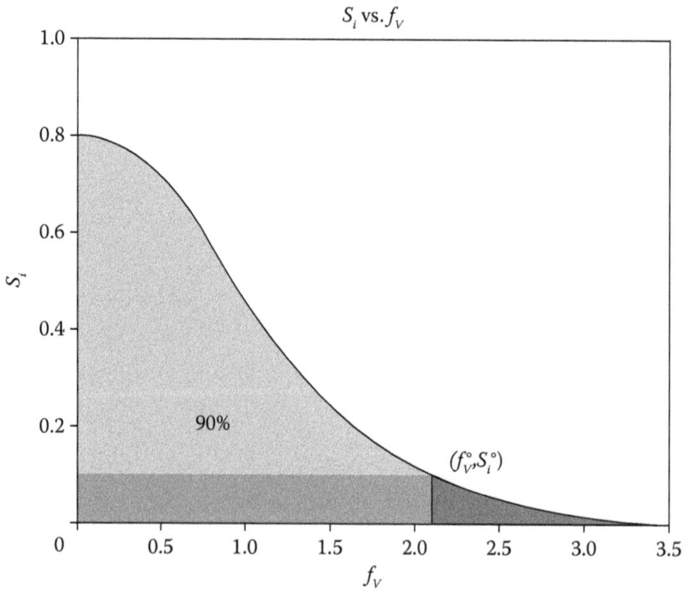

FIGURE 6.19 Integration of the injected fluid saturation $S_i(f_V)$.

SUMMARY

This chapter presented very simple descriptions of porosity–pressure waves and saturation waves. However, this work exposes a weakness in the theory presented in the earlier chapters. It was observed that although the theory based on volume averaging is on firm physical grounds when describing two phases mixed at the macroscale (pore scale), it does not generalize to fluids mixed at multiple scales or more than two phases. The details of how this limitation is overcome will be presented in subsequent chapters.

An important experimental observation is the coupling between porosity waves and saturation waves. These coupled processes were observed to suppress viscous fingering, enhance dispersion and thus enhance sweep efficiency both in the lab and in reservoirs. Based on these observations, very simplistic theoretical approximations were investigated, subject to the observation that saturation waves are strongly coupled to porosity–pressure waves. It was shown how to calculate the velocity of a porosity wave as a function of frequency, permeability, porosity and the physical properties of the fluid and solid. As well, calculations of the velocity of saturation waves were presented. The calculated velocities illustrate the frequency, viscosity, saturation and permeability dependence of the saturation wave velocities.

A numerical illustration of the saturation wave velocity was presented. Frontal dispersion of pulsed fluid injection into a well was modelled by showing how the permeability parameters may be modified to describe dispersion instead of fingering. Finally, the radius of influence of a pulsed fluid surrounding an injection well was calculated.

REFERENCES

Bear, J., 1972, *Dynamics of Fluids in Porous Media*, Dover, New York.

Bouzidi, Y., 2003, The Acoustic Reflectivity and Transmissivity of Liquid Saturated Porous Media: Experimental Tests of Theoretical Concepts, Ph.D. Dissertation in Physics, University of Alberta.

Brooks, R.H., and Corey, A.T., 1964, Hydraulic Properties of Porous Media, Hydrology Papers 3, Colorado State University, 1–37.

Davidson, B.C., Dusseault, M.B., Geilikman, M.B., Hayes, K.W., and Spanos, T.J.T., United States Patent #6,241,019, June 5, 2001.

Davidson, B.C., Dusseault, M.B., Geilikman, M.B., M.B., Hayes, K.W., and Spanos, T.J.T., United States Patent #6,405,797, June 18, 2002a.

Davidson, B.C., Dusseault, M.B, Geilikman, M.B., Hayes, K.W., and Spanos, T.J.T., United Kingdom Patent #GB232819, January 2, 2002b.

Davidson, B.C., Dusseault, M.B., Geilikman, M.B., Hayes, K.W., and Spanos, T.J.T., Enhancement of flow rates through porous media Canadian Patent #CA2232948, June 7, 2005.

Davidson, B.C., Pringle, R.E., Samaroo, M., and Spanos, T.J.T., Borehole seismic pulse generation using rapid-opening valve United States Patent #8,061,421, November 22, 2011

Davidson, B.C., Spanos, T.J.T., and Dusseault, M., 1999, Experiments on pressure pulse flow enhancement in porous media, Proceedings of the CIM Petroleum Society of Regina Annual Meeting, Saskatchewan, October.

de la Cruz, V., and Spanos, T.J.T., 1989, Thermomechanical coupling during seismic wave propagation in a porous medium, J. Geophys. Res., 94, 637–642.

de la Cruz, V., Spanos, T.J.T., and Yang, D., 1995, Macroscopic capillary pressure, *Trans. Porous Media*, 19, 67–77.

Dusseault, M.B., Davidson, B., and Spanos, T., 2000a, Removing mechanical skin in heavy oil wells, SPE 58718, 2000 SPE Symposium on Formation Damage, Lafayette, Louisiana, 2000a.

Dusseault, M.B., Davidson, B.C., and Spanos, T.J.T., 2000b, Pressure pulsing: The ups and downs of starting a new technology, *JCPT*, 39(2), 13–17.

Dusseault, M.B., Gall, C., Shand, D., Davidson, B.C., and Hayes, K., 2001, Rehabilitating heavy oil wells using pulsing workovers to place treatment chemicals, Proceedings of the CIM Petroleum Society 52nd Annual Technical Meeting, Paper 2001-57, Calgary, 2001.

Dusseault, M.B., Geilikman, M.B., and Spanos T.J.T., 1998, Heavy oil production from unconsolidated sandstones using sand production and SAGD, *JPT*, 50(9), 92–94.

Dusseault, M.B., Hayes, K.C., Kremer, M., and Wallin, C., 2000c, Workover strategies in CHOP wells, Proceedings of the CIM Petroleum Society 51st Annual Technical Meeting, Paper 2000-69, Calgary, 2000.

Dusseault, M.B., Shand, D., and Davidson, B.C., 2002a, Pressure pulse workovers in heavy oil, SPE/Petroleum Society of CIM/CHOA 79033, SPE International Thermal Operations and Heavy Oil Symposium and International Horizontal Well Technology Conference, Calgary, AB, 4–7 November.

Dusseault, M.B., Shand, D., Meling, T., Spanos, T.J.T., and Davidson, B.C., 2002b, *Field Applications of Pressure Pulsing in Porous Media*, Proc, 2nd Biot Conference on Poromechanics, Grenoble France, Balkema, Rotterdam, 639–645.

Dusseault, M.B., Spanos, T., and Davidson, B.C., 1999, A new workover approach for oil wells based on high amplitude pressure pulsing, Proceedings of the 50th CIM Petroleum Society Annual Technical Meeting, Calgary, AB, June 1999.

Geilikman, M.B., Spanos, T.J.T., and Nyland, E., 1993, Porosity diffusion in fluid-saturated media, *Tectonophysics*, 217(1/2), 111–115, 1993.

Grey, P., Davidson B.C., and Macdonald, A., 2001, Dramatic NLAPL recover at an Ontario manufacturing facility, *Environ. Sci. Eng.*, 22–24 January.

Groenenboom, J., Wong, S., Meling, T., Zschuppe, R., Davidson, B.C., 2003, SPE Paper 84856, SPE International Improved Oil Recovery Conference in Asia Pacific held in Kuala Lumpur, Malaysia, 20–21 October 2003.

Hickey, C.J., 1994, Mechanics of porous media, PhD dissertation, University of Alberta.

Samaroo, M., 1999, Pressure pulse enhancement: Report on the First Reservoir Scale experiment conducted by PE-TECH inc in section 36 of Wascana Energy inc's Morgan Field Lease, MSc thesis, University of Waterloo, 1999.

Spanos, T.J.T., 2002, The Thermophysics of Porous Media, Chapman & Hall/CRC Press, Monographs and Surveys in Pure and Applied Mathematics series. 212 pages.

Spanos, T.J.T., Davidson, B.C., Dusseault, M., and Samaroo, M., 1999, Pressure pulsing at the reservoir scale: A new IOR approach, Proceedings of the CIM Petroleum Society 50th ATM, Calgary, pp. 99–11.

Spanos, T.J.T., de la Cruz, V., and Hube, J., 1988, An Analysis of the Foundations of Relative Permeability Curves, AOSTRA Journal of Research, 4, 1–10.

Talwani, P., 2000, Seismogenic properties of the crust inferred from recent studies of reservoir-induced Seismicity-Application to Koyna, *Curr. Sci.*, 79, 9.

Talwani, P., and Acree, S., 1984, Pore pressure diffusion and the mechanism of reservoir-induced seismicity, *Pure Appl. Geophys.*, 122, 947–965.

Udey, N., 2009, Dispersion waves of two fluids in a porous Medium, *Trans. Porous Media*, 79, 107–115.

Udey, N., 2012, Coupled porosity and saturation waves in porous media, *Mathematical and Numerical Modeling in Porous Media: Applications in Geosciences*, edited by M.A. Diaz Viera, P. Sahay, M. Coronado, and A.O. Tapia, CRC Press, Boca Raton, FL., 313–343.

van Genabeek, O., and Rothman D.H., 1996, Macroscopic Manifestations of Microscopic Flows through Porous Media: Phenomenology from Simulation, Annual Review of Earth and Planetary Sciences, 24, 63–87.

Wang, J., Dusseault, M.B., Davidson, B.C., and Spanos, T.J.T., 1998, Fluid enhancement under liquid pressure pulsing at low frequency. Proceedings of the 7th UNITAR International Conference on Heavy Crude and Tar Sands, Beijing, October 1998.

CHAPTER 7

Miscible Fluid Flow in Porous Media (Part 1)

OBJECTIVE OF THIS CHAPTER

A system of equations for miscible displacement in porous media was constructed (Udey and Spanos. 1993) by considering the limiting case of immiscible flow with zero surface tension. Yang et al. (1998) demonstrated that for miscible displacement the megascopic pressure difference between phases must be non-zero due to the effect of an average pressure drop between the displacing and displaced phases. The theory presented by Udey et al. (1993) is reviewed and expanded upon. The effect that a dynamic capillary pressure has on this theory is discussed. The resulting equations differ from the theory that has been the most standard description of miscible displacement (e.g. Greenkorn 1983; Bear 1988; Dullien 1992). In the standard description the dynamics of the fluid is described by introducing a dispersion tensor in analogy to Fick's law for diffusion. When this description is combined with the equation of continuity (conservation of mass), the dynamics is embodied in the well-known convection–diffusion equation. This approach is based on the equilibrium thermodynamics of molecular mixtures, which does not apply to porous media. The correct description for such systems is given in Chapter 8. In the present discussion, it is shown that in the proper description of miscible flow with negligible molecular diffusion may be obtained from the equations of motion for immiscible flow and setting surface tension to zero. In the slow flow limit, the flow is shown to be described by a Fokker–Planck equation.

EQUATIONS OF MOTION

In physical descriptions of nature, one of the most common tests of a theory is to check to see if it agrees with established theory in various limits. If they do not agree, then the only possibilities are that one of the theories is incorrect or the limit was not taken correctly. In the case of immiscible flow of two fluids, the equations have been rigorously derived and the limiting case where it must agree with miscible flow is simply the case where surface tension becomes zero. Physically, such scenarios where the two theories must agree are fairly straightforward to construct. For example, consider a porous matrix where the pores are completely filled with one fluid (e.g. glycerine). Now allow this fluid to be displaced by another fluid that is miscible with it (e.g. water). Assume that the time to sweep out the original fluid by the displacing fluid is smaller than the characteristic time for molecular diffusion to occur. Therefore, throughout the displacement, the two fluids remain distinct and occupy their own volume but, because the fluids are miscible, the interface between the fluids has zero interfacial tension. This scenario looks exactly like immiscible flow with zero interfacial tension.

The use of the equations of immiscible flow to describe miscible flow is not a new approach. It has been used by, for example, Dougherty (1963), Koval (1963), Todd and Longstaff (1972) and Jankovic (1986). Stalkup (1982) noted that this approach is often used in the numerical modelling of miscible flows, in contrast to using simulators, which solve the convection–diffusion equations. This application of the immiscible theory was based on a Buckley–Leverett type of solution to Muscat's equations. The fractional flow in this solution is represented by suitable algebraic expressions; the specific expression is determined by matching the theory to actual breakthrough curves or production curves obtained from miscible flow experiments. A generalization of Darcy's equation to two-phase flow was rigorously constructed by de la Cruz and Spanos (1983). This generalization was an improvement over Muscat's equations, which described two-phase flow in terms of simple linear equations and a zeroth order constraint on the pressure difference between the fluid phases. This new theory showed that physical consistency required that multiphase flow in porous media be described by a non-linear field theory. In this description, the fluids now interact with each other and an additional degree of freedom and thus an additional equation is required to account for saturation changes.

In the case of immiscible displacements, it was observed in Chapter 6 that dispersion and viscous fingering must be accounted for and it must be determined which process dominates. An understanding of both phenomena and their interaction is of paramount importance in determining the behaviour of both miscible and immiscible displacements.

The technique presented here of seeking a common description for miscible and immiscible displacement in the absence of surface tension and molecular diffusion has been employed by Cyr et al. (1988) in the analysis of polymer flooding.

Recall that the equations of motion for the steady-state flow of two incompressible immiscible fluids is given by

$$\frac{\partial(P_1 - P_2)}{\partial t} = \beta_2 \frac{\partial^2 S_1}{\partial t^2} - \beta_1 \frac{\partial S_1}{\partial t} \tag{7.1}$$

$$Q_{11}\vec{q}_1 - Q_{12}\vec{q}_2 = -\vec{\nabla} P_1 - \rho_1 \vec{g} \tag{7.2}$$

$$Q_{22}\vec{q}_2 - Q_{21}\vec{q}_1 = -\vec{\nabla} P_2 - \rho_2 \vec{g} \tag{7.3}$$

de la Cruz et al. (1995) showed that β_1 is proportional to surface tension and thus the term $\beta_1 \frac{\partial S_1}{\partial t}$ is 0 in the present case. This system of equations can be used to describe miscible flow processes dominated by the flow velocities.

One case already mentioned in Chapter 5 that may be described by these equations and easily observed in the lab is to displace clear water with dyed water. In this experiment silica sand is packed with about 1-darcy permeability and 33% porosity. Then the dyed water is injected with, say, a 1-meter head. Here, the saturation contours accelerate until they reach a certain velocity, with the lower saturations of injected water moving faster than the higher saturations as predicted by the equations in Chapter 6. The experiment occurs over a sufficiently short period of time that molecular diffusion is negligible.

The purpose of this chapter is to rewrite the equations of immiscible flow in terms of mass fractions and to consider the slow flow limit, which is the case normally considered in miscible displacement processes in the earth.

EQUATION OF CONTINUITY

Consider a three-dimensional homogeneous and isotropic porous medium with porosity η, which is spatially constant and time independent. Fluid 1 with density ρ_1 and viscosity μ_1 initially fills the pores and is being displaced by Fluid 2 with density ρ_2 and viscosity μ_2. It is assumed that each fluid is incompressible.

Since the molecular diffusion is negligible, each fluid occupies its own volume. Furthermore, each fluid obeys its own equation of continuity and the dynamics of the fluid flow is governed by the Navier–Stokes equations. To obtain the megascopic equations of flow, volume averaging is applied to the pore-scale equations, as is described in de la Cruz and Spanos (1983).

Let V represent a volume over which volume averaging may be performed. The fluid volume V_f in V is the sum of the two fluid volumes:

$$V_f = \eta V = V_1 + V_2 \tag{7.4}$$

The fractional volume of fluid i is defined by

$$\eta_i = \frac{V_i}{V} \tag{7.5}$$

which, by virtue of Equation 7.4, implies that

$$\eta = \eta_1 + \eta_2 \tag{7.6}$$

The saturation (proportion by volume) of each fluid is given by

$$S_i = \frac{V_i}{V} = \frac{\eta_i}{\eta},\ i = 1..2 \tag{7.7}$$

Equation 7.6 may be rewritten via Equation 7.7 as

$$1 = S_1 + S_2 \tag{7.8}$$

Using Equation 7.8, S_2 may be replaced by $(1-S_1)$ whenever it appears in the equations. S_1 alone may characterize the composition of the fluids during miscible flow.

The mass of a fluid component, M_i, contained in the volume V is

$$M_i = \rho_i V_i,\ i = 1..2 \tag{7.9}$$

The mass concentration of a fluid component (Fried and Combarnous 1971) is the mass of that component in a volume divided by the volume; thus the concentration is

$$c_i = \frac{M_i}{V} = \rho_i S_i,\ i = 1..2 \tag{7.10}$$

The mass density of the fluid is the total fluid mass in V divided by V

$$\rho = \frac{M_f}{V},\ M_f = M_1 + M_2 \tag{7.11}$$

which leads to the expression

$$\rho = c_1 + c_2 = \rho_1 S_1 + \rho_2 S_2 = \rho_2 + (\rho_1 - \rho_2) S_1 \tag{7.12}$$

This result is rather important in that it serves as a guide to the construction of the continuity equation.

A quantity related to the mass concentration is the fractional mass concentration

$$c_{fi} = \frac{c_i}{\rho},\ i = 1,2 \tag{7.13}$$

Equations 7.12 and 7.13 now imply

$$1 = c_{f1} + c_{f2} \tag{7.14}$$

Equations 7.10, 7.12 and 7.13 may be used to express c_{f2} in terms of S_1

$$c_{f2} = \frac{\rho_2 (1 - S_1)}{\rho_2 + (\rho_1 - \rho_2) S_1} \tag{7.15}$$

which in turn may be rearranged to express the saturation in terms of the fractional concentration:

$$S_1 = \frac{\rho_2 (1 - c_{f2})}{\rho_2 + (\rho_1 - \rho_2) c_{f2}} \tag{7.16}$$

The present description of miscible flow is formulated in terms of saturation; S_1 is the independent variable that describes the composition of the fluid. However, the composition of the fluids during miscible flows is often described by the fractional concentration of the displacing fluid, namely c_{f2}. If necessary, Equation 7.16 permits these results to be expressed in terms of concentration.

When volume averaging is applied to the pore-scale equations of continuity for each fluid the following set of equations is obtained (de la Cruz and Spanos 1983):

$$\frac{\partial \eta_i}{\partial t} + \vec{\nabla} \cdot \vec{q}_i = 0 \quad i = 1,2 \tag{7.17}$$

Here, $\vec{q}_i$ are the Darcy velocities for each fluid. Simply adding these equations together results in

$$\frac{\partial \eta}{\partial t} + \vec{\nabla} \cdot \vec{q} = 0 \tag{7.18}$$

where the total Darcy flow for the fluid as a whole is

$$\vec{q} = \vec{q}_1 + \vec{q}_2 \tag{7.19}$$

Since the porosity is a constant in time equation, Equation 7.18 immediately yields the equation of incompressibility:

$$\vec{\nabla} \cdot \vec{q} = 0 \tag{7.20}$$

An alternative form of Equation 7.17 is obtained by simply dividing it by the porosity:

$$\frac{\partial S_i}{\partial t} + \vec{\nabla} \cdot \left(\frac{\vec{q}}{\eta} \right) = 0 \quad i = 1,2 \tag{7.21}$$

If this equation is now multiplied by the fluid density ρ_i, the equation of continuity for fluid component i is obtained, namely

$$\frac{\partial c_i}{\partial t} + \vec{\nabla} \cdot (c_i \mathrm{v}_i) = 0 \quad i = 1,2 \tag{7.22}$$

where the velocity of fluid i is given by

$$\vec{\mathrm{v}}_i = \frac{1}{S_i} \frac{\vec{q}_i}{\eta}, \; i = 1..2 \tag{7.23}$$

and may be seen to be the momentum per unit mass for fluid component i.

Now Equation 7.12 suggests that the continuity equations of the fluid components, Equation 7.22, should be added to obtain the continuity equation for the whole fluid. This operation yields

$$\frac{\partial \rho}{\partial t} + \vec{\nabla} \cdot (c_1 \vec{v}_1 + c_2 \vec{v}_2) = 0 \tag{7.24}$$

Here the equation of continuity of a multicomponent fluid serves to define the fluid velocity as the total momentum per unit mass of fluid (Landau and Lifshitz 1975). Therefore, Equation 7.24 may be expressed as

$$\frac{\partial \rho}{\partial t} + \vec{\nabla} \cdot (\rho \vec{v}) = 0 \tag{7.25}$$

so that the fluid velocity $\vec{V}$ is defined by

$$\vec{v} = \frac{c_1 \vec{v}_1 + c_2 \vec{v}_2}{\rho}$$

$$= \frac{\rho_1 \vec{q}_1 + \rho_2 \vec{q}_2}{\eta \rho} \tag{7.26}$$

Note that when the two fluids have the same density

$$\rho = \rho_1 = \rho_2, \; c_{f2} = 1 - S \tag{7.27}$$

so the fluid velocity becomes

$$\vec{v} = \vec{u} \tag{7.28}$$

where the average interstitial fluid velocity is given by

$$\vec{u} = \frac{\vec{q}}{\eta} \tag{7.29}$$

Consequently, the equation of continuity (Equation 7.25) reduces to the equation of incompressibility (Equation 7.20) in this case. Equation 7.28 is the Dupuit–Forcheimer assumption (Scheidegger 1974) that is normally used in the convection–diffusion theory of miscible displacement. Here, it is valid only for equal density fluids, but this is often the situation in laboratory experiments (e.g. Brigham et al. 1961).

THE DYNAMICAL EQUATIONS OF TWO-PHASE IMMISCIBLE FLOW

In addition to the equations of continuity and compressibility, equations that describe the dynamics of the miscible flow are obtained by volume-averaging the Navier–Stokes equations. For steady-state flow, these equations are as follows (de la Cruz and Spanos 1983):

$$Q_{11}\vec{q}_1 - Q_{12}\vec{q}_2 = -\vec{\nabla} P_1 + \rho_1 \vec{g} \tag{7.30}$$

$$Q_{22}\vec{q}_2 - Q_{21}\vec{q}_1 = -\vec{\nabla} P_2 + \rho_2 \vec{g} \tag{7.31}$$

Neglecting gravity, these component flows may also be expressed in terms of the fractional flow as follows (Udey and Spanos 1993):

$$\vec{q}_1 = f_1(S_1)\vec{q} \tag{7.32}$$

$$\vec{q}_2 = [1 - f_1(S_1)]\vec{q} \tag{7.33}$$

The fractional flow is often the starting point for discussions of miscible flow using the limiting form of the immiscible equations (Dougherty 1963; Koval 1963; Jankovic 1986).

Combining Equations 7.32 and 7.33 with the continuity equation in the form of Equation 7.21 yields

$$\frac{\partial S_1}{\partial t} + \vec{u}_s \cdot \vec{\nabla} S_1 = 0 \tag{7.34}$$

where the isosaturation velocity $\vec{u}_s$ is defined by

$$\vec{u}_s = v_q(S_1)\vec{u} \tag{7.35}$$

$$v_q(S_1) = \frac{df_1}{dS_1} \tag{7.36}$$

The left-hand side of Equation 7.34 is the material derivative of the saturation. Therefore Equation 7.34 may be written as

$$\frac{dS_1}{dt} = 0 \tag{7.37}$$

which is the equation of motion of a surface of constant saturation.

Equation 7.12 states that density is solely a function of saturation. Therefore the time derivative of density is

$$\frac{\partial \rho}{\partial t} = \frac{\partial \rho}{\partial S_1}\frac{\partial S_1}{\partial t} = (\rho_1 - \rho_2)\frac{\partial S_1}{\partial t} \tag{7.38}$$

Now substituting Equations 7.29 and 7.30 into the definition of the fluid velocity equation (Equation 7.26) yields

$$\vec{v} = \frac{\rho_2 + f_1(\rho_1 - \rho_2)}{\rho}\vec{u} \tag{7.39}$$

This yields an expression for the divergence of the mass flux:

$$\nabla \cdot (\rho\vec{v}) = (\rho_1 - \rho_2)\vec{u}_s \cdot \nabla S_1 \tag{7.40}$$

Therefore the equation of continuity becomes

$$(\rho_1 - \rho_2)\left(\frac{\partial S_1}{\partial t} + \vec{u}_s \cdot \nabla S_1\right) = 0 \tag{7.41}$$

which leads to Equation 7.34 when the densities are different and is trivially satisfied when the densities are the same.

DISPERSION

For a miscible flow process, in general, the flow is experiencing dispersion. Furthermore when frontal displacement occurs, physical consistency requires that a dynamic pressure difference exists between the phases. Here, the nature of this dispersion will be examined and compared with the experiment. For the remainder of this discussion it will be assumed that gravity is negligible, $\vec{g} = 0$. Note that the equations describing miscible flow in one dimension become

$$\frac{\partial p_1 - p_2}{\partial x} = (Q_{11} + Q_{21} + Q_{22} + Q_{12})q_1 - (Q_{22} + Q_{12})$$

$$+\rho_1 \frac{\partial v_1}{\partial t} - \rho_2 \frac{\partial v_2}{\partial t} \tag{7.42}$$

$$\frac{\partial p_1 - p_2}{\partial x} = -\beta' \frac{\partial^2 S_1}{\partial t^2} \tag{7.43}$$

$$\frac{\partial S_1}{\partial t} + u_s \frac{\partial S_1}{\partial x} = 0 \tag{7.44}$$

in the frontal region.

A simple change of variable using the chain rule yields Equation 7.44 as

$$\frac{\partial c_{f2}}{\partial t} + u_s \frac{\partial c_{f2}}{\partial x} = 0 \tag{7.45}$$

Here, $\vec{u}_s$ is referred to as the *isoconcentration velocity.* It may also be thought of as referring to a saturation profile, a saturation distribution, or a displacement front. There are three equivalent ways of viewing the flow. The saturation may be described at a given position as a function of time, $S_1 = S_1(x, t)$; it may also be described by specifying the saturation for a given time and finding where in space the saturation is, $x = x(S_1, t)$, or finding the time for a given position to have a specified saturation, $t = t(S_1, x)$. Observe that

$$\frac{\partial}{\partial t} x(S_1, t) = u_s(S_1, t) \tag{7.46}$$

Consider two saturation values, S_L and S_H (low and high), where $S_H > S_L$. A large value of saturation, S_1, corresponds to a small value of concentration, c_{f2}. The distance between these two saturations is given by

$$w(S_L, S_H, t) = x(S_H, t) - x(S_L, t) > 0 \quad (7.47)$$

This may be called the width of the displacement front. The overall width of the front, or, as it is often called, the length of the mixing zone, is defined by specifying a pair of values for S_L and S_H. Fried and Combarnous (1971) specify the pair ($S_H = 0.84$, $S_L = 0.16$) that is most appropriate for the error function solution of the convection–diffusion theory. Another frequently used pair is (0.9, 0.1) (e.g. Brigham 1974).

In laboratory experiments on miscible flows, it is observed that the distance between different saturation values always increases, that is, dispersion, and this is expressed mathematically as

$$\frac{\partial}{\partial t} w(S_L, S_H) > 0, \ \forall(S_L, S_H), \ S_L < S_H \quad (7.48)$$

This condition prohibits the Buckley–Leverett paradox from occurring.

The dispersion of immiscible flows may be classified into three categories according to the second time derivative of the width:

$$\frac{\partial^2}{\partial t^2} w(S_L, S_H, t) \left\langle \begin{array}{l} < 0 \text{ dispersion decelerates} \\ = 0 \text{ dispersion is constant} \\ > 0 \text{ dispersion accelerates} \end{array} \right\rangle \quad (7.49)$$

Arya et al. (1988) also classify dispersion in three ways according to the time behaviour of the width; however, two of their classes fall into the constant dispersion class given here, and their third is the same as the decelerating dispersion class. Here, the classification equation (Equation 7.49) is used to compare and contrast the conventional convection–diffusion theory with the results of this chapter.

Now examine the nature of the dispersion in the present theory. Equations 7.48 and 7.46 yield

$$\frac{\partial}{\partial t} w(S_L, S_H, t) = u_S(S_H, t) - u_S(S_L, t) \quad (7.50)$$

which now implies, from Equation 7.44, that

$$u_S(S_L, t) < u_S(S_H, t), \ \ \forall(S_L, S_H), \ S_L < S_H \quad (7.51)$$

Thus the isosaturation speed must be a monotonically increasing function of saturation. In the analysis of Udey and Spanos (1993), the isosaturation speed had no explicit dependence upon time; the time derivative of Equation 7.50 gave

$$\frac{\partial^2}{\partial t^2} w(S_L, S_H, t) = \frac{\partial}{\partial t} u_S(S_H) - \frac{\partial}{\partial t} u_S(S_L) = 0 \tag{7.52}$$

and it was observed that the dispersion is constant. This type of dispersion, arising from a constant isosaturation velocity term, is called *flux-induced dispersion* (Arya et al. 1988). The effect of the megascopic dynamic capillary pressure equation is to allow for a variable dispersion rate.

The analysis above shows that Equation 7.34 or 7.45 will describe dispersion in the three-dimensional case, provided the isosaturation speed is a monotonically increasing function of saturation. Equation 7.34 also implies that a three-dimensional flow experiences lateral dispersion in addition to longitudinal dispersion. The actual motion of an isosaturation front is along the saturation gradient with an apparent speed given by the dot product between the saturation gradient and the Darcy velocity. This motion may be thought of as having a component along the Darcy velocity (longitudinal dispersion) and a component perpendicular to the Darcy velocity (lateral dispersion).

Equations 7.34 and 7.45 produce longitudinal and lateral dispersion in the following manner. In Equation 7.22 for c_2 write

$$c_2 \vec{v}_2 = \rho_2 \frac{\vec{q}_2}{\eta} = c_2 \vec{v} + \vec{j}_2 \tag{7.53}$$

so that Equation 7.22 for c_2 becomes

$$\frac{\partial c_2}{\partial t} + \vec{\nabla} \cdot (c_2 \vec{v}) + \vec{\nabla} \cdot (\vec{j}_2) = 0 \tag{7.54}$$

The definition of the fractional concentration and the equation of continuity yield

$$\frac{\partial c_2}{\partial t} + \vec{\nabla} \cdot (c_2 \vec{v}) = \rho \left(\frac{\partial c_{f2}}{\partial t} + \vec{v} \cdot \vec{\nabla} c_{f2} \right) \tag{7.55}$$

When this expression is employed in Equation 7.54, it yields

$$\frac{\partial c_{f2}}{\partial t} + \vec{v} \cdot \vec{\nabla} c_{f2} = -\frac{1}{\rho} \vec{\nabla} \cdot \vec{j}_2 \tag{7.56}$$

A comparison of this result with Equation 7.45 shows that

$$\frac{1}{\rho} \vec{\nabla} \cdot \vec{j}_2 = -\vec{v} \cdot \vec{\nabla} c_{f2} + \vec{u}_S \cdot \vec{\nabla} c_{f2} \tag{7.57}$$

Equation 7.53 may be viewed as the definition for a deviation mass flux $\vec{j}_2$. This flux may be written, using Equations 7.33 and 7.39 with zero gravity, as

$$\vec{j}_2 = F\left(c_{f2}\right)\vec{u} \tag{7.58}$$

where $F(c_{f2})$ is given by

$$F\left(c_{f2}\right) = \rho_2\left(1 - c_{f2}\right) - \left[\rho_2 + \left(\rho_1 - \rho_2\right)c_{f2}\right]f_1 \tag{7.59}$$

Equation 7.58 also leads to the result expressed by Equation 7.57.

Now define an apparent mechanical dispersion tensor, $\mathbf{D}_m$, by writing

$$\vec{j}_2 = -\rho\mathbf{D}_m \cdot \vec{\nabla}c_{f2} \tag{7.60}$$

in analogy with Fick's law (Fried and Combarnous 1971). Equation 7.57 now becomes

$$\vec{v} \cdot \vec{\nabla}c_{f2} - \frac{1}{\rho}\vec{\nabla} \cdot \left(\rho\mathbf{D}_m \cdot \vec{\nabla}c_{f2}\right) = \vec{u}_S \cdot \vec{\nabla}c_{f2} \tag{7.61}$$

It is important to realize that, unlike the dispersion tensor in the convection–diffusion theory, this apparent dispersion is described by a Fokker–Plank equation and is not a constant. It varies in value from point to point in the flow. This is evident by comparing Equations 7.60 and 7.58; this leads to the relationship

$$\rho\mathbf{D}_m \cdot \vec{\nabla}c_{f2} = -F(c_{f2})\vec{u} \tag{7.62}$$

A formal solution to Equation 7.62 for $\mathbf{D}_m$ is

$$\mathbf{D}_m = -\frac{F_1(c_{f2})}{\rho}\vec{u} \otimes \frac{\vec{\nabla}c_{f2}}{\left|\vec{\nabla}c_{f2}\right|^2} \tag{7.63}$$

where ⊗ is the outer product operator. Thus the Fokker–Plank equation describing the motion is given by

$$\frac{\partial c_{f2}}{\partial t} + \mathbf{v} \cdot \nabla c_{f2} = \frac{1}{\rho}\nabla \cdot (\rho\mathbf{D}_m \cdot \nabla c_{f2}) \tag{7.64}$$

This expression shows the explicit dependence of the apparent dispersion tensor upon the concentration, the concentration gradient and the flow rate. This construction of an apparent mechanical dispersion tensor illustrates that the present equations describe longitudinal and lateral dispersion.

COMPARISON WITH EXPERIMENT

Koval (1963) noted that L. Handy, in some unreported experiments using radioactive tracers in miscible fluids, found that a given solvent saturation moved at constant speed. This is consistent with the visual experiment described previously of dyed water displacing clear water. This property was illustrated in Figure 1 of Koval's paper; the figure shows the movement of solvent saturation through a linear Alhambra sandstone. This figure also shows that low solvent saturations move more quickly than higher ones.

These results are consistent with the theoretical description for the isosaturation velocity given in Equations 7.41 and 7.45.

Another comparison may be made by analysing data from typical laboratory experiments. The typical experiment consists of displacing one fluid in a slim tube or core by another miscible fluid. The tube or core is completely filled by the original fluid and a second fluid is injected at one end at a constant volumetric rate. The purpose of the experiment is to measure the effluent concentration at the emitting end as a function of time, that is, a breakthrough curve.

If the tube or core has a length L and a cross-sectional area A, then the pore volume is given by

$$V_p = \eta AL \tag{7.65}$$

At time t, the volume injected is

$$V_i = qAt = \eta uAt \tag{7.66}$$

Thus

$$\frac{V_i}{V_p} = t\frac{u}{L} \tag{7.67}$$

Thus any breakthrough time may be expressed in terms of a ratio of volume injected to pore volume.

The effluent concentration c_e may be defined as the ratio of the mass of displacing fluid emitted per unit time to the total mass emitted per unit time:

$$c_e = \frac{c_2\vec{v}_2 \cdot \vec{n}A}{\rho\vec{v} \cdot \vec{n}A} \tag{7.68}$$

where $\vec{n}$ is the surface normal of the emitting surface A. Equation 7.68 may be rewritten using Equations 7.10, 7.12, 7.23 and 7.26 as

$$c_e = \frac{\rho_2\vec{q}_2 \cdot \vec{n}}{\rho_1\vec{q}_1 \cdot \vec{n} + \rho_2\vec{q}_2 \cdot \vec{n}} \tag{7.69}$$

Substituting Equations 7.32 and 7.33 into Equation 7.69 yields

$$c_e = \frac{\rho_2(1-f_1)}{\rho_1 f_1 + \rho_2(1-f_1)} \tag{7.70}$$

Using this expression, the fractional flow may be determined from the effluent concentration

$$f_1 = \frac{\rho_2(1-c_e)}{\rho_1 c_e + \rho_2(1-c_e)} \tag{7.71}$$

When the densities of the two fluids are the same, Equations 7.70 and 7.71 simplify to

$$f_1 = 1 - c_e \tag{7.72}$$

It is conventional to display breakthrough curves as a graph of c_e versus V_i/V_p.

Figure 7.1 is a typical breakthrough curve, which will be used for theoretical comparisons. This data was presented by Brigham (1974) and appears in Figure 5 of his paper. Figure 7.1 represents a best-fit convection–diffusion solution for the effluent concentration.

This solution is given by the following (see Udey and Spanos 1993):

$$c_e = P\left(\frac{ut - x}{\sqrt{2D_L t}}\right) \tag{7.73}$$

where P(x) is the normal distribution function (Abramowitz and Stegun 1965) and $x_o = 0$, $t_o = 0$. In the regime being considered, the contribution of molecular diffusion to the longitudinal dispersion coefficient is negligible; the convection–diffusion theory is only

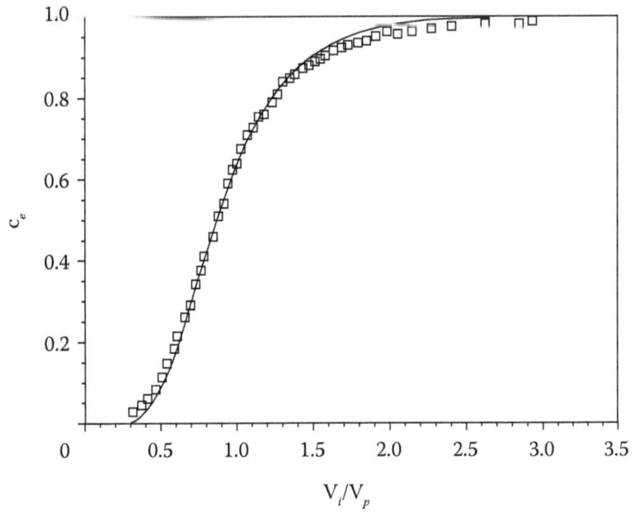

FIGURE 7.1 Breakthrough curve of Brigham's (1974) Zama Lake core data plus convection diffusion best fit. (From Udey, N., and Spanos, T.J.T., *Transp. Porous Media*, 10, 1–41, 1993.)

describing mechanical dispersion in this case. Brigham et al. (1961) introduced a modified volume function, which was changed slightly by Perkins and Johnston (1963). In the present discussion, this will be referred to as the *effluent volume.*

$$V_e = \frac{V_i/V_p - 1}{\sqrt{V_i/V_p}} \tag{7.74}$$

Equations 6.67 and 7.74 permit Equation 7.73 to be written as

$$c_e = P\left(\sqrt{\frac{Lu}{2D_L}} V_e\right) \tag{7.75}$$

or equivalently

$$V_e = \sqrt{\frac{2D_L}{Lu}} P^{-1}(c_e). \tag{7.76}$$

The convection diffusion theory predicts that a plot of a breakthrough curve on a V_e versus $P^{-1}(c_e)$ graph should be a straight line. The data in Figure 7.1 is presented as a plot of V_e versus $P^{-1}(c_e)$ in Figure 7.2. The best-fit straight line through the data from the values of c_e between 0.84 and 0.16 represents the best-fit convection–diffusion solution (Brigham 1974); in fact, this best fit was used in that paper to generate the solid curve in Figure 7.1.

In Figures 7.1 and 7.2 it is observed that the miscible displacement deviates from the convection–diffusion solution at low and high concentration values. This is a typical observation of many experimental breakthrough curves (Brigham et al. 1961; Fried and Combarnous 1971; Brigham 1974; Dullien 1992). The deviation at low concentration values corresponds to a breakthrough that occurs sooner than predicted by the

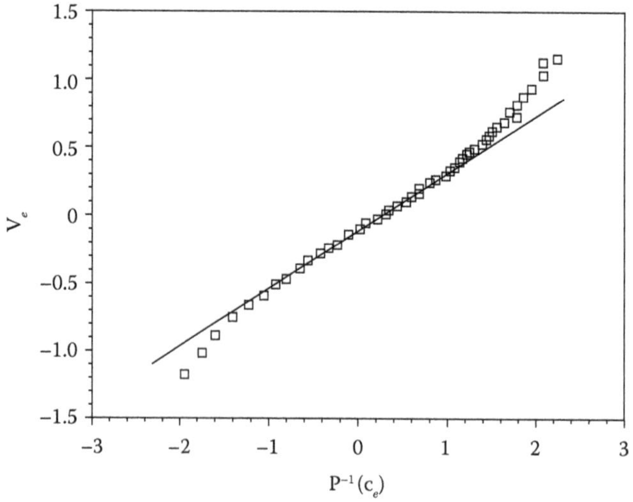

FIGURE 7.2 Probability plot of Zama Lake core plus the convection–diffusion best fit. (From Udey, N., and Spanos, T.J.T., *Transp. Porous Media*, 10, 1–41, 1993.)

convection–diffusion theory, and the high concentration value deviations corresponds to a delayed breakthrough. The dead-end pore model was proposed to explain these deviations (Coats and Smith 1964). In particular, the long tailing of Figure 7.1 at high concentration values ($V_e > 0.5$ in Figure 7.2) was attributed to a continuous production of fluid via the mechanism of diffusion between the flowing fraction of fluid and the fluid in the dead-end pores. However, Brigham (1974) has pointed out that no combination of the dead-end pore model parameters can account for the data's deviations at low concentration values.

Now consider the predictions of the theory derived in this chapter. Consider some concentration value $c_e < 0.5$ on the leading edge of the displacement. In the theory constructed in this chapter, the values of the isoconcentration speed stays constant while the convection–diffusion theory says it should slow down. Thus it is predicted that c_e will break through sooner than expected by the convection–diffusion theory. Similarly, a concentration value $c_e > 0.5$ breaks through later than by the convection-diffusion theory. This is precisely what the data displays in Figure 7.1. Thus there is a single, simple explanation for the deviations at both high and low concentration values. Furthermore, miscible displacements often show asymmetry in the breakthrough curves. This implies that the time-independent isoconcentration velocities are not symmetric around the displacement velocity u, as is the case in the convection–diffusion theory.

One of the goals of laboratory miscible displacements is to obtain a longitudinal dispersion coefficient. This is obtained by assuming that the flow is described by the convection–diffusion theory. In such a case, the dispersion constant is given by

$$D_L = \frac{1}{8}\left\{\frac{L - ut_{0.16}}{\sqrt{t_{0.16}}} \frac{L - ut_{0.84}}{\sqrt{t_{0.84}}}\right\}^2 \tag{7.77}$$

where $x_o = 0$, $t_o = 0$, $x = L$ (see Udey and Spanos 1993). However, if the theory constructed in this chapter is used with time-independent isoconcentration velocities, then the breakthrough times appearing in Equation 7.76 can be expressed as

$$t_{0.16} = \frac{L}{uv_q(0.16)} \quad t_{0.84} = \frac{L}{uv_q(0.84)} \tag{7.78}$$

However, Equations 7.77 and 7.78 together imply the longitudinal dispersion coefficient is not a constant at all but

$$D_L = D_L^o Lu \tag{7.79}$$

where D_L^o is given by

$$D_L^o = \frac{1}{8}\left\{\left(\sqrt{v_q(0.16)} - \frac{1}{\sqrt{v_q(0.16)}}\right) - \left(\sqrt{v_q(0.84)} - \frac{1}{\sqrt{v_q(0.84)}}\right)\right\}^2 \tag{7.80}$$

Therefore, according to the theory constructed here, experimental attempts to measure the dispersion coefficient for the convection diffusion theory will observe that the dispersion constant is not a constant but is proportional to both the flow velocity and the length of the core.

The result that the dispersion coefficient was proportional to the flow velocity for cores of constant length was noted by Koval (1963) and observed experimentally (Perkins and Johnston 1963; Fried and Combarnous 1971; Bear 1988).

In experiments on miscible displacement, the dispersion constant is normally considered independent of length, so experimentalists have not specifically looked for length dependence. However, several authors have reported observations of some length dependence. Stalkup (1970) and Baker (1977) noted that the dead-end pore model in long cores resembles the convection–diffusion theory error function solution with a much larger dispersion coefficient. A similar point by Pickens and Grisak (1981) is that the dispersion coefficients that arise in computer models of existing groundwater contamination zones are much larger than those than those obtained from laboratory breakthrough experiments. Bretz and Orr (1987) and Han et al. (1985) noted that measured dispersion coefficients increase with the length scale of a miscible displacement. Arya et al. (1988) examined the length dependence of the megascopic dispersivity

$$\alpha_{ME} = \frac{\eta D_L}{u} \tag{7.81}$$

which, using Equation 7.79, becomes

$$\alpha_{ME} = \eta D_L^o L \tag{7.82}$$

In an examination of the data of Lellemand-Barres and Peaudecerf (1978) and Pickens and Grisak (1981), they noted a trend for α_{ME} to increase with increasing length. Their log–log fit to the field data produces

$$\alpha_{ME} = 0.229 L^{0.775} \tag{7.83}$$

and for all the data they obtain

$$\alpha_{ME} = 0.044 L^{1.13} \tag{7.84}$$

These results are consistent with Equation 7.82 but are not conclusive because of the large scatter in the data. Strictly speaking, Equation 7.82 is only valid for a single type of porous medium; the term ηD_L^o will be different for various types of porous media.

When the results for disparate types of porous media are plotted together, a great deal of scatter should be expected, which should obscure the length dependence of α_{ME}.

Now let us return to the data of Brigham (1974), which was presented in Figures 7.1 and 7.2. Note the data in Figure 7.2 strongly resembles a cubic function, so fitting this data with a cubic polynomial using the method of least squares (Curtis and Gerald 1978), the best-fit cubic is

$$V_e = -0.1092 + 0.3966x + 0.0003x^2 + 0.0376x^3 x = P^{-1}(c_e) \tag{7.85}$$

This function is used to generate a best-fit set of points $[V_{ei}, P^{-1}(c_e)_i]$, $i = 1 \ldots N$; with $N = 100$. The data in Figure 7.2 are replotted in Figure 7.3 along with the best-fit cubic, which is drawn by simply connecting the best-fit points $[V_{ei}, P^{-1}(c_e)_i]$ with straight line segments. These best-fit points can be mapped from V_e versus $P^{-1}(c_e)$ coordinates into c_e versus V_i/V_p coordinates. This generates a set of best-fit points $[c_{ei}, (V_i/V_p)_i]$, which are used to draw the best-fit breakthrough curve plotted in Figure 7.4.

The breakthrough time for a given saturation may be expressed as

$$t = \frac{L}{u v_q(S_1)} \tag{7.86}$$

Equations 7.67 and 7.86 combine to yield

$$\frac{V_i}{V_p} = \frac{1}{v_q(S_1)} \tag{7.87}$$

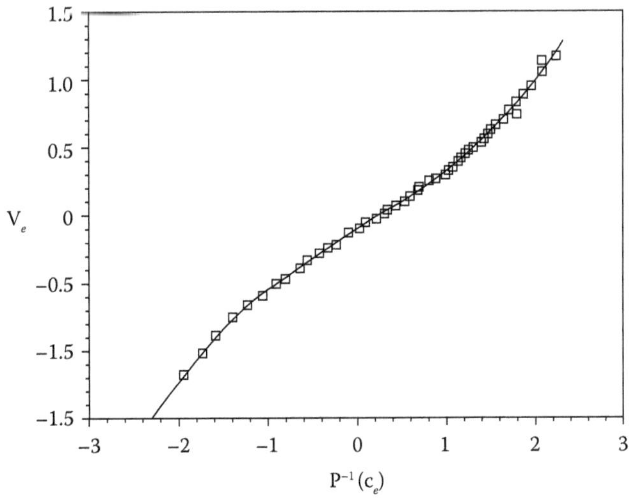

FIGURE 7.3 Probability plot of Zama Lake core data plus best-fit cubic polynomial. (From Udey, N., and Spanos, T.J.T., *Transp. Porous Media*, 10, 1–41, 1993.)

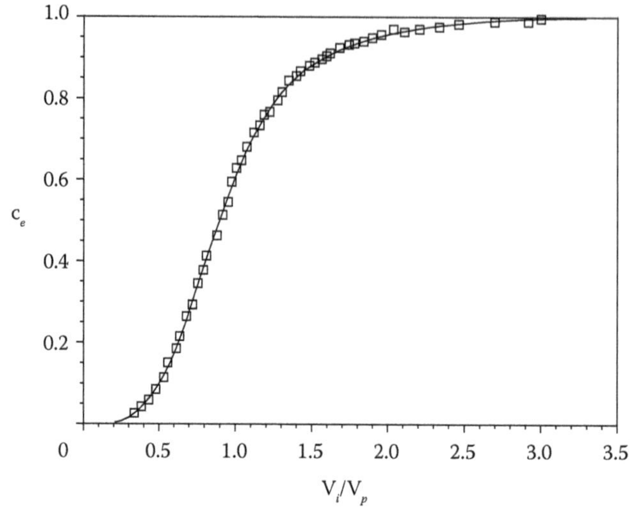

FIGURE 7.4 Breakthrough curve: data and best fit. (From Udey, N., and Spanos, T.J.T., *Transp. Porous Media*, 10, 1–41, 1993.)

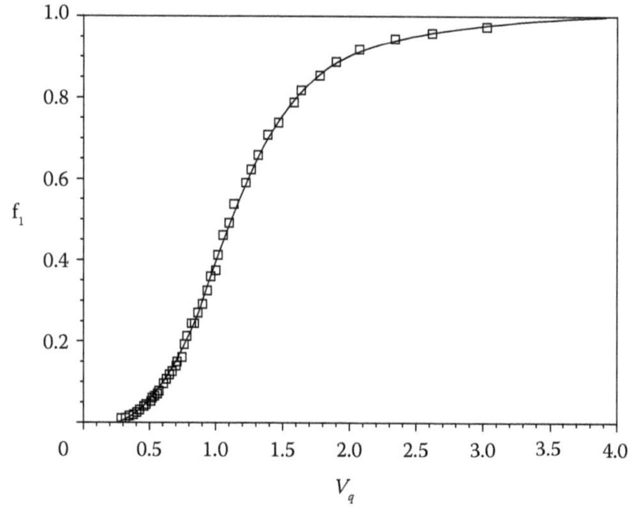

FIGURE 7.5 Fractional flow versus isosaturation speed. Zama Lake data and derived best fit. (From Udey, N., and Spanos, T.J.T., *Transp. Porous Media*, 10, 1–41, 1993.)

Therefore, the breakthrough curve may be transformed using Equations 7.72 and 7.87 into a graph of the fractional flow versus normalized isosaturation speed. The application of this transformation and best-fit points in Figure 7.4 produces the graph shown in Figure 7.5. In Figure 7.4 it can be observed that

$$\lim_{v_i/v_p \to \infty} c_e = 1 \tag{7.88}$$

which is equivalent to

$$\lim_{v_q \to \infty} f_1 = 1 \tag{7.89}$$

by virtue of Equations 7.72 and 7.87. The trend in Figure 7.5 is consistent with this limit and is realized by the extrapolation to the origin.

Now observe that the curve in Figure 7.4 flattens out as V_i/V_p approaches infinity

$$\lim_{V_i/V_p \to \infty} \frac{dc_e}{d(V_i/V_p)} = 0 \tag{7.90}$$

The equivalent statement for fractional flow and isosaturation velocity is

$$\lim_{v_q \to \infty} v_q^2 \frac{df_1}{dv_q} = 0 \tag{7.91}$$

Now, construct a natural cubic spline (Curtis and Gerald 1978) to the best-fit points (f_{1i}, v_{qi}) in Figure 7.5 and employ it to compute theoretical values of the slope df_1/dv_q. The derivative of dS_1/dv_q may be computed by

$$\frac{dS_1}{dv_q} = \frac{dS_1}{df_1}\frac{df_1}{dv_q} = \frac{1}{v_q}\frac{df_1}{dv_q} \tag{7.92}$$

A set of best-fit points are obtained, which are used to draw the solid curve in Figure 7.6.

A graph of S_1 versus v_q may now be obtained by integrating the dS_1/dv_q curve. A cubic spline may be computed from the best-fit points $[(dS_1/dv_q)_i, v_{qi}]$. Consequently, for each interval (i, i+1) a representation of the dS_1/dv_q curve is obtained in the form

$$\frac{dS_1}{dv_q}(v_q) = \left(\frac{dS_1}{dv_q}\right)_i + a_1(v_q - v_{qi}) + a_2(v_q - v_{qi})^2 + a_3(v_q - v_{qi})^3, v_{qi} \le v_q \le v_{qi+1} \tag{7.93}$$

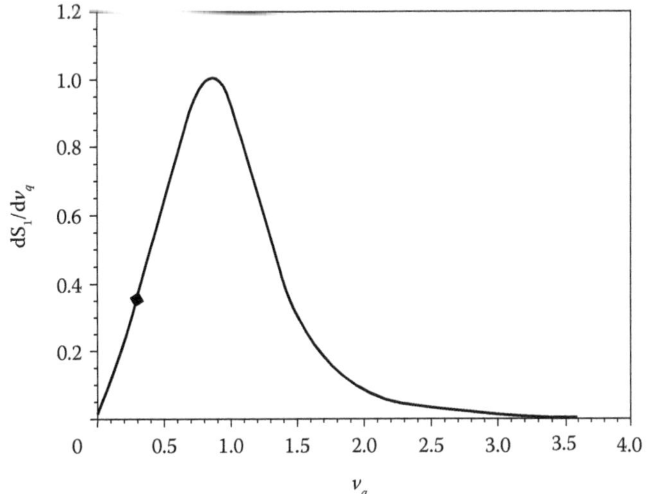

FIGURE 7.6 Derived best fit for dS_1/dv_q versus v_q. (From Udey, N., and Spanos, T.J.T., *Transp. Porous Media*, 10, 1–41, 1993.)

where the coefficients a_1, a_2 and a_3 are known values. The indefinite integral of the dS_1/dv_q cubic spline generates a piecewise quartic. In the interval (i, i+1) it has the form

$$S_1(v_q)=S_{1i}+\left(\frac{dS_1}{dv_q}\right)_i(v_q-v_{qi})+\frac{a_1}{2}(v_q-v_{qi})^2$$

$$+\frac{a_2}{3}(v_q-v_{qi})^3+\frac{a_3}{4}(v_q-v_{qi})^4,\ v_{qi}\le v_q\le v_{qi+1}, \tag{7.94}$$

where the constants of integration S_{1i} are unknown. These may be evaluated as follows.

The largest value of v_q, namely, v_{qN}, has associated with it the value of f_{1N}. When S_1 is close to 1, f_1 should also be close to 1. This suggests that $S_{1N} = f_{1N}$ may be taken as a starting value. Some error in the saturation values will exist initially but may be corrected. Therefore, starting from i = N − 1 and working down to i = 1, S_{1i} may be evaluated from the formula

$$S_{1i}=S_{1i+1}-\left(\frac{dS_1}{dv_q}\right)_i(v_{qi+1}-v_{qi})$$

$$-\frac{a_1}{2}(v_{qi+1}-v_{qi})^2-\frac{a_2}{3}(v_{qi+1}-v_{qi})^3-\frac{a_3}{4}(v_{qi+1}-v_{qN})^4 \tag{7.95}$$

The result of this process is a set of best-fit points (S_{1i}, v_{qi}), which delineate the function S_1 versus v_q.

The next step is to obtain the curve relating S_1 and f_1. Rewriting Equation 7.36 as

$$\int_{f_{1i}}^{f_{1i+1}} df_1=\int_{S_{1i}}^{S_{1i+1}} v_q\, dS_1$$

$$=S_{1i+1}v_{qi+1}-S_{1i}v_{qi}-\int_{v_{qi}}^{v_{qi+1}} S_1\, dv_q \tag{7.96}$$

Now Equation 7.94 may be substituted into the integrand in Equation 7.96. Taking the integral yields

$$f_{1i+1}-f_{1i}=S_{1i+1}v_{qi+1}-S_{1i}v_{qi}-S_{1i}(v_{qi+1}-v_{qi})-\frac{1}{2}\left(\frac{dS_1}{dv_q}\right)_i(v_{qi+1}-v_{qi})^2$$

$$-\frac{a_1}{6}(v_{qi+1}-v_{qi})^3-\frac{a_2}{12}(v_{qi+1}-v_{qi})^4-\frac{a_3}{20}(v_{qi+1}-v_{qN})^5 \tag{7.97}$$

This equation may be used to calculate the value of f_{1i} from the value of f_{1i+1}. For i = N − 1 the starting values are f_{1N}, S_{1N} and v_{qN}. Thus, a set of best-fit points (S_{1i}, f_{1i}) are generated.

The next step is to correct for the error in the saturation values in the (S_1, f_1) data sets. A natural cubic spline to the best-fit points (S_{1i}, f_{1i}) is computed. The interval (N−1, N) lies just below the value $f_1 = 1$. Thus the cubic spline is used in this interval to find the saturation value for $f_1 = 1$. The difference between this saturation value and the true value, 1, is the error, say ΔS_1, in all the saturation values. All these values in the (S_1, v_q) and (S_1, f_1) data sets are corrected by letting $S_{1i} \rightarrow S_{1i} + \Delta S_1$. It can be verified that (S_1, v_q) and (S_1, f_1) curves are correct by noting that, from a point (S_{1i}, f_{1i}) on the S_1 versus f_1 curve, and the corresponding point (S_{1i}, f_{1i}) on the S_1 versus v_q curve, a point (f_1, v_q) may be constructed. This point must agree with the corresponding point on the f_1 versus v_q curve. In the calculations, this correspondence is found to be quite accurate. Figure 7.7 is a plot of the best-fit points (S_{1i}, f_{qi}) as a curve of v_q versus S_1; and Figure 7.8 shows the (S_{1i}, f_{1i}) points as a curve of f_1 versus S_1.

Udey and Spanos (1993) defined an effective viscosity of the two fluids by

$$\frac{1}{\mu(S_1)} = \frac{A(S_1)}{\mu_1} + \frac{B(S_1)}{\mu_2} \tag{7.98}$$

and an effective flow density by

$$\rho_q = \mu(S_1)\left(\frac{A(S_1)}{\mu_1}\rho_1 + \frac{B(S_1)}{\mu_2}\rho_2\right) \tag{7.99}$$

This yields a relation with the fractional flow given by

$$f_1(S_1) = S_1 A(S_1)\frac{\mu(S_1)}{\mu_1} \tag{7.100}$$

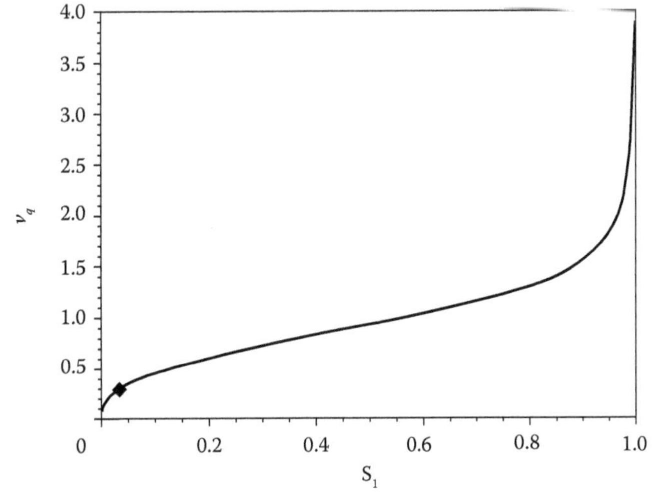

FIGURE 7.7 Derived isosaturation speed versus saturation curve. (From Udey, N., and Spanos, T.J.T., *Transp. Porous Media*, 10, 1–41, 1993.)

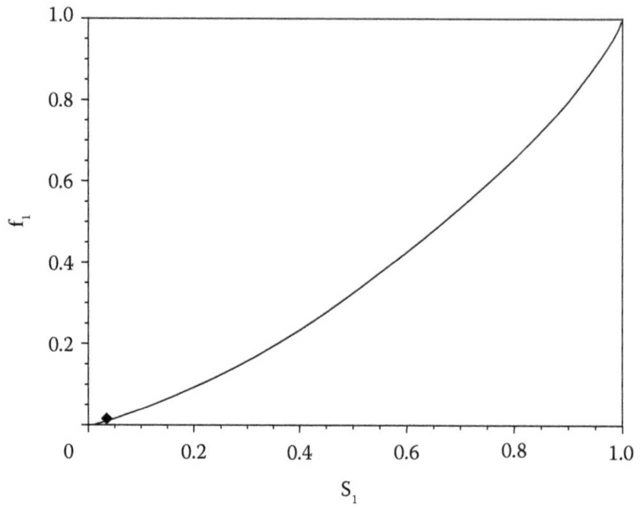

FIGURE 7.8 Derived fractional flow versus saturation curve. (From Udey, N., and Spanos, T.J.T., *Transp. Porous Media*, 10, 1–41, 1993.)

This equation means that A (S_1) may be calculated from the fractional flow and effective viscosity by

$$A(S_1)=\frac{f_1(S_1)/S_1}{\mu(S_1)/\mu_1} \tag{7.101}$$

They showed that the appropriate extrapolation of the curve S_1 versus $f_1(S_1)/S_1$ can be represented as a simple cubic of the form

$$S_1 = a_2\left(\frac{f_1}{S_1}\right)^2 + a_3\left(\frac{f_1}{S_1}\right)^3 \tag{7.102}$$

where the coefficients a_2 and a_3 must be positive. These coefficients are determined by requiring that the resultant curve and its derivative be continuous at the joining point between the extrapolation curve and the derived curve. The resultant composite curve is shown in Figure 7.9.

In studies on the stability of miscible displacement, the functional form of the effective viscosity is usually specified as an exponential (Tan and Homsey 1986). Therefore the effective viscosity is represented here as

$$\frac{\mu(S_1)}{\mu_1} = \frac{\mu_2}{\mu_1}\exp\left(-S_1\ln\left(\frac{\mu_2}{\mu_1}\right)\right) \tag{7.103}$$

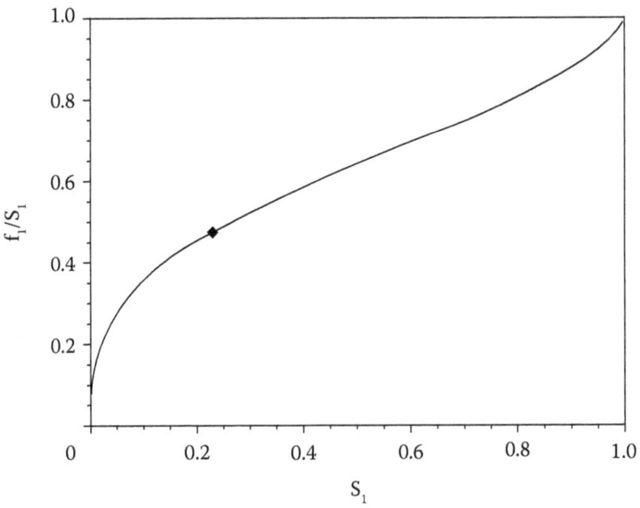

FIGURE 7.9 Derived $f_1(S_1)/S_1$ versus S_1(From Udey, N., and Spanos, T.J.T., *Transp. Porous Media*, 10, 1–41, 1993.).

Using this effective viscosity function, it is now possible to compute $A(S_1)$. However, Brigham did not specify the viscosities of the fluids that were used to produce the data that is being analysed here. It is therefore assumed that $\mu_2/\mu_1 = 2$. Using Equation 7.101, a plot of A (S_1) versus S_1 is given in Figure 7.10.

Equation 7.95 may now be rearranged to determine B (S_1). This plot is shown in Figure 7.11.

It is hoped that this analysis will encourage more experimental investigations.

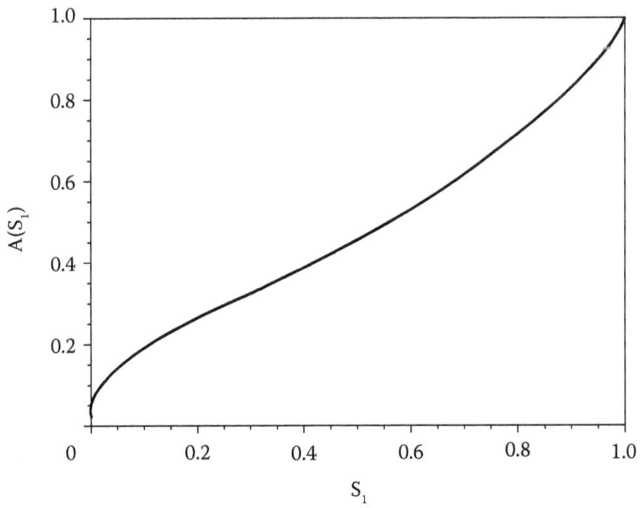

FIGURE 7.10 Derived $A(S_1)$ versus S_1. (From Udey, N., and Spanos, T.J.T., *Transp. Porous Media*, 10, 1–41, 1993.)

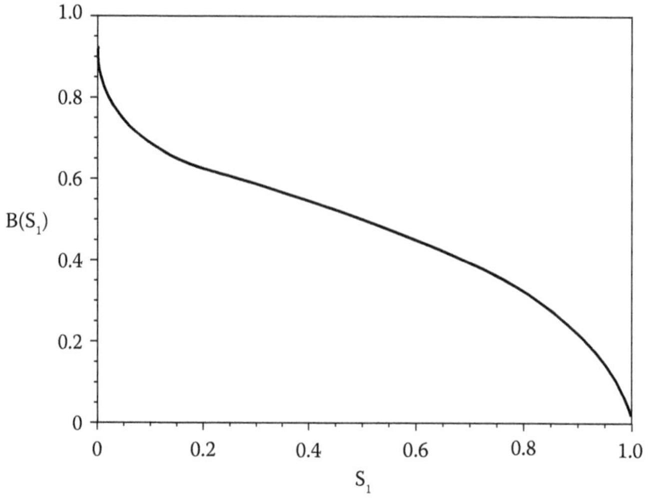

FIGURE 7.11 Derived $B(S_1)$ versus S_1. (From Udey, N., and Spanos, T.J.T., *Transp. Porous Media*, 10, 1–41, 1993.)

SUMMARY

The equations of two-phase miscible flow with negligible molecular diffusion were derived from the equations of two-phase immiscible flow in the limit of zero interfacial tension. A solution to these equations was obtained and was shown to give rise to three-dimensional dispersion. The dispersion can be decomposed into longitudinal and transverse components with respect to the Darcy velocity.

The dispersion that arises from the theory constructed here permits predictions of several features that arise in miscible displacement experiments. It is predicted that laboratory breakthrough curves should display systematic deviations from the expected breakthrough curves predicted by the convection diffusion theory; these deviations are early breakthrough and long tailing in the breakthrough curve. In slim tubes and cores, it is predicted that the longitudinal dispersion coefficient should be proportional to the average fluid velocity and the length of the tube or core; this second feature shows that the dispersion coefficient is scale dependent. All of these features were observed.

The standard laboratory experiments that only measure the effluent concentration emitted from a slim tube or core during a miscible displacement are inadequate. They do not provide all the information that is needed to characterize a miscible flow. Future experiments should monitor, throughout a slim tube or core, the evolution of pressure, average fluid velocity, and the kinematics of the surfaces of constant saturation. An understanding of the time development of the isosaturation velocity is vital in classifying the type of dispersion the flow is experiencing; furthermore, this information is important in untangling the relative contribution to the flow of mechanical dispersion and molecular diffusion.

REFERENCES

Abramowitz, M., and Stegun, I.A., 1965, *Handbook of Mathematical Functions*, Dover, New York.

Arya, A., Hewett, T.A., Larson, R.G., and Lake, L.W., 1988, Dispersion and reservoir heterogeneity, *SPE Reservoir Eng.*, 3, 139–148.

Baker, L.E., 1977, Effects of dispersion and dead-end pore volume in miscible flooding, *Soc. Petrol. Eng. J.*, 17, 219–227.

Bear, J., 1988, *Dynamics of Fluids in Porous Media*, 2nd edn., Dover, New York.

Bretz, R.E., and Orr, F.M., Jr., 1987, Interpretation of miscible displacements in laboratory cores, *SPE Reservoir Eng.*, 2, 492–500.

Brigham, W.E., 1974, Mixing equations in short laboratory cores, *Soc. Petrol. Eng. J.*, 14, 91–99.

Brigham, W.E., Reed, P.W., and Dew, J.N., 1961, Experiments on mixing during miscible displacement in porous media, *Soc. Petrol. Eng. J.*, I, 1–8.

Coats, K.H., and Smith, B.D., 1964, Dead-end pore volume and dispersion in porous media, *Soc. Petrol. Eng. J.*, 4, 73–84.

Curtis, F., and Gerald, C.F., 1978, *Applied Numerical Analysis*, 2nd edn., Addison-Wesley, Reading, MA.

Cyr, T.J., de la Cruz, V., and Spanos, T.J.T., 1988, An analysis of the viability of polymer flooding as an enhanced oil recovery technology, *Trans. Porous Media*, 3, 591–618.

de la Cruz, V., and Spanos, T.J.T., 1983, Mobilization of oil ganglia, *AIChE J.*, 29, 854–858.

de la Cruz, V., Spanos, T.J.T., and Yang, D.S., 1995, Macroscopic capillary pressure, *Trans. Porous Media*, 19, 67–77, 1995.

Dougherty, E.L., 1963, Mathematical model of an unstable miscible displacement, *Soc. Petrol. Eng. J.*, 3, 155–163.

Fried, J.J., and Combarnous, M.A., 1971, Dispersion in porous media, *Adv. Hydrosci.*, 7, 169–282.

Dullien, F.A.L., 1992, *Porous Media, Fluid Transport and Pore Structure*, 2nd edn., Academic Press, Toronto.

Greenkorn, R.A., 1983, *Flow Phenomena in Porous Media*, Marcel Dekker, New York.

Han, N., Bhakta, J., and Carbonell, R.G., 1985, Longitudinal and lateral dispersion in packed beds: Effect of column length and particle size distribution, *AIChE J.*, 31, 277–288.

Jankovic, M.S., 1986, Analytical miscible relative permeability curves and their usage with compositional and pseudo-miscible simulators, *J. Can. Petrol. Technol.*, 25, 55–65.

Koval, E.J., 1963, A method for predicting the performance of unstable miscible displacement in heterogeneous media, *Soc. Petrol. Eng. J.*, 3, 145–154.

Lallemand-Barres, A., and Peaudecerf, P., 1978, Recherche des relations entre le valeur de Ia dispersivite macroscopique d'un milieu aquifere, ses autres characteristiques et les conditions de mesure, *Bull. BRGM*, 2e Serie, Sec. III. No. 4, 277–284.

Landau, L.D., and Lifshitz, E.M., 1959, *Fluid Mechanics, Course of Theoretical Physics*, vol. 6, Pergamon Press, Oxford.

Perkins, T.K., and Johnston, O.C., 1963, A review of diffusion and dispersion in porous media, *Soc. Petrol. Eng. J.*, 3, 70–84.

Pickens, J.F., and Grisak, G.E., 1981, Scale-dependent dispersion in a stratified granular aquifer, *Water Resour. Res.*, 17, 1191–1211.

Scheidegger, A.E., 1974, *The Physics of Flow Through Porous Media*, 3rd edn., University of Toronto Press, Toronto.

Spanos, T.J.T., de la Cruz, V., and Hube, J., 1988, An analysis of the theoretical foundations of relative permeability curves, *AOSTRA J. Res.*, 4, 181–192.

Stalkup F.I., Jr., 1970, Displacement of oil by solvent at high water saturation, *Soc. Petrol. Eng. J.*, 10, 337–348.

Stalkup F.I., Jr., 1982, *Miscible Displacement*, Monograph 8, Henry L. Doherty Memorial Fund of AIME, and Society of Petroleum Engineers of AIME, New York.

Tan, C.T., and Homsey, G.M., 1986, Stability of miscible displacements in porous media: Rectilinear flow, *Phys. Fluids*, 29, 3549–3556.

Todd, M.R., and Longstaff, W.J., 1972, The development, testing, and application of a numerical simulator for predicting miscible flood performance, *J. Petrol. Technol.*, 24, 874–882.

Udey, N., and Spanos, T.J.T., 1993, The equations of miscible flow with negligible molecular diffusion, *Trans. Porous Media*, 10, 1–41, 1993.

Yang, D., Udey, N., and Spanos, T.J.T., 1998, Automata Simulations of Dispersion in Porous Media, Transport in Porous Media, 32, 187–198.

CHAPTER 8

Miscible Fluid Flow in Porous Media (Part 2)

OBJECTIVE OF THIS CHAPTER

A system of equations for miscible displacement in porous media was constructed (Udey and Spanos 1993) by considering the limiting case of immiscible flow with zero surface tension. This construction was summarized in Chapter 7. In that construction it was shown that miscible flow could be described by a Fokker–Plank equation. At this point, the assumption of negligible molecular diffusion is discarded. Here, it is observed that the analysis involving volume averaging in the previous chapters can no longer be used. As well, the concept of using component velocities and volume fractions must be discarded. In the previous chapters these quantities only appeared to be thermomechanical and thermodynamic variables because they were proportional to the actual thermodynamic variables, which are momentum fluxes and mass fractions that will be described in the next three chapters. This previous work in terms of volume averaging only applies to two phases that are mixed at one scale, the pore scale.

EQUATION OF CONTINUITY

Thermodynamics is based on conservation of energy and the statement that the entropy of an isolated system can never decrease (Fermi 1956; Yourgrau et al. 1982; De Groot and Mazer 1984; Haase 1990). Consider a system composed of n components. Here, the mass concentration of a fluid component is

$$c_k = \frac{M_k}{V_f} - k = 1, \ldots, n \tag{8.1}$$

The mass density of the total fluid mass in V_f divided by the volume V_f is

$$\rho = \frac{M_f}{V_f} - M_f = M_1 + \ldots + M_n \tag{8.2}$$

which leads to the expression

$$\rho = c_1 + \ldots + c_n \tag{8.3}$$

The fractional mass concentration is defined by

$$\hat{c}_{fk} = \frac{c_k}{\rho} \tag{8.4}$$

Conservation of total mass states

$$\frac{\partial \rho}{\partial t} = -\nabla \cdot (\rho \mathbf{v}) \tag{8.5}$$

where ρ is the total density and $\mathbf{v}$ is the velocity of momentum flux. Note that the various components may be mixed at the molecular scale through chemical reactions or at the pore scale through mechanical work. In order to include the mechanical work at the pore scale for a two-phase system, an additional velocity is required, the momentum flux of the interactions $\mathbf{v}_i$. Here, each macroscopically segregated phase at the pore scale has its own distinct conservation properties at that scale, and this information may not be lost by specifying a single continuum model at the megascale. Here, the density associated with $\mathbf{v}_i$ is

$$\rho^{(i)} = (c_1 - c_2) \tag{8.6}$$

and the continuity equation for out-of-phase motions is

$$\frac{\partial \rho^{(\mathrm{i})}}{\partial t} = -\nabla \cdot (\rho^{(\mathrm{i})} \mathbf{v}^{(i)}) + \nabla \cdot \mathbf{J}^{(i)} \tag{8.7}$$

Note for the component phases, there are two types of flow: the diffusion flow or molecular-scale mixing of the phases and the pore-scale mixing of the phases.

Two mass fractions are now obtained that are required to describe the mixing at both the molecular scale and the macroscale, the second describing macroscopic mixing.

$$\hat{c}^{(\mathrm{i})} = \rho^{(\mathrm{i})} / \rho \tag{8.8}$$

Note that if the phases are only mixed at the molecular scale this second continuity equation yields no information, that is, it becomes unnecessary.

In this case, the macroscale and megascale descriptions are the same. If mixing only occurs at the macroscale, then two equations of motion are firmly established at the macroscale and the interactions of the phases at the macroscale must be retained when describing the motions at the megascale. Here, two equations of motion are required, as well as the description of how the volume fractions are changing, as was discussed in the previous description of immiscible flow. If mixing is occurring at various scales, then two equations of motion are required, as well as descriptions of how the above two mass fractions are changing in time.

In the most general case, equations of motion for the component phases have little meaning. In this case, the motions must be described in terms of the total momentum flux, the momentum flux of the interactions, and two degrees of freedom in the changing of the mass fractions, since the mass fractions may change due to molecular mixing or due to mechanical work at the macroscale. Thus there is an additional degree of freedom that was not present in the description of immiscible flow. It is also clear the volume fraction only appeared to be a thermodynamic variable because it was proportional to the mass fraction in the previous descriptions.

EQUATIONS OF MOTION

$$\rho\frac{\partial v_j}{\partial t} = \sigma_{jk,k} + \rho_k F_{jk} \tag{8.9}$$

$$\rho^{(i)}\frac{\partial v_j^{(i)}}{\partial t} = \sigma_{jk,k}^{(i)} + \rho_k^{(i)} F_{jk}^{(i)} \tag{8.10}$$

$$\frac{\partial \hat{c}_{fk}}{\partial t} = \delta_m \nabla\cdot\mathbf{v} + \delta_i \nabla\cdot\mathbf{v}^{(i)} \tag{8.11}$$

$$\frac{\partial \hat{c}^{(\mathrm{i})}}{\partial t} = \beta_m \nabla\cdot\mathbf{v} - \beta_i \nabla\cdot\mathbf{v}^{(i)} \tag{8.12}$$

Now assume that molecular diffusion is negligible; then the first three equations are described by Udey and Spanos (1993).

In this case, the equation

$$\frac{\partial \hat{c}_{fk}}{\partial t} = \delta_m \nabla\cdot\mathbf{v} + \delta_i \nabla\cdot\mathbf{v}^{(i)} \tag{8.13}$$

becomes

$$\kappa\frac{\partial S_1}{\partial t} = \delta_m \nabla\cdot[S_1\rho_1\mathbf{v}_1 + (1-S_1)\rho_2\mathbf{v}_2] + \delta_i \nabla\cdot[S_1\rho_1\mathbf{v}_1 - (1-S_1)\rho_2\mathbf{v}_2] \tag{8.14}$$

which may be written as

$$\frac{\partial S_1}{\partial t} = \frac{S_1\rho_1}{\kappa}(\delta_m + \delta_i)\nabla\cdot\mathbf{v}_1 - \frac{(1-S_1)\rho_2}{\kappa}(\delta_i - \delta_m)\nabla\cdot\mathbf{v}_2 \tag{8.15}$$

which is the familiar equation

$$\frac{\partial S_1}{\partial t} = \delta_1 \nabla \cdot \mathbf{v}_1 - \delta_2 \nabla \cdot \mathbf{v}_2 \tag{8.16}$$

The final equation

$$\frac{\partial \hat{c}^{(i)}}{\partial t} = \beta_m \nabla \cdot \mathbf{v} - \beta_i \nabla \cdot \mathbf{v}^{(i)} \tag{8.17}$$

is no longer independent in this limit and thus provides no additional information.

Thus its role in general is to provide information about the coupling between molecular-scale and macroscale mixing.

This may be seen as follows:

$$\frac{\partial c_1}{\partial t} = -c_1 \nabla \cdot \mathbf{v}_1 \tag{8.18}$$

$$\frac{\partial c_2}{\partial t} = -c_2 \nabla \cdot \mathbf{v}_2 \tag{8.19}$$

$$\mathbf{v} = \frac{c_1 \mathbf{v}_1 + c_2 \mathbf{v}_2}{\rho} \tag{8.20}$$

$$\mathbf{v}^{(i)} = \frac{c_1 \mathbf{v}_1 - c_2 \mathbf{v}_2}{\rho} \tag{8.21}$$

upon substitution

$$\frac{\partial c_1}{\partial t} - \frac{\partial c_2}{\partial t} = \beta_m \left(-\frac{\partial c_1}{\partial t} - \frac{\partial c_2}{\partial t} \right) - \beta_i \left(-\frac{\partial c_1}{\partial t} + \frac{\partial c_2}{\partial t} \right) \tag{8.22}$$

thus

$$\frac{\partial c_1}{\partial t} - \frac{\partial c_2}{\partial t} = (\beta_i - \beta_m) \frac{\partial c_1}{\partial t} - (\beta_i + \beta_m) \frac{\partial c_2}{\partial t} \tag{8.23}$$

and

$$\beta_i = 1 \tag{8.24}$$

$$\beta_m = 0 \tag{8.25}$$

Now consider the material derivatives of the megascopic concentrations (the mass fraction obtained from both molecular scale and macroscale mixing) $\hat{c}_{fk}$ and $\hat{c}^{(i)}$ (here, the definition of $\mathbf{J}^{(i)}$ is changed by a constant)

$$\frac{d\hat{c}_{fk}}{dt} = \nabla \cdot \mathbf{J}_{fk} \tag{8.26}$$

$$\frac{d\hat{c}^{(i)}}{dt} = \nabla \cdot \mathbf{J}^{(i)} \tag{8.27}$$

which yield

$$\frac{\partial \hat{c}_{fk}}{\partial t} + \mathbf{v} \cdot \nabla \hat{c}_{fk} = \nabla \cdot \mathbf{J}_{fk} \tag{8.28}$$

and

$$\frac{\partial \hat{c}^{(i)}}{\partial t} + \mathbf{v} \cdot \nabla \hat{c}^{(i)} = \nabla \cdot \mathbf{J}^{(i)} \tag{8.29}$$

Here

$$\nabla \cdot \mathbf{J}_{fk} = -\rho \mathbf{D}_m \cdot \nabla \hat{c}_{fk} \tag{8.30}$$

and

$$\nabla \cdot \mathbf{J}^{(i)} = -\rho^{(i)} \mathbf{D}_{(i)} \cdot \nabla \hat{c}^{(i)} \tag{8.31}$$

$$\mathbf{v} \cdot \nabla \hat{c}_{fk} - \frac{1}{\rho} \nabla \cdot (\rho \mathbf{D}_m \cdot \nabla \hat{c}_{fk}) = \mathbf{v}_c \cdot \nabla \hat{c}_{fk} \tag{8.32}$$

$$\mathbf{v} \cdot \nabla \hat{c}^{(i)} - \frac{1}{\rho} \nabla \cdot (\rho^{(i)} \mathbf{D}_{(i)} \cdot \nabla \hat{c}^{(i)}) = \mathbf{v}^{(i)} \cdot \nabla \hat{c}^{(i)} \tag{8.33}$$

Here, $\mathbf{v}_c$ is the isoconcentration velocity and $\mathbf{v}^{(i)}$ isodiffusion velocity (which is a composite of molecular diffusion and macroscopic mixing). This leads to two coupled Fokker–Planck equations. This implies that there is an additional physical process, likely a coupling between diffusion and dispersion.

The equations

$$\frac{\partial \hat{c}_{fk}}{\partial t} + \mathbf{v} \cdot \nabla \hat{c}_{fk} = \frac{1}{\rho} \nabla \cdot (\rho \mathbf{D}_m \cdot \nabla \hat{c}_{fk}) \tag{8.34}$$

and

$$\frac{\partial \hat{c}^{(i)}}{\partial t} + \mathbf{v} \cdot \nabla \hat{c}^{(i)} = \frac{1}{\rho} \nabla \cdot (\rho^{(i)} \mathbf{D}_{(i)} \cdot \nabla \hat{c}^{(i)}) \tag{8.35}$$

are two coupled Fokker–Planck equations. In the limit, as molecular diffusion becomes negligible, Equation 8.34 becomes identical to Equation 7.61 and Equation 8.35 is trivially satisfied.

THREE-FLUID INTERACTIONS

As discussed previously, in going from two- to three-phase interactions, the components may now interact in four different ways. In terms of the mass, there is still the total mass

$$\rho = c_1 + c_2 + c_3 \tag{8.36}$$

Conservation of total mass again is stated by

$$\frac{\partial \rho}{\partial t} = -\nabla \cdot (\rho \mathbf{v}) \tag{8.37}$$

However, in the absence of diffusion there are now four ways the masses may interact at the two different scales (one in-phase motion and three out-of-phase motions associated with mechanical interactions of the phases at the macroscale).

$$\rho_{12(3)}^{(i)} = c_1 + c_2 - c_3 \tag{8.38}$$

$$\rho_{1(2)3}^{(i)} = c_1 - c_2 + c_3 \tag{8.39}$$

$$\rho_{(1)23}^{(i)} = -c_1 + c_2 + c_3 \tag{8.40}$$

The continuity equations for these out-of-phase motions are given by

$$\frac{\partial \rho_{12(3)}^{(i)}}{\partial t} = -\nabla \cdot \rho_{12(3)}^{(i)} \mathbf{v}_{12(3)}^{(i)} + \mathbf{J}_{12(3)}^{(i)} \tag{8.41}$$

$$\frac{\partial \rho_{1(2)3}^{(i)}}{\partial t} = -\nabla \cdot \rho_{1(2)3}^{(i)} \mathbf{v}_{1(2)3}^{(i)} + \mathbf{J}_{1(2)3}^{(i)} \tag{8.42}$$

$$\frac{\partial \rho_{(1)23}^{(i)}}{\partial t} = -\nabla \cdot \rho_{(1)23}^{(i)} \mathbf{v}_{(1)23}^{(i)} + \mathbf{J}_{(1)23}^{(i)} \tag{8.43}$$

where $v^{(i)}_{(1)23}$, $v^{(i)}_{1(2)3}$ and $v^{(i)}_{12(3)}$ are the velocities of momentum flux of the out-of-phase motions. As in Equation 8.21

$$v^{(i)}_{(1)23} = \frac{-\rho_1 v_1 + \rho_2 v_2 + \rho_3 v_3}{\rho} \tag{8.44}$$

$$v^{(i)}_{1(2)3} = \frac{\rho_1 v_1 - \rho_2 v_2 + \rho_3 v_3}{\rho} \tag{8.45}$$

$$v^{(i)}_{12(3)} = \frac{\rho_1 v_1 + \rho_2 v_2 - \rho_3 v_3}{\rho} \tag{8.46}$$

However, when molecular diffusion is also allowed, there now five different coupled Fokker–Planck equations obtained—two equations associated with the values for $\hat{c}_{fk}$ and three associated with the values for $\hat{c}^{(i)}$.

SUMMARY

As mentioned in the previous chapter, dispersion that arises from this solution permits a prediction of several features that arise in miscible displacement experiments. It was predicted that laboratory breakthrough curves should display systematic deviations from the expected breakthrough curves predicted by the convection diffusion theory; these deviations are early breakthrough and long tailing in the breakthrough curve. In slim tubes and cores, it was predicted that the longitudinal dispersion coefficient should be proportional to the average fluid velocity and the length of the tube or core; this second feature shows that the dispersion coefficient is scale dependent. All of these features have been observed.

A flux equation such as diffusional flow at the molecular scale is described by a diffusion equation. Here, it was shown that if the flow is mixed at the macroscale as well, then a megascopic description of that flow requires two coupled Fokker–Planck equations. In the case where molecular scale mixing becomes negligible, this description reduces to one Fokker–Planck equation. Of course, it is straightforward to present identical constructions for other macroscale diffusion processes occurring at the pore scale in porous media.

Going from two components to three components without molecular diffusion requires that four momenta and four mass fractions be considered associated with one in-phase motion and three out-of-phase motions. The three components are so strongly coupled that they essentially lose their independent identities. The momenta are, however, defined in terms of the component phases, and the four momenta may be expressed in terms of the three components. When molecular diffusion is included, the component phases no longer have meaning and all equations may only be expressed in terms of momenta and mass fractions. However, molecular diffusion is also observed to introduce an additional degree of freedom into the description of the motion. As a result, four coupled Fokker–Planck equations are observed when molecular diffusion is not included and five coupled Fokker–Planck equations are observed when molecular diffusion is included.

REFERENCES

De Groot, S.R., and Mazur, P., 1984, *Non-Equilibrium Thermodynamics*, Dover, New York.

Fermi, E., 1956, *Thermodynamics*, Dover, New York.

Haase, R., 1990, *Thermodynamics of Irreversible Processes*, Dover, New York.

Udey, N., and Spanos, T.J.T., 1993, The equations of miscible flow with negligible molecular diffusion, *Trans. Porous Media*, 10, 1–41, 1993.

Yourgrau, W., van der Merwe, A., and Raw, G., 1982, *Treatise on Irreversible and Statistical Thermophysics; An Introduction to Nonclassical Thermodynamics*, Dover, New York.

CHAPTER 9

Non-Equilibrium Thermodynamics

Fluid Flow and Granular Motions

THE OBJECTIVES OF THIS CHAPTER

Thermodynamics generally describes transformations between heat and mechanical work (Planck 1917; Fermi 1956; Schrodinger 1952; Landau and Lifshitz 1958; Reiss 1965; Pauli 1973; Callen 1985). In the present discussion the materials are also mixed at different scales, bringing additional processes into the discussion. As well, for dynamic processes, non-equilibrium thermodynamics needs to be considered (Jammer 1964; Yourgrau et al. 1982; de Groot and Mazer 1984; Haase 1990). For porous materials and other composite media, an important departure in the thermodynamic status of porosity and saturation described in Chapter 4 occurs when irreversible deformations are allowed. In these cases, it is observed that the volume fractions and component velocities are not thermodynamic variables. This is in stark contrast to the previous discussion of equilibrium thermodynamics, where they appeared to be thermodynamic variables. Of course, in hindsight it is observed that they were proportional to the actual thermodynamic variables in the cases considered. In the case of porosity, it is observed that specific mass fractions replace porosity and saturation (volume fractions) as the thermodynamic variables, and a new fundamental quantity appears which is associated with the mass fractions, the order (or structure) of the medium. Of particular importance to non-equilibrium thermodynamics are the modifications momentum flux, etc., make to the Clapeyron equation and the new information about dispersion, porosity waves, fluid flow and their interactions obtained from the new Onsager's relations.

Fluid transport in porous media (Slattery 1967; Whitaker 1967; Scheidegger 1974; Marle 1982; de la Cruz and Spanos 1983; Entov et al. 1990; Bear and Bachmat 1990, 1991; Dullien 1992; de la Cruz et al. 1995; Cushman 1997), the coupling to the motions of the medium (Burridge and Keller 1981; de la Cruz and Spanos 1985, 1989; Geilikman et al. 1993; Geilikman 1999; Sahay et al. 2001; Spanos 2002) and the associated thermodynamics

(Hassanizadeh and Grey 1990, 1993; de la Cruz et al. 1993; Udey et al. 1999; Bennethum and Cushman 2002a, 2002b) have been considered by a number of authors.

A description of the thermodynamics of air, water vapour, gas (a mixture of air and vapour), and liquid water in a rigid porous matrix was described by Fremond and Nicolas (1990). The thermodynamics of phase transitions in porous media was considered by de Boer (1995) and de la Cruz et al. (1985). A discussion of the effect larger scale interactions have on Onsager's relations was presented by Gyftopoulos (2004). Extensions of Onsager's relations to non-linear fields and far from equilibrium processes were discussed by Garcia-Conlin and Uribe (1991) and Velasco et al. (2011). A description of how Fokker–Planck equations apply to many physical systems including porous media was presented by Chavanis (2006, 2008). Green (1952) showed the H'-theorem and the Onsager's reciprocal relations could be constructed in the framework of Fokker–Planck equations. Hasegawa (1977) showed that important concepts in thermodynamics could be incorporated into the statistical mechanical framework by means of Fokker–Planck equations. A review of multiscale simulations was given by Hyodo (2002).

The position taken in this analysis is that reciprocity, which is referred to here as the *Onsager's relations*, is fundamental. That is, these relations not only apply to molecular mixtures in which fluxes are described by diffusion equations but also at the megascale and for non-linear processes where fluxes are described by Fokker–Planck equations. This appears to follow from the result that reciprocity cannot be constructed from the component equations for a three-component system or a system mixed at multiple scales. As well, it is observed that a proper description of the observed physics (e.g. the coupling between porosity and saturation waves or the mixing of fluids at both the molecular scale and macroscale) also cannot be constructed from the component equations in these cases. However, when the theory is constructed in terms of momentum and the associated mass fractions, both constructions (the description of observed motions and reciprocity) follow.

THE COMPONENT PHASES AND VOLUME FRACTIONS

As a starting point in this discussion, consider a single fluid and an elastic matrix. This case was considered for equilibrium thermodynamics, where it was observed that the thermodynamics could be written in terms of the components and the volume fractions. Allowing for fluid flow and/or granular motion, irreversible processes occur and equilibrium thermodynamics can no longer be used. The only reasonable cases in which equilibrium thermodynamics can be used to describe fluid-filled porous media is for statics or dynamic motions of a perfectly elastic medium (Landau and Lifshitz 1975). All other dynamic processes involve irreversible motions. The various dynamic processes that can occur have different relations between fluid flow, porosity change and the dilational motions of the phases (e.g. porosity wave, seismic wave propagation, porosity soliton). In order to specify a process, the time rate of change of the porosity and the strain rate must be specified. The relation between porosity and dilational motions is once again given by

$$\frac{\partial \eta}{\partial t} = \delta_s \nabla \cdot \mathrm{v}_s - \delta_f \nabla \cdot \mathrm{v}_f \tag{9.1}$$

This equation relates the time derivative of porosity to the dilational velocities of the fluid and solid. In the context of a non-equilibrium process, displacement vectors and porosity do not have clear physical meaning, since it is the relationship between their time derivatives that specifies the process under consideration. This is a major change from the previous description of equilibrium thermodynamics, where these quantities played an important role. Also note the assumption of local equilibrium, which is usually imposed on single continuum models, is no longer acceptable. In the present case, it is the dynamic relationship (Equation 9.1) that defines the process under consideration. Mathematically, one may take the time integral of this equation once a process has been chosen (a soliton, say) and formally write local relations in the form of equilibrium thermodynamics. In the case of the soliton (and the porosity wave), δ_s and δ_f are parameters that describe viscous dissipation within the fluid, frictional sliding of the grains and the shear elastic properties of the matrix, since compressional motions of the component phases are negligible. In the case of a P wave, δ_s and δ_f are parameters that primarily describe the compressional motions of the fluid and solid and the associated flow of the fluid.

In order to better study these parameters, the fluid and solid velocity may be eliminated from the equations of motion, yielding first for the fluid

$$\frac{\rho_{\eta f}}{\eta_o}\left(\frac{\partial^2\eta}{\partial t^2}+2a_{\eta f}\frac{\partial\eta}{\partial t}-2b_{\eta f}\nabla^2\frac{\partial\eta}{\partial t}\right)= \tag{9.2}$$

$$-\frac{\rho_{pf}}{K_f}\left(\frac{\partial^2 p_f}{\partial t^2}+2a_{pf}\frac{\partial p_f}{\partial t}-2b_{pf}\nabla^2\frac{\partial p_f}{\partial t}-v_{pf}^2\nabla^2 p_f\right)$$

The stress tensor for the solid must now account for both reversible and irreversible motions of the matrix. Recall that for the elastic matrix, the porosity is given by $\eta = \eta_o + \eta'$, where η_o is the unperturbed porosity and η' is a reversible porosity perturbation. In the case of frictional sliding of the grains, we now have both reversible and irreversible motions of the grains. This results in a matrix whose porosity is given by $\eta = \eta_o^i + (\eta_o^f - \eta_o^i)n'$, where η_o^i is the initial porosity, η_o^f is the final porosity after the granular motion, $(\eta_o^f - \eta_o^i)$ is the irreversible porosity change and η' is the reversible change in porosity. The equation of motion for the solid becomes

$$\frac{\alpha_1\rho_{\eta s}}{\eta_o}\left(\frac{\partial^2\eta}{\partial t^2}+2a_{\eta s}\frac{\partial\eta}{\partial t}-2b_{\eta s}\nabla^2\frac{\partial\eta_o^f}{\partial t}-v_{\eta s}^2\nabla^2\eta'\right)= \tag{9.3}$$

$$-\frac{\alpha_2\rho_{ps}}{K_f}\left(\frac{\partial^2 p_f}{\partial t^2}+2a_{ps}\frac{\partial p_f}{\partial t}-2b_{ps}\nabla^2\frac{\partial p_f}{\partial t}-v_{ps}^2\nabla^2 p_f\right)$$

where the parameters in Equation 9.1 specify the dynamic process under consideration. However, Equations 9.2 and 9.3 indicate that Equation 9.1 should now be written as

$$\frac{\partial(\eta_o^f+\eta_o')}{\partial t}=\delta_s\nabla\cdot\mathrm{v}_s-\delta_f\nabla\cdot\mathrm{v}_f \tag{9.4}$$

which may be broken up into the reversible and irreversible parts. For now, it will be assumed that the irreversible motions are small in comparison to the reversible motions. Now writing the reversible and irreversible motions separately yields

$$\frac{\partial \eta^r}{\partial t} = \delta_s^r \nabla \cdot \mathrm{v}_s - \delta_f^r \nabla \cdot \mathrm{v}_f \tag{9.5}$$

and

$$\frac{\partial \eta_o^f}{\partial t} = \delta_s^i \nabla \cdot \mathrm{v}_s - \delta_f^i \nabla \cdot \mathrm{v}_\mathrm{f} \tag{9.6}$$

In terms of the solid fraction, the quantities $\phi_o^s = 1 - \eta_o^f$ and $\phi^r = 1 - \eta^r$ are defined to beconsistent with the notation in Chapter 5.

Now note that Equation 9.5 defines such reversible porosity changes (reversible changes in the elastic matrix) as static compressions, seismic wave propagation and low amplitude porosity wave propagation. The values of. δ_s^r and δ_f^r would be slightly different for the description of the first two processes due to the effect of fluid flow during seismic wave propagation. These values are substantially different in the case of porosity wave propagation due to the negligible amount of compressional motions and the net flux of fluid in the direction of the wave motion. Equation 9.6 accounts for irreversible motions of the solid and in particular allows the equations of motion to predict a soliton wave, which enhances the porosity wave by allowing it to extract mechanical energy from the stress field of the matrix and move fluid through the matrix essentially without any net attenuation of the wave.

In this case, the entropy change may result from molecular-scale processes, yielding a heat flux at the macroscale, and it may also change due to work being done at the macroscale, yielding a change in structure at the megascale.

The entropy is composed of

$$\begin{bmatrix} \text{rate of increase of} \\ \text{entropy inside dV} \end{bmatrix} + \begin{bmatrix} \text{outward flux of} \\ \text{entropy from dV} \end{bmatrix} = \begin{bmatrix} \text{entropy production} \\ \text{inside dV} \end{bmatrix}$$

$$\frac{dS}{dt} - \frac{1}{T}\frac{dQ}{dt} - \frac{1}{c_s}\frac{dO}{dt} = \int_V \sigma \, dV \tag{9.7}$$

Here, for a closed system it may exchange heat with its surroundings or the matrix may undergo an irreversible change in structure, causing it to exchange mass with its surroundings. Here, the exchange in heat has its origins from the molecular scale, while the exchange in mass occurs because of mechanical work altering the stress on the matrix at the pore scale. From the second law of thermodynamics, the right-hand side is greater than or equal to 0. Therefore

$$\frac{dS}{dt} \geq \frac{1}{T}\frac{dQ}{dt} + \frac{1}{c_s}\frac{dO}{dt} \tag{9.8}$$

Here, $c_s = \phi$ is the mass (volume) fraction of the solid inside V and O is the order (how the structure and stress of the matrix is changing) of the system. This term allows for the propagating source term and the soliton observed in Equation 9.3.

This allows for a construction of non-equilibrium thermodynamics in terms of the component phases and volume fractions in this special case. Thus, a direct generalization of the description in Chapter 5 may be constructed here.

$$\frac{dQ_s}{dt} = \frac{\partial U_s}{\partial T_s}\frac{\partial T_s}{\partial t} \tag{9.9}$$

$$\frac{dQ_f}{dt} = \frac{\partial U_f}{\partial T_f}\frac{\partial T_f}{\partial t} \tag{9.10}$$

$$\frac{dO_s}{dt} = \frac{\partial O_s}{\partial \phi}\frac{\partial \phi}{\partial t} \tag{9.11}$$

$$Q = Q_s + Q_f \tag{9.12}$$

$$O = O_s \tag{9.13}$$

The internal energy is now given by

$$dU_s = \tau_{ik}^s du'_{ik} + T_s dS_s^T + \phi dS_s^\phi \tag{9.14}$$

where

$$S_s = S_s^T + S_s^\phi \tag{9.15}$$

Here, S_s^T is the portion of the entropy associated with thermal effects and S_s^ϕ is the portion of the entropy associated with the structure of the medium.

$$S_s^T = \frac{1}{T}\frac{dQ}{dt} \tag{9.16}$$

$$S_s^\phi = \frac{1}{\phi}\frac{dO}{dt} \tag{9.17}$$

The solid stress is

$$\tau_{ik}^s = \phi_o K_s \alpha_s \left(T^s - T_o\right)\delta_{ik} + 2\mu_M\left[u_{ik}'^s - \tfrac{1}{3}\delta_{ik}u_{jj}'^s\right] + \phi_o K_s \delta_{ik} u_{jj}'^s + \upsilon\left(\phi_o^f - \phi_o^i\right)\delta_{ik} \tag{9.18}$$

where the solid strain is given by

$$u_{ik}'^{s} = u_{ik}^{s} + \frac{1}{3}\frac{\phi' - \phi_o}{\phi_o}\delta_{ik} \tag{9.19}$$

The solid internal energy is given by

$$\begin{aligned} U_s(S, u_{ik}'^{s}) &= \phi_o U_o^s(S_s) - \frac{T_o^s K_s \alpha_s}{c_v^s}(S_s^T - S_o)u_{jj}'^{s} - \Omega(S_s^{\phi} - S_o)u_{jj}'^{s} \\ &+ \mu_M\left[u_{ik}'^{s} - \tfrac{1}{3}\delta_{ik}u_{jj}'^{s}\right]^2 + \frac{1}{2}\phi_o K_{ad}^s u_{jj}'^{s2} \end{aligned} \tag{9.20}$$

The entropy balance of the solid component is now given by

$$S^S - S_o^S = \phi_o c_v^s (T^s - T_o)/T_o + \phi_o K_s \alpha_s u_{jj}'^{s} + c_\beta\left(\phi_o^f - \phi_o^i\right) \tag{9.21}$$

$$S_o^S = \phi_o \frac{dF_o(T_o)}{dT_o} + \phi_o \frac{dF_o(\phi_o)}{d\phi_o^i} \tag{9.22}$$

$$S^S = \phi_o \frac{dF_o(T_o)}{dT} + \phi_o \frac{dF_o(\phi_o)}{d\phi_o^f} \tag{9.23}$$

$$c_v^s = \phi_o \frac{d^2F_o(T_o)}{dT_o^2} \tag{9.24}$$

$$c_\beta = \frac{d^2F_o(\phi_o)}{d\phi_o^2} \tag{9.25}$$

$$\mu_1(P_f, T) = \mu_2(P_f, T) \tag{9.26}$$

$$\mu_1(P_s, \phi) = \mu_2(P_s, \phi) \tag{9.27}$$

$$\frac{\partial \mu_1}{\partial T} + \frac{\partial \mu_1}{\partial P_f}\frac{\partial P_f}{\partial T} = \frac{\partial \mu_2}{\partial T} + \frac{\partial \mu_2}{\partial P_f}\frac{\partial P_f}{\partial T} \tag{9.28}$$

The Clapeyron equation for a steam–water phase boundary in a rigid porous medium was discussed in Chapter 6. Here, the transition from an intact porous medium containing a fluid to a fluidized mixture of solid and fluid is considered. Note at the megascale a change in density is observed due to a different proportion of fluid and solid on each side of the boundary. In the previous case, a change in density was observed due to the steam–water phase transition.

$$\frac{\partial \mu_1}{\partial \phi} + \frac{\partial \mu_1}{\partial P_s}\frac{\partial P_s}{\partial \phi} = \frac{\partial \mu_2}{\partial \phi} + \frac{\partial \mu_2}{\partial P_s}\frac{\partial P_s}{\partial \phi} \tag{9.29}$$

This yields

$$-S_{s1}^{\phi} + V_{s1}\frac{\partial P_s}{\partial \phi} = -S_{s2}^{\phi} + V_{s2}\frac{\partial P_s}{\partial \phi} \tag{9.30}$$

and thus

$$\frac{\partial P_s}{\partial \phi} = \frac{S_{s1}^{\phi} - S_{s2}^{\phi}}{V_{s1} - V_{s2}} \tag{9.31}$$

These entropies originate from the structure at the macroscale and the volumes describe the volume occupied by equivalent masses of solid on each side of the boundary.

Note that in the porous medium $P_s > P_f$ and in the fluidized mixture $P_s = P_f$. Another type of phase transition may also be observed under dynamic conditions. If a porosity wave passes through a granular porous medium and the magnitude of the dilatational stress exceeds the yield strength of the medium, then the medium will be fluidized. This is a second-order phase transition that may be easily observed in the lab and in shallow sand and soil observations in the earth. Wavefront Technology Solutions has demonstrated this process in the earth using its environmental pulsing tool.

Here it should be noted that Equation 9.31 gives rise to a phase transition called *liquefaction* (Geilikman 1999). The transition from a porous medium containing a fluid to solid suspensions in a fluid occurs across this boundary. In heavy oil reservoirs, this process is enhanced by gas coming out of solution as the pressure is reduced, causing foamy oil to drive the oil–sand mixture into production wells. A thermodynamic description of this process, however, cannot be obtained from the current formulation, since it also includes microscale processes. This would prohibit the use of volumes and component phases in the thermodynamic description. Thus, a generalization of the description in the following sections to include solid motions would be required.

THE NECESSITY OF A CONSTRUCTION IN TERMS OF MOMENTA AND MASS FRACTIONS

At this point, it is important to observe that Equations 9.1, 9.4, 9.5 and 9.6 have served their purpose in describing the current processes. However, when the thermodynamic relations such as Onsager's relations and the Clapeyron equation are constructed for more complex systems, it is observed that volume fraction is not a thermodynamic variable. Note that these equations are required to complete Newton's second law. In the following discussion, the main objective is to describe the coupled porosity saturation waves so the additional complexity of irreversible solid motions will be neglected. These equations in a more general form state

$$\frac{\partial c_{fk}}{\partial t} = \delta_m \nabla \cdot \mathbf{v} + \delta_i \nabla \cdot \mathbf{v}^{(i)} \tag{9.32}$$

$$\frac{\partial c^{(i)}}{\partial t} = \beta_m \nabla \cdot \mathbf{v} - \beta_i \nabla \cdot \mathbf{v}^{(i)} \tag{9.33}$$

Now note that the strain and strain rate tensors for more complex systems no longer have thermodynamic meaning. The component velocities no longer have clear physical meaning in the context of the processes being considered (since they describe momentum fluxes), so the deformations that are related to stress must now be specified in terms of the velocity of momentum flux and the velocity of momentum flux of the interactions. This removes all reference to the scale at which the phases are mixed.

This introduces the new strain tensors

$$\varepsilon_{ij} = \frac{1}{2}\left(\frac{\partial \mathrm{v}_i}{\partial x_j} + \frac{\partial \mathrm{v}_j}{\partial x_i}\right) \tag{9.34}$$

and

$$\varepsilon_{ij}^{(i)} = \frac{1}{2}\left(\frac{\partial \mathrm{v}_i^{(i)}}{\partial x_j} + \frac{\partial \mathrm{v}_j^{(i)}}{\partial x_i}\right) \tag{9.35}$$

$$\vec{\mathrm{v}}^i = \nabla \cdot \varphi \tag{9.36}$$

and $\vec{\mathrm{v}}^i$ is the velocity of the momentum of interaction between the phases.

Also

$$\vec{\mathrm{v}} = \nabla \cdot \theta \tag{9.37}$$

and $\vec{\mathrm{v}}$ the velocity of the total momentum of the phases. Note that, in the case of two immiscible fluid phases in a rigid porous medium or if there is only one fluid phase in a deformable porous medium, Equation 9.33 yields no new information and Equation 9.32 becomes equivalent to Equation 9.1. However, if the phases become miscible or if the matrix motions are coupled to the fluid motions of the two immiscible phases, then Equations 9.32 and 9.33 are required to describe the thermodynamics. As a result, they are also required to correctly describe the thermomechanics.

The total internal energy is given by

$$\mathrm{dU} = \tau_{ij}^{m} \mathrm{d}\varepsilon_{ij} + \tau_{ij}^{(i)} \mathrm{d}\varepsilon_{ij}^{(i)} + \mathrm{TdS} \tag{9.38}$$

The energy momentum tensor for the total momentum flux is

$$T_{\mu\nu}^{(t)} = \begin{bmatrix} \tau_{11}^{(t)} & \tau_{12}^{(t)} & \tau_{13}^{(t)} & m_1^{(t)} \\ \tau_{21}^{(t)} & \tau_{22}^{(t)} & \tau_{23}^{(t)} & m_2^{(t)} \\ \tau_{31}^{(t)} & \tau_{32}^{(t)} & \tau_{33}^{(t)} & m_3^{(t)} \\ m_1^{(t)} & m_2^{(t)} & m_3^{(t)} & \varepsilon^{(t)} \end{bmatrix} \tag{9.39}$$

The internal energy for the total momentum flux is

$$dU^{(m)} = \tau_{ij}^{(m)} d\varepsilon_{ij}^{(m)} + TdS^{(m)} \tag{9.40}$$

The energy momentum tensor for the interaction between the two phases is

$$T_{\mu\nu}^{(i)} = \begin{bmatrix} \tau_{11}^{(i)} & \tau_{12}^{(i)} & \tau_{13}^{(i)} & m_1^{(i)} \\ \tau_{21}^{(i)} & \tau_{22}^{(i)} & \tau_{23}^{(i)} & m_2^{(i)} \\ \tau_{31}^{(i)} & \tau_{32}^{(i)} & \tau_{33}^{(i)} & m_3^{(i)} \\ m_1^{(i)} & m_2^{(i)} & m_3^{(i)} & \varepsilon^{(i)} \end{bmatrix} \tag{9.41}$$

The symmetry of these tensors is required for angular momentum to be conserved. The internal energy for the interaction is

$$dU^{(i)} = \tau_{ij}^{(i)} d\varepsilon_{ij}^{(i)} + TdS^{(i)} \tag{9.42}$$

The equations of motion (i.e. the field equations) are obtained in accordance with the principle of least action

$$\begin{aligned} \delta S &= \int \left(\frac{\partial \Lambda}{\partial \theta} \delta\theta + \frac{\partial \Lambda}{\partial v_i^{(t)}} \delta v_i^{(t)} \right) d\Omega \\ &= \int \left(\frac{\partial \Lambda}{\partial \theta} \delta\theta + \frac{\partial}{\partial x_i} \left(\frac{\partial \Lambda}{\partial v_i^{(t)}} \delta\theta \right) - \delta\theta \frac{\partial}{\partial x_i} \frac{\partial \Lambda}{\partial v_i^{(t)}} \right) d\Omega = 0 \end{aligned} \tag{9.43}$$

which yields

$$\frac{\partial}{\partial x_i} \frac{\partial \Lambda}{\partial v_i^{(t)}} - \frac{\partial \Lambda}{\partial \theta} = 0 \tag{9.44}$$

$$T_{\mu\nu}^{(t)} = \delta_{\mu\nu} \frac{\partial \Lambda}{\partial \theta} - v_\mu^{(t)} \frac{\partial \Lambda}{\partial v_\nu^{(t)}} \tag{9.45}$$

$$\delta S = \int \left(\frac{\partial \Lambda}{\partial \theta} \delta \varphi + \frac{\partial \Lambda}{\partial v_j^{(i)}} \delta v_j^{(i)} \right) d\Omega \tag{9.46}$$

Similarly

$$\frac{\partial}{\partial x_j} \frac{\partial \Lambda}{\partial v_j^{(i)}} - \frac{\partial \Lambda}{\partial \theta} = 0 \tag{9.47}$$

$$T_{\mu\nu}^{(i)} = \delta_{\mu\nu} \frac{\partial \Lambda}{\partial \theta} - v_\mu^{(i)} \frac{\partial \Lambda}{\partial v_\nu^{(i)}} \tag{9.48}$$

THE CONSTRUCTION OF THE THERMODYNAMIC POTENTIALS

The change in the Gibb's free energies* is given by

$$\mathrm{d}G^{(m)} = -S^{(m)}\mathrm{d}T + \varepsilon_{ij}^{(m)} d\tau_{ij}^{(m)} + \Pi^{(m)}\, \mathrm{d}c_{fk} \tag{9.49}$$

$$\mathrm{d}G^{(i)} = -S^{(i)}\mathrm{d}T + \varepsilon_{ij}^{(i)} d\tau_{ij}^{(i)} + \Pi^{(i)}\, \mathrm{d}c^{(i)} \tag{9.50}$$

and

$$\mathrm{d}G = -S\mathrm{d}T + \varepsilon_{ij}^{(m)} d\tau_{ij}^{(m)} + \varepsilon_{ij}^{(i)} d\tau_{ij}^{(i)} + \Pi^{(m)}\, \mathrm{d}c_{fk} + \Pi^{(i)}\, \mathrm{d}c^{(i)} \tag{9.51}$$

where $\Pi^{(m)}$ and $\Pi^{(i)}$ are chemical potentials.

The change in the internal energies is now

$$\mathrm{d}U^{(m)} = \sigma_{ik}^{(m)} \mathrm{d}\varepsilon_{ik}^{(m)} + T\mathrm{d}S^{(m)} + \Pi^{(m)}\, \mathrm{d}c_{fk} \tag{9.52}$$

$$\mathrm{d}U^{(i)} = \sigma_{ik}^{(i)} \mathrm{d}\varepsilon_{ik}^{(i)} + T\mathrm{d}S^{(i)} + \Pi^{(i)}\, \mathrm{d}c^{(i)} \tag{9.53}$$

$$\mathrm{d}U = \sigma_{ik}^{(m)} \mathrm{d}\varepsilon_{ik}^{(m)} + \sigma_{ik}^{(i)} \mathrm{d}\varepsilon_{ik}^{(i)} + T\mathrm{d}S + \Pi^{(m)}\, \mathrm{d}c_{fk} + \Pi^{(i)}\, \mathrm{d}c^{(i)} \tag{9.54}$$

Note in the static limit of a fluid and solid mixed only at the macroscale, Equations 9.27 and 9.28 may be rewritten in the form of Equations 4.1 and 4.14.

The Gibbs–Duhem equations are now formally

$$0 = \mathrm{u}_{ik}^{(m)} \mathrm{d}\sigma_{ik}^{(m)} + S^{(m)}\mathrm{d}T + c_{fk}\mathrm{d}\Pi^{(m)} \tag{9.55}$$

* Note the final term in these equations did not occur in Chapter 4 because there was no irreversible change in volume fractions, so the changes in volume fractions were incorporated into the definition of *strain*.

$$0 = u_{ik}^{(i)} d\sigma_{ik}^{(i)} + S^{(i)} dT + c^{(i)} d\Pi^{(i)} \quad (9.56)$$

$$0 = u_{ik}^{(m)} d\sigma_{ik}^{(m)} + u_{ik}^{(i)} d\sigma_{ik}^{(i)} + SdT + c^{(i)} d\Pi^{(i)} + c_{fk} d\Pi^{(m)} \quad (9.57)$$

The change in the Helmholtz free energy is formally given by

$$dF^{(m)} = \sigma_{ik}^{(m)} d\varepsilon_{ik}^{(m)} - S^{(m)} dT + \Pi^{(m)} dc_{fk} \quad (9.58)$$

$$dF^{(i)} = \sigma_{ik}^{(i)} d\varepsilon_{ik}^{(i)} - S^{(i)} dT + \Pi^{(i)} dc^{(i)} \quad (9.59)$$

$$dF = \sigma_{ik}^{(m)} d\varepsilon_{ik}^{(m)} + \sigma_{ik}^{(i)} d\varepsilon_{ik}^{(i)} - SdT + \Pi^{(m)} dc_{fk} + \Pi^{(i)} dc^{(i)} \quad (9.60)$$

The change in the enthalpy is formally given by

$$dH^{(m)} = -u_{ik}^{(m)} d\sigma_{ik}^{(m)} + TdS^{(m)} + \Pi^{(m)} dc_{fk} \quad (9.61)$$

$$dH^{(i)} = -u_{ik}^{(i)} d\sigma_{ik}^{(i)} + TdS^{(i)} + \Pi^{(i)} dc^{(i)} \quad (9.62)$$

$$dH = -u_{ik}^{(m)} d\sigma_{ik}^{(m)} - u_{ik}^{(i)} d\sigma_{ik}^{(i)} + TdS + \Pi^{(m)} dc_{fk} + \Pi^{(i)} dc^{(i)} \quad (9.63)$$

In this current construction, there is a one-to-one relationship between the mass fractions and the volume fractions. As well, there is a one-to-one relationship between the component velocities and the momentum fluxes. These relations hold when there are two components mixing at the macroscale. In Chapter 7 it was observed that these relations break down when mixing also occurs at the molecular scale. In that case, the concept of volume fraction or a component velocity does not have clear meaning. As a result, the description of miscible flow with molecular diffusion required that mass fractions and momentum fluxes be used. In the case where the components are a fluid and a solid only mixing at the macroscale and equilibrium processes are considered, these relations may be reduced to the form constructed in Chapter 5.

NON-EQUILIBRIUM THERMODYNAMICS INCLUDING GRANULAR MOTION AND TWO FLUIDS

In the case of two immiscible fluids, it was observed that saturation became a dynamic variable just as porosity was in the case of a fluid and a solid. In the case of miscible flow, it was observed that volume fractions and component velocities only appeared to be thermodynamic variables because they were proportional to the actual thermodynamic variables' mass fraction and the momentum fluxes. Attempts to use volume fractions (porosity and saturation) and the equations of motion for the component velocities to describe thermodynamic processes for three components fail to describe reciprocity. The problem that arises is that the three-component equations of motion may be combined in four different ways to describe the total momentum and the momentums of the interactions between the components. As well, if interactions are occurring at various scales, the volume fractions must be replaced by statements about the mass fractions. Here, the momenta and mass fractions that must be used in the case of three phases are as follows.

The velocity of momentum flux:

$$\mathrm{v}_{\mathrm{m}} = \frac{\eta_s \rho_s \mathrm{v}_s + \eta_{f1} \rho_{f1} \mathrm{v}_{f1} + \eta_{f2} \rho_{f2} \mathrm{v}_{f2}}{\eta_s \rho_s + \eta_{f1} \rho_{f1} + \eta_{f2} \rho_{f2}} \tag{9.64}$$

The total mass:

$$\rho^{(m)} = \eta_s \rho_s + \eta_{f1} \rho_{f1} + \eta_{f2} \rho_{f2} \tag{9.65}$$

The total mass fraction is given by

$$c^{(m)} = \frac{\rho^{(m)}}{\rho_o^{(m)}} \tag{9.66}$$

The velocity of momentum flux of the interaction between the solid and the two fluids (i.e. the two fluids moving in phase with each other but out of phase with the solid; it should be noted that, as was the case for two phases, the phase angles yield almost in-phase and almost out-of-phase motions):

$$\mathrm{v}_{s(12)} = \frac{\eta_s \rho_s \mathrm{v}_s - \eta_{f1} \rho_{f1} \mathrm{v}_{f1} - \eta_{f2} \rho_{f2} \mathrm{v}_{f2}}{\eta_s \rho_s + \eta_{f1} \rho_{f1} + \eta_{f2} \rho_{f2}} \tag{9.67}$$

The velocity of momentum flux of the interaction that occurs when the solid and Fluid 1 move in phase with each other but out of phase with Fluid 2:

$$\mathrm{v}_{s1(2)} = \frac{\eta_s \rho_s \mathrm{v}_s + \eta_{f1} \rho_{f1} \mathrm{v}_{f1} - \eta_{f2} \rho_{f2} \mathrm{v}_{f2}}{\eta_s \rho_s + \eta_{f1} \rho_{f1} + \eta_{f2} \rho_{f2}} \tag{9.68}$$

The velocity of momentum flux of the interaction that occurs when the solid and Fluid 2 move in phase with each other but out of phase with Fluid 1:

$$\mathrm{v}_{s2(1)} = \frac{\eta_s \rho_s \mathrm{v}_s - \eta_{f1} \rho_{f1} \mathrm{v}_{f1} + \eta_{f2} \rho_{f2} \mathrm{v}_{f2}}{\eta_s \rho_s + \eta_{f1} \rho_{f1} + \eta_{f2} \rho_{f2}} \tag{9.69}$$

The density of interaction when the solid moves out of phase with the two fluids:

$$\rho_{s(12)}^{(i)} = \eta_s \rho_s - \eta_{f1} \rho_{f1} - \eta_{f2} \rho_{f2} \tag{9.70}$$

The density of interaction when the solid and Fluid 1 move out of phase with Fluid 2:

$$\rho_{s1(2)}^{(i)} = \eta_s\rho_s + \eta_{f1}\rho_{f1} - \eta_{f2}\rho_{f2} \tag{9.71}$$

The density of interaction when the solid and Fluid 2 move out of phase with Fluid 1:

$$\rho_{s2(1)}^{(i)} = \eta_s\rho_s - \eta_{f1}\rho_{f1} + \eta_{f2}\rho_{f2} \tag{9.72}$$

The change in the Gibb's free energies are given by

$$\mathrm{d}G^{(m)} = -S^{(m)}\mathrm{d}T + \varepsilon_{ij}^{(m)} d\tau_{ij}^{(m)} \tag{9.73}$$

$$\mathrm{d}G^{(is(12))} = -S^{(is(12))}\mathrm{d}T + \varepsilon_{ij}^{(is(12))} d\tau_{ij}^{(is(12))} \tag{9.74}$$

$$\mathrm{d}G^{(is1(2))} = -S^{(is1(2))}\mathrm{d}T + \varepsilon_{ij}^{(is1(2))} d\tau_{ij}^{(is1(2))} \tag{9.75}$$

$$\mathrm{d}G^{(is2(1))} = -S^{(is2(1))}\mathrm{d}T + \varepsilon_{ij}^{(is2(1))} d\tau_{ij}^{(is2(1))} \tag{9.76}$$

and

$$\mathrm{d}G = -S\mathrm{d}T + \varepsilon_{ij}^{(m)} d\tau_{ij}^{(m)} + \varepsilon_{ij}^{(is(12))} d\tau_{ij}^{(is(12))} + \varepsilon_{ij}^{(is1(2))} d\tau_{ij}^{(is1(2))} + \varepsilon_{ij}^{(is2(1))} d\tau_{ij}^{(is2(1))} \tag{9.77}$$

The change in internal energies is given by

$$\mathrm{d}U^{(m)} = \tau_{ik}^{(m)}\mathrm{d}\varepsilon_{ik}^{(m)} + T\mathrm{d}S^{(m)} + \prod{}^{(m)}\mathrm{d}c^{(m)} \tag{9.78}$$

$$\mathrm{d}U^{(is(12))} = \tau_{ik}^{(is(12))}\mathrm{d}\varepsilon_{ik}^{(is(12))} + T\mathrm{d}S^{(is(12))} + \prod{}^{(is(12))}\mathrm{d}c^{(is(12))} \tag{9.79}$$

$$\mathrm{d}U^{(is1(2))} = \tau_{ik}^{(is1(2))}\mathrm{d}\varepsilon_{ik}^{(is1(2))} + T\mathrm{d}S^{(is1(2))} + \prod{}^{(is1(2))}\mathrm{d}c^{(is1(2))} \tag{9.80}$$

$$\mathrm{d}U^{(is2(1))} = \tau_{ik}^{(is2(1))}\mathrm{d}\varepsilon_{ik}^{(is2(1))} + T\mathrm{d}S^{(is2(1))} + \prod{}^{(is2(1))}\mathrm{d}c^{(is2(1))} \tag{9.81}$$

$$\begin{aligned}\mathrm{d}U = -T\mathrm{d}S &+ \tau_{ik}^{(m)}\mathrm{d}\varepsilon_{ik}^{(m)} + \tau_{ik}^{(is(12))}\mathrm{d}\varepsilon_{ik}^{(is(12))} + \tau_{ik}^{(is1(2))}\mathrm{d}\varepsilon_{ik}^{(is1(2))} + \tau_{ik}^{(is2(1))}\mathrm{d}\varepsilon_{ik}^{(is2(1))} \\ &+ \prod{}^{(m)}\mathrm{d}c^{(m)} + \prod{}^{(is(12))}\mathrm{d}c^{(is(12))} + \prod{}^{(is1(2))}\mathrm{d}c^{(is1(2))} + \prod{}^{(is2(1))}\mathrm{d}c^{(is2(1))}\end{aligned} \tag{9.82}$$

The Gibbs–Duhem equations are now

$$0 = u_{ik}^{(m)}\mathrm{d}\tau_{ik}^{(m)} + S^{(m)}\mathrm{d}T + c^{(m)}\mathrm{d}\prod{}^{(m)} \tag{9.83}$$

$$0 = u_{ik}^{(is(12))} d\tau_{ik}^{(is(12))} + S^{(is(12))} dT + c^{(is(12))} d\prod^{(is(12))} \tag{9.84}$$

$$0 = u_{ik}^{(is1(2))} d\tau_{ik}^{(is1(2))} + S^{(is1(2))} dT + c^{(is1(2))} d\prod^{(is1(2))} \tag{9.85}$$

$$0 = u_{ik}^{(is2(1))} d\tau_{ik}^{(is2(1))} + S^{(is2(1))} dT + c^{(is2(1))} d\prod^{(is2(1))} \tag{9.86}$$

$$\begin{aligned} 0 = {} & u_{ik}^{(m)} d\tau_{ik}^{(m)} + u_{ik}^{(is(12))} d\tau_{ik}^{(is(12))} + u_{ik}^{(is1(2))} d\tau_{ik}^{(is1(2))} + u_{ik}^{(is2(1))} d\tau_{ik}^{(is2(1))} + SdT \\ & + c^{(m)} d\prod^{(m)} + c^{(is(12))} d\prod^{(is(12))} + c^{(is1(2))} d\prod^{(is1(2))} + c^{(is2(1))} d\prod^{(is2(1))} \end{aligned} \tag{9.87}$$

The change in the Helmholtz free energy is given by

$$dF^{(m)} = \sigma_{ik}^{(m)} d\varepsilon_{ik}^{(m)} - S^{(m)} dT + \prod^{(m)} dc^{(m)} \tag{9.88}$$

$$dF^{(is(12))} = \sigma_{ik}^{(is(12))} d\varepsilon_{ik}^{(is(12))} - S^{(is(12))} dT + \prod^{(is(12))} dc^{(is(12))} \tag{9.89}$$

$$dF^{(is1(2))} = \sigma_{ik}^{(is1(2))} d\varepsilon_{ik}^{(is1(2))} - S^{(is1(2))} dT + \prod^{(is1(2))} dc^{(is1(2))} \tag{9.90}$$

$$dF^{(is2(1))} = \sigma_{ik}^{(is2(1))} d\varepsilon_{ik}^{(is2(1))} - S^{(is2(1))} dT + \prod^{(is2(1))} dc^{(is2(1))} \tag{9.91}$$

$$\begin{aligned} dF = {} & -SdT + \sigma_{ik}^{(m)} d\varepsilon_{ik}^{(m)} + \sigma_{ik}^{(is(12))} d\varepsilon_{ik}^{(is(12))} + \sigma_{ik}^{(is1(2))} d\varepsilon_{ik}^{(is1(2))} + \sigma_{ik}^{(is2(1))} d\varepsilon_{ik}^{(is2(1))} \\ & + \prod^{(m)} dc^{(m)} + \prod^{(is(12))} dc^{(is(12))} + \prod^{(is1(2))} dc^{(is1(2))} + \prod^{(is2(1))} dc^{(is2(1))} \end{aligned} \tag{9.92}$$

The change in the enthalpy is given by

$$dH^{(m)} = -u_{ik}^{(m)} d\sigma_{ik}^{(m)} + TdS^{(m)} + \prod^{(m)} dc^{(m)} \tag{9.93}$$

$$dH^{(is(12))} = \varepsilon_{ik}^{(is(12))} d\tau_{ik}^{(is(12))} + TdS^{(is(12))} + \prod^{(is(12))} dc^{(is(12))} \tag{9.94}$$

$$dH^{(is1(2))} = \varepsilon_{ik}^{(is1(2))} d\tau_{ik}^{(is1(2))} + TdS^{(is1(2))} + \prod^{(is1(2))} dc^{(is1(2))} \tag{9.95}$$

$$dH^{(is2(1))} = \varepsilon_{ik}^{(is2(1))} d\tau_{ik}^{(is2(1))} + TdS^{(is2(1))} + \prod^{(is2(1))} dc^{(is2(1))} \tag{9.96}$$

$$\begin{aligned} dH = {} & -TdS + \varepsilon_{ik}^{(m)} d\tau_{ik}^{(m)} + \varepsilon_{ik}^{(is(12))} d\tau_{ik}^{(is(12))} + \varepsilon_{ik}^{(is1(2))} d\tau_{ik}^{(is1(2))} + \varepsilon_{ik}^{(is2(1))} d\tau_{ik}^{(is2(1))} \\ & + \prod^{(m)} dc^{(m)} + \prod^{(is(12))} dc^{(is(12))} + \prod^{(is1(2))} dc^{(is1(2))} + \prod^{(is2(1))} dc^{(is2(1))} \end{aligned} \tag{9.97}$$

RECIPROCITY

Onsager's relations describe flux relations, heat, chemical, electromagnetic and flow. However, in the analysis of composite materials, it is observed that, as the complexity increases, these fluxes are no longer described by diffusion equations. The description of mixing with negligible diffusion required a Fokker–Plank equation, and when molecular mixing was allowed the description required coupled Fokker–Plank equations. As the number of physical processes increases in the interactions (e.g. the number of components at the pore scale), the number of coupled Fokker–Planck equations increases. This causes additional fluxes, thus additional variables and additional constraints to be imposed.

Here, the ultimate goal as is the case for any physical theory is to provide a set of equations that make predictions subject to experimental scrutiny. As in Chapter 5, the expressions will be given in terms of heat, mass flux (as expressed by the above concentrations) and momentum (as expressed by the total momentum and momenta of interactions that correlate with these concentrations). Thus, there are a number of additional constraints and volume fractions no longer appear in the Onsager's relations.

Temperature (heat):

$$\vec{J}_Q = -\kappa \nabla T \tag{9.98}$$

In order to simplify the analysis, temperature is assumed to be macroscopically uniform among the component phases. Thus the macroscopic and megascopic temperatures are assumed to be the same for the processes under consideration.

Mass variations (megascopic chemical potential or order parameter) associated with deformations enter the thermodynamics through the mass variations

$$\mathrm{J}_{\Pi_m} = -\mathrm{D}_m \nabla c^{(m)} \tag{9.99}$$

$$\mathrm{J}_{\Pi_{is(12)}} = -\mathrm{D}_{is(12)} \nabla c^{(is(12))} \tag{9.100}$$

$$\mathrm{J}_{\Pi_{is(1)2}} = -\mathrm{D}_{is(1)2} \nabla c^{(is(1)2)} \tag{9.101}$$

$$\mathrm{J}_{\Pi_{is1(2)}} = -\mathrm{D}_{is1(2)} \nabla c^{(is1(2))} \tag{9.102}$$

Velocity of momentum flux:

$$\mathrm{J}_M = -\mathrm{D}_M \nabla \varphi_m \tag{9.103}$$

$$\mathrm{v}_m = \nabla \varphi_m \tag{9.104}$$

$$\mathrm{J}_{M_{s(12)}} = -\mathrm{D}_{M_{s(12)}} \nabla \varphi_{m_{s(12)}} \tag{9.105}$$

$$\mathrm{v}_{m_{s(12)}} = \nabla \varphi_{m_{s(12)}} \tag{9.106}$$

$$\mathrm{J}_{M_{s1(2)}} = -\mathrm{D}_{M_{s1(2)}} \nabla \varphi_{m_{s1(2)}} \tag{9.107}$$

$$\mathrm{v}_{m_{s1(2)}} = \nabla\varphi_{m_{s1(2)}} \tag{9.108}$$

$$\mathrm{J}_{M_{s2(1)}} = -\mathrm{D}_{M_{s2(1)}}\nabla\varphi_{m_{s2(1)}} \tag{9.109}$$

$$\mathrm{v}_{m_{s2(1)}} = \nabla\varphi_{m_{s2(1)}} \tag{9.110}$$

Note that the diffusion coefficients in these flux equations are associated with coupled Fokker–Plank equations, unlike equilibrium thermodynamics (or no-equilibrium thermodynamics of molecular mixtures), where they represent diffusion equations.

These quantities may also be expressed as

$$\begin{aligned}\mathrm{J}_Q = \mathrm{L}_{QQ}\left(-\frac{\nabla\mathrm{T}}{\mathrm{T}}\right) - \mathrm{L}_{Q\Pi_s}\nabla c^{(m)} - \mathrm{L}_{Q\Pi_{is(12)}}\nabla c^{(is(12))} - \mathrm{L}_{Q\Pi_{is1(2)}}\nabla c^{(is1(2))} - \mathrm{L}_{Q\Pi_{is2(1)}}\nabla c^{(is2(1))} \\ -\mathrm{L}_{QM}\nabla\varphi_m - \mathrm{L}_{QM_{s(12)}}\nabla\varphi_{m_{s(12)}} - \mathrm{L}_{QM_{s1(2)}}\nabla\varphi_{m_{s1(2)}} - \mathrm{L}_{QM_{s2(1)}}\nabla\varphi_{m_{s2(1)}}\end{aligned} \tag{9.111}$$

$$\begin{aligned}\mathrm{J}_{\Pi_s} = \mathrm{L}_{\Pi_s Q}\left(-\frac{\nabla\mathrm{T}}{\mathrm{T}}\right) - \mathrm{L}_{\Pi_s\Pi_s}\nabla c_{fk} - \mathrm{L}_{\Pi_s\Pi_{is(12)}}\nabla c^{(is(12))} - \mathrm{L}_{\Pi_s\Pi_{is1(2)}}\nabla c^{(is1(2))} - \mathrm{L}_{\Pi_s\Pi_{is2(1)}}\nabla c^{(is2(1))} \\ -\mathrm{L}_{\Pi_s M}\nabla\varphi_m - \mathrm{L}_{\Pi_s M_{s(12)}}\nabla\varphi_{m_{s(12)}} - \mathrm{L}_{\Pi_s M_{s1(2)}}\nabla\varphi_{m_{s1(2)}} - \mathrm{L}_{\Pi_s M_{s2(1)}}\nabla\varphi_{m_{s2(1)}}\end{aligned} \tag{9.112}$$

$$\begin{aligned}\mathrm{J}_{\Pi_{is(12)}} = \mathrm{L}_{\Pi_{is(12)}Q}\left(-\frac{\nabla\mathrm{T}}{\mathrm{T}}\right) - \mathrm{L}_{\Pi_{is(12)}\Pi_s}\nabla c_{fk} - \mathrm{L}_{\Pi_{is(12)}\Pi_{is(12)}}\nabla c^{(is(12))} - \mathrm{L}_{\Pi_{is(12)}\Pi_{is1(2)}}\nabla c^{(is1(2))} \\ -\mathrm{L}_{\Pi_{is(12)}\Pi_{is2(1)}}\nabla c^{(is2(1))} - \mathrm{L}_{\Pi_{is(12)}M}\nabla\varphi_m - \mathrm{L}_{\Pi_{is(12)}M_{s(12)}}\nabla\varphi_{m_{s(12)}} \\ -\mathrm{L}_{\Pi_{is(12)}M_{s1(2)}}\nabla\varphi_{m_{s1(2)}} - \mathrm{L}_{\Pi_{is(12)}M_{s2(1)}}\nabla\varphi_{m_{s2(1)}}\end{aligned} \tag{9.113}$$

$$\begin{aligned}\mathrm{J}_{\Pi_{is1(2)}} = \mathrm{L}_{\Pi_{is1(2)}Q}\left(-\frac{\nabla\mathrm{T}}{\mathrm{T}}\right) - \mathrm{L}_{\Pi_{is1(2)}\Pi_s}\nabla c_{fk} - \mathrm{L}_{\Pi_{is1(2)}\Pi_{is(12)}}\nabla c^{(is(12))} - \mathrm{L}_{\Pi_{is1(2)}\Pi_{is1(2)}}\nabla c^{(is1(2))} \\ -\mathrm{L}_{\Pi_{is1(2)}\Pi_{is2(1)}}\nabla c^{(is2(1))} - \mathrm{L}_{\Pi_{is1(2)}M}\nabla\varphi_m - \mathrm{L}_{\Pi_{is1(2)}M_{s(12)}}\nabla\varphi_{m_{s(12)}} \\ -\mathrm{L}_{\Pi_{is1(2)}M_{s1(2)}}\nabla\varphi_{m_{s1(2)}} - \mathrm{L}_{\Pi_{is1(2)}M_{s2(1)}}\nabla\varphi_{m_{s2(1)}}\end{aligned} \tag{9.114}$$

$$\begin{aligned}\mathrm{J}_{\Pi_{is2(1)}} = \mathrm{L}_{\Pi_{is2(1)}Q}\left(-\frac{\nabla\mathrm{T}}{\mathrm{T}}\right) - \mathrm{L}_{\Pi_{is2(1)}\Pi_s}\nabla c_{fk} - \mathrm{L}_{\Pi_{is2(1)}\Pi_{is(12)}}\nabla c^{(is(12))} - \mathrm{L}_{\Pi_{is2(1)}\Pi_{is1(2)}}\nabla c^{(is1(2))} \\ -\mathrm{L}_{\Pi_{is2(1)}\Pi_{is2(1)}}\nabla c^{(is2(1))} - \mathrm{L}_{\Pi_{is2(1)}M}\nabla\varphi_m - \mathrm{L}_{\Pi_{is2(1)}M_{s(12)}}\nabla\varphi_{m_{s(12)}} \\ -\mathrm{L}_{\Pi_{is2(1)}M_{s1(2)}}\nabla\varphi_{m_{s1(2)}} - \mathrm{L}_{\Pi_{is2(1)}M_{s2(1)}}\nabla\varphi_{m_{s2(1)}}\end{aligned} \tag{9.115}$$

$$\mathrm{J}_M = \mathrm{L}_{MQ}\left(-\frac{\nabla \mathrm{T}}{\mathrm{T}}\right) - \mathrm{L}_{M\Pi_s}\nabla c_{fk} - \mathrm{L}_{M\Pi_{is(12)}}\nabla c^{(is(12))} - \mathrm{L}_{M\Pi_{is1(2)}}\nabla c^{(is1(2))} - \mathrm{L}_{M\Pi_{is2(1)}}\nabla c^{(is2(1))}$$
$$-\mathrm{L}_{MM}\nabla\varphi_m - \mathrm{L}_{MM_{s(12)}}\nabla\varphi_{m_{s(12)}} - \mathrm{L}_{MM_{s1(2)}}\nabla\varphi_{m_{s1(2)}} - \mathrm{L}_{MM_{s2(1)}}\nabla\varphi_{m_{s2(1)}} \quad (9.116)$$

$$\mathrm{J}_{M_{s(12)}} = \mathrm{L}_{M_{s(12)}Q}\left(-\frac{\nabla \mathrm{T}}{\mathrm{T}}\right) - \mathrm{L}_{M_{s(12)}\Pi_s}\nabla c_{fk} - \mathrm{L}_{M_{s(12)}\Pi_{is(12)}}\nabla c^{(is(12))} - \mathrm{L}_{M_{s(12)}\Pi_{is1(2)}}\nabla c^{(is1(2))}$$
$$-\mathrm{L}_{M_{s(12)}\Pi_{is2(1)}}\nabla c^{(is2(1))} - \mathrm{L}_{M_{s(12)}M}\nabla\varphi_m - \mathrm{L}_{M_{s(12)}M_{s(12)}}\nabla\varphi_{m_{s(12)}} \quad (9.117)$$
$$-\mathrm{L}_{M_{s(12)}M_{s1(2)}}\nabla\varphi_{m_{s1(2)}} - \mathrm{L}_{M_{s(12)}M_{s2(1)}}\nabla\varphi_{m_{s2(1)}}$$

$$\mathrm{J}_{M_{s1(2)}} = \mathrm{L}_{M_{s1(2)}Q}\left(-\frac{\nabla \mathrm{T}}{\mathrm{T}}\right) - \mathrm{L}_{M_{s1(2)}\Pi_s}\nabla c_{fk} - \mathrm{L}_{M_{s1(2)}\Pi_{is(12)}}\nabla c^{(is(12))} - \mathrm{L}_{M_{s1(2)}\Pi_{is1(2)}}\nabla c^{(is1(2))}$$
$$-\mathrm{L}_{M_{s1(2)}\Pi_{is2(1)}}\nabla c^{(is2(1))} - \mathrm{L}_{M_{s1(2)}M}\nabla\varphi_m - \mathrm{L}_{M_{s1(2)}M_{s(12)}}\nabla\varphi_{m_{s(12)}} \quad (9.118)$$
$$-\mathrm{L}_{M_{s1(2)}M_{s1(2)}}\nabla\varphi_{m_{s1(2)}} - \mathrm{L}_{M_{s1(2)}M_{s2(1)}}\nabla\varphi_{m_{s2(1)}}$$

$$\mathrm{J}_{M_{s2(1)}} = \mathrm{L}_{M_{s2(1)}Q}\left(-\frac{\nabla \mathrm{T}}{\mathrm{T}}\right) - \mathrm{L}_{M_{s2(1)}\Pi_s}\nabla c_{fk} - \mathrm{L}_{M_{s2(1)}\Pi_{is(12)}}\nabla c^{(is(12))} - \mathrm{L}_{M_{s2(1)}\Pi_{is1(2)}}\nabla c^{(is1(2))}$$
$$-\mathrm{L}_{M_{s2(1)}\Pi_{is2(1)}}\nabla c^{(is2(1))} - \mathrm{L}_{M_{s2(1)}M}\nabla\varphi_m - \mathrm{L}_{M_{s2(1)}M_{s(12)}}\nabla\varphi_{m_{s(12)}} \quad (9.119)$$
$$-\mathrm{L}_{M_{s2(1)}M_{s1(2)}}\nabla\varphi_{m_{s1(2)}} - \mathrm{L}_{M_{s2(1)}M_{s2(1)}}\nabla\varphi_{m_{s2(1)}}$$

Onsager's relations now require that $\mathrm{L}_{ab} = \mathrm{L}_{ba}$.

SUMMARY

When irreversible deformations are allowed, it is observed that volume fractions and component velocities are not thermodynamic variables. This is in stark contrast to the previous discussion of equilibrium thermodynamics, where they appeared to be thermodynamic variables. In hindsight, it is observed that they were proportional to the actual thermodynamic variables in the cases considered. Of particular importance to non-equilibrium thermodynamics are the modifications momentum flux, etc., make to the Clapeyron equation and the new information about dispersion, porosity waves, fluid flow and their interactions obtained from the new Onsager's relations.

The thermodynamic potentials were constructed in the case of two components and three components. These potentials are expressed in terms of the total momentum, the total mass, the momenta of interaction between the various phases and the mass fractions associated with these interactions.

Onsager's relations describe flux relations, heat, chemical, electromagnetic and flow. However, in the analysis of composite materials, it is observed that as the complexity increases these fluxes are no longer described by diffusion equations. As the number of physical processes increases in the interactions (e.g. the number of components at the pore scale), the number of coupled Fokker–Planck equations increases. This causes additional fluxes, thus additional variables and additional constraints to be imposed.

REFERENCES

Bear, J., and Bachmat, Y., 1990, *Theory and Applications of Transport in Porous Media: Introduction to Modeling of Transport Phenomena in Porous Media*, Kluwer Academic, Dordrecht.

Bear, J., and Bachmat, Y., 1991, *Introduction to Modeling of Transport Phenomena in Porous Media (Theory and Applications of Transport in Porous Media, Vol 4)*, Kluwer Academic, London.

Bennethum, L.S., and Cushman, J.H., 2002a, Multicomponent, multiphase thermodynamics of swelling porous media with electroquasistatics: I. Macroscale field equations, *Trans. Porous Media,* 47(3), 309–336.

Bennethum, L.S., and Cushman, J.H., 2002b, Multicomponent, multiphase thermodynamics of swelling porous media with electroquasistatics: II. Constitutive theory, *Trans. Porous Media,* 47(3), 337–362.

Burridge, R., and Keller, J.B., 1981, Poroelasticity equations derived from microstructure, *J. Acoust. Soc. Am.*, 70, 1140–1146.

Callen, H.B., 1985, *Thermodynamics and an Introduction to Thermostatistics*, Wiley, Singapore.

Chavanis, P., 2006, Nonlinear mean field Fokker-Planck equations and their applications in physics astrophysics and biology, *C. R. Physique* 7, 318–330.

Chavanis, P., 2008, Nonlinear mean field Fokker-Planck equations. Applications in physics the chemotaxis of biological populations, *Eur. Phys. J. B*, 62, 179.

Cushman, J., 1997, *The Physics of Fluids in Hierarchical Porous Media: Angstroms to Miles (Theory and Applications of Transport in Porous Media Volume 10)*, Kluwer Academic, London.

De Boer, R., 1995, Thermodynamics of phase transitions in porous media, *Appl. Mech. Rev.*, 48(10), 613–622.

De Groot, S.R., and Mazur, P., 1984, *Non-Equilibrium Thermodynamics*, Dover, New York.

de la Cruz, V., Sahay, P.N., and Spanos, T.J.T., 1993, Thermodynamics of porous media, *Proc. Roy. Soc. Lond. A*, 433, 247–255.

de la Cruz, V., and Spanos, T.J.T., 1983, Mobilization of oil ganglia, *AIChE J.*, 29(7), 854–858.

de la Cruz, V., and Spanos, T.J.T., 1985, Seismic wave propagation in a porous medium, *Geophysics*, 50(10), 1556–1565.

de la Cruz, V., and Spanos, T.J.T., 1989, Thermo-mechanical coupling during seismic wave propagation in a porous medium, *J. Geophys. Res.*, 94, 637–642.

de la Cruz, V., Spanos, T.J.T., and Sharma, R.C., 1985, The stability of a steam water front in a porous medium, *Can. J. Chem. Eng.*, 63, 735–745.

de la Cruz, V., Spanos, T.J.T., and Yang, D.S., 1995, Macroscopic capillary pressure, *Trans. Porous Media*, 19, 67–77.

Dullien, F.A.L., 1992, *Porous Media: Fluid Transport and Pore Structure*, Academic Press, San Diego, CA.

Entov, V.M., Ryzhik, V.M., Barenblatt, G.I., 1990, *Theory of Fluid Flows Through Natural Rocks, (Theory and Applications of Transport in Porous Media, Volume 3)*, Kluwer Academic, London.

Fermi, E., 1956, *Thermodynamics*, Dover.

Fremond, M., and Nicolas, P., 1990, Macroscopic thermodynamics of porous media, *Continuum Mech. Thermodyn.*, 2, 119–139.

Garcia-Conlin, L.S., and Uribe, 1991, Extended irreversible thermodynamics beyond the linear regime: A critical overview, *J. Non Equilib. Thermodyn.*, 16, 89–128.

Geilikman, M.B., 1999, Sand production caused by foamy oil flow, *Trans. Porous Media*, 35(2), 259–272.

Geilikman, M.B., Spanos, T.J.T., and Nyland, E., 1993, Porosity diffusion in fluid—Saturated media, *Tectonophysics*, 217, 111–115.

Green, M.S., 1952, Markoff random processes and the statistical mechanics of time-dependent phenomena, *J. Chem. Phys.*, 20, 1281–1295.

Gyftopoulos, E.P., 2004, Thermodynamic derivation of reciprocal relations, Proceedings of ASME International Conference ECOS2004 on Efficiency, Costs, Optimization, Simulation and Environmental Impact of Energy Systems, In: *Energy-Efficient, Cost-Effective, and Environmentally-Sustainable Systems and Processes*, edited by R. Rivero, L. Monroy, R. Pulido, and G. Tsatsaronis, Guanajuato, Mexico, pp. 735–751.

Haase, R., 1990, *Thermodynamics of Irreversible Processes*, Dover, New York.

Hasegawa, H., 1977, Thermodynamic properties of non-equilibrium states subject to Fokker-Planck equations, *Prog. Theor. Phys.*, 57(5), 1523–1537.

Hassanizadeh, S.M., and Grey W.G., 1990, Mechanics and thermodynamics of multiphase flow in porous media including interphase boundaries, *Water Resour. Res.*, 13(4), 169–186.

Hassanizadeh, S.M., and Grey W.G., 1993, Thermodynamic basis of capillary pressure in porous media, *Water Resour. Res.*, 29(10), 3389–3405.

Hyodo, S., 2002, Multiscale simulations for materials (Conspectus), *R&D Rev. Toyota CRDL*, 38(1), 1–9.

Jammer, M., 1964, *Concepts of Mass in Classical and Modern Physics*, Dover. Mineola, New York.

Landau, L.D., and Lifshitz, E.M., 1958, *Statistical Physics*, Pergamon, Toronto.

Landau, L.D., and Lifshitz, E.M., 1975, *Theory of Elasticity*, Pergamon, Toronto.

Marle, C.M., 1982, On macroscopic equations governing multiphase flow with diffusion and chemical reactions in porous media, *Int. J. Eng. Sci.*, 20(5), 643–662.

Pauli, W., 1973, *Thermodynamics and the Kinetic Theory of Gases: Volume 3 of Pauli Lectures on Physics*, Dover. Mineola, New York.

Planck, M., 1917, *Treatise on Thermodynamics*, Dover, New York.

Reiss, H., 1965, *Methods of Thermodynamics*, Dover, Mineola, New York.

Sahay, P.N., Spanos, T.J.T., and de la Cruz, V., 2001, Seismic wave propagation in inhomogeneous and anisotropic porous media, *Geophys. J. Int.*, 145, 209–223.

Scheidegger, A.E., 1974, *The Physics of Flow Through Porous Media*, University of Toronto Press, Toronto.

Schrodinger, E., 1952, *Statistical Thermodynamics*, Dover, New York.

Slattery, J.C., 1967, Flow of viscoelastic fluids through porous media, *AIChE J.*, 13, 1066–1071.

Spanos, T.J.T., 2002, *The Thermophysics of Porous Media*, Monographs and Surveys in Pure and Applied Mathematics Series, Chapman & Hall, Boca Raton, FL.

Udey, N., Shim D., and Spanos, T.J.T., 1999, A Lorentz invariant thermal Lattice Gas Model, *Proc. Roy. Soc. Lond. A*, 455, 3565–3587.

Velasco, R.M., Garcia-Conlin, L.S., and Uribe, F.J., 2011, Entropy production: Its role in non-equilibrium thermodynamics, *Entropy*, 13, 82–116.

Whitaker, S., 1967, Diffusion and dispersion in porous media, *AIChE J.*, 13, 420–427.

Yourgrau, W., van der Merwe, A., and Raw, G., 1982, *Treatise on Irreversible and Statistical Thermophysics*, Dover, New York.

CHAPTER 10

Coupled Porosity and Saturation Waves in Porous Media

OBJECTIVE OF THIS CHAPTER

The equations for dilatational waves in a porous medium consisting of a solid and two almost incompressible and immiscible fluids are constructed. As seen in the previous chapter, when three phases are present the description of the motions must be given in terms of mass fractions and momenta. The solutions of these wave equations appear to be coupled porosity and saturation waves. However, this is just an illusion because the equations of motion are expressing statements about momenta and mass fractions. Since the components are segregated at the macroscale (pore scale), these statements in terms of mass fractions and momentum may be reformulated in terms of the component phases and the volume fractions.

The porosity wave described previously is an almost incompressible fluid flow process where a dynamic change in the porosity of a porous medium is coupled to a dynamic pressure change in the fluid, which in turn induces the fluid to flow. For this wave to occur, the solid matrix of the porous medium must be deformable, and the pressure changes must be slow enough for the fluid to behave in an almost incompressible fashion, that is, the fluid is forced to move instead of just compressing. Therefore, in this process, a damped travelling wave of porosity, pressure and fluid is obtained. The theoretical description of porosity waves was first presented by Spanos (2002). In the diffusional limit, the inertial terms appearing in the theory are negligible, and the porosity wave becomes a porosity diffusion process. This diffusion process was first recognized and analysed by Geilikman et al. (1993).

The saturation wave is a fluid flow process where a dynamic change occurs in the proportion of two incompressible fluids in a porous medium, that is, the saturation is coupled to the pressure in each fluid. Since the fluid pressure in each fluid changes, so does their difference, which is defined as the megascopic capillary pressure. Saturation waves are inherently dispersional. Dispersion is a pore-scale process where a displacing fluid tends to bypass a displaced fluid. It is a fundamental characteristic of fluid flow in porous media. Many of the essential features of dispersion were presented in the classic book by Bear (1972). A theory of saturation waves and their dispersional nature was recently presented by Udey (2009).

Saturation waves are strongly coupled to porosity waves as they pass. Laboratory studies (Wang et al., 1998; Zschuppe, 2001) have demonstrated that porosity waves can enhance fluid flow and displacement; furthermore, they can generate saturation waves, which are highly dispersional and can suppress viscous fingering. Consequently, when oil is being displaced by water using a porosity wave, more oil can be accessed and swept out by the saturation wave. At the field scale, these processes have generated an observed increase in total fluid flux and increased oil production (Dusseault et al., 2002; Groenenboom et al., 2003; Spanos et al., 2003).

In order to implement reservoir simulations and enhanced recovery estimation tools, a deep understanding of the physics of porosity waves, saturation waves and the coupling between them must be attained. The theory of porosity waves and the theory of saturation waves have been developed separately. In the research presented here, those theories will be used to guide the construction and analysis of the wave equations for coupled porosity and saturation waves in porous media. Again, it is emphasized that this description is only possible because the components are distinct separated phases at the pore scale and is a special case of the more general description in terms of momenta and mass fractions.

THE GOVERNING EQUATIONS

Variables and Definitions

Consider a porous medium consisting of a solid and two immiscible and incompressible fluids. The properties of these materials are labelled by the subscripts A = s, 1, 2. The densities of the solid and two fluids are ρ_A (A = s, 1, 2), respectively. The bulk moduli are K_A (A = s, 1, 2). The solid's shear modulus is μs and the fluids' shear and bulk viscosities are μ_A (A = 1, 2) and ξ_A (A = 1, 2). The solid's displacement is u_s, while the solid's velocity is $v_s = \frac{\partial u_s}{\partial t}$. The fluid velocities are v_A (A = 1, 2).

The proportion of the volume occupied by the two fluids is the porosity η, while the proportion occupied by the solid is the solidosity $\eta_s = -\eta$. The proportion of each fluid is its fractional porosity η_A (A = 1, 2). The saturation of each fluid S_A (A = 1, 2) is defined by writing $\eta_A = \eta S_A$ (A = 1, 2). Since $\eta = \eta_1 + \eta_2$, then one has $S_1 + S_2 = 1$. The irreducible saturations of Fluid 1 and Fluid 2 are denoted by S_{1r} and S_{2r} respectively. The permeability of the porous medium is denoted by K.

The governing equations and physical quantities at the megascopic scale of the porous medium are obtained from their corresponding pore-scale equations and quantities via the technique of volume averaging (Slattery 1967; Whitaker 1967; de la Cruz and Spanos 1985; Spanos et al. 1988). At the megascopic scale, the governing equations are the equations of continuity for the fluids, the equations of motion for the solid and the two fluids, the porosity equation and the saturation equation. Since temperature is not being incorporated into the description, the system of equations is isothermal. Also, the effects of gravity are being excluded in the analysis and one lets g = 0.

At the megascopic scale, the temporal and spatial variations of volume-averaged quantities are considered to be of first order. For example, the porosity may be split into a zeroth and first-order part by writing $\eta = \eta_o + \delta\eta$. Then one has $\delta_t\eta = \delta_t\delta\eta$, so no distinction is necessary between a derivative of η and its first-order part $\delta\eta$. The velocities are first order since they represent a deviation from a background state of no flow.

The Equations of Continuity

The megascopic equations of continuity (Spanos 2002) are

$$\frac{(\rho_s-\rho_s^o)}{\rho_s^o}-\frac{(\eta-\eta_o)}{(1-\eta_o)}+\nabla\cdot\vec{u}_s=0 \tag{10.1}$$

$$\partial_t(\rho_1\eta S_1)+\rho_1\eta S_1\nabla\cdot\vec{v}_1=0 \tag{10.2}$$

$$\partial_t(\rho_2\eta S_2)+\rho_2\eta S_2\nabla\cdot\vec{v}_2=0 \tag{10.3}$$

Now solve for the fluid divergences $\nabla\cdot\vec{u}_s$, $\nabla\cdot\vec{v}_1$ and $\nabla\cdot\vec{v}_2$. This yields

$$\nabla\cdot\vec{u}_s=-\frac{(\rho_s-\rho_s^o)}{\rho_s^o}+\frac{(\eta-\eta_o)}{(1-\eta_o)} \tag{10.4}$$

$$\nabla\cdot\vec{v}_1=-\frac{1}{\rho_1}\frac{\partial\rho_1}{\partial t}-\frac{1}{\eta}\frac{\partial\eta}{\partial t}-\frac{1}{S_1}\frac{\partial S_1}{\partial t} \tag{10.5}$$

$$\nabla\cdot\vec{v}_2=-\frac{1}{\rho_2}\frac{\partial\rho_2}{\partial t}-\frac{1}{\eta}\frac{\partial\eta}{\partial t}-\frac{1}{S_2}\frac{\partial S_2}{\partial t} \tag{10.6}$$

Since the system of equations is isothermal, the fluid densities depend only on pressure. Thus

$$\frac{(\rho_s-\rho_s^o)}{\rho_s^o}=\frac{1}{K_s}(P_s-P_o) \tag{10.7}$$

$$\frac{1}{\rho_1}\frac{\partial\rho_1}{\partial t}=\frac{1}{K_1}\frac{\partial P_1}{\partial t} \tag{10.8}$$

$$\frac{1}{\rho_2}\frac{\partial\rho_2}{\partial t}=\frac{1}{K_2}\frac{\partial P_2}{\partial t} \tag{10.9}$$

where the bulk modulus of each fluid is K_1 and K_2, respectively.

Substituting the pressure equations into the solid and fluid divergence equations yields

$$\nabla \cdot \vec{u}_s = -\frac{1}{K_s}(P_s - P_o) + \frac{(\eta - \eta_o)}{(1-\eta_o)} \tag{10.10}$$

$$\nabla \cdot \vec{v}_1 = -\frac{1}{\eta}\frac{\partial \eta}{\partial t} - \frac{1}{S_1}\frac{\partial S_1}{\partial t} - \frac{1}{K_1}\frac{\partial P_1}{\partial t} \tag{10.11}$$

$$\nabla \cdot \vec{v}_2 = -\frac{1}{\eta}\frac{\partial \eta}{\partial t} - \frac{1}{S_2}\frac{\partial S_2}{\partial t} - \frac{1}{K_2}\frac{\partial P_2}{\partial t} \tag{10.12}$$

The volumetric flow for each fluid is

$$\vec{q}_1 = \eta_1 \vec{v}_1 \tag{10.13}$$

$$\vec{q}_2 = \eta_2 \vec{v}_2 \tag{10.14}$$

and the total volumetric flow is

$$\vec{q} = \vec{q}_1 + \vec{q}_2 \tag{10.15}$$

The volumetric filter velocity $\nabla \cdot \vec{v}_q$ is defined by

$$\nabla \cdot \vec{v}_q = \frac{1}{\eta}\vec{q} = S_1 \vec{v}_1 + S_2 \vec{v}_2 \tag{10.16}$$

since the divergences of the volumetric flows are

$$\nabla \cdot \vec{q}_1 = -S_1 \partial_t \eta - \eta \partial_t S_1 - \frac{\eta S_1}{K_1}\partial_t P_1 \tag{10.17}$$

$$\nabla \cdot \vec{q}_2 = -S_2 \partial_t \eta + \eta \partial_t S_1 - \frac{\eta S_2}{K_2}\partial_t P_2 \tag{10.18}$$

The Equations of Motion

In the absence of gravity, the megascopic equations of motion for the solid and each fluid (Hickey 1994; Spanos 2002) are

$$\begin{aligned}\eta_s^o \rho_s^o \partial_t \mathrm{v}_s &= \eta_s^o K_s \nabla(\nabla \cdot \mathrm{u}_s) + \eta_s^o \mu_{Mss}\left(\nabla^2 \mathrm{u}_s + \frac{1}{3}\nabla(\nabla \cdot \mathrm{u}_s)\right) - K_s \nabla \eta \\ &\quad + Q_{s1}(\mathrm{v}_1 - \mathrm{v}_s) + Q_{s2}(\mathrm{v}_2 - \mathrm{v}_s) + \rho_{s1}\partial_t(\mathrm{v}_1 - \mathrm{v}_s) + \rho_{s2}\partial_t(\mathrm{v}_2 - \mathrm{v}_s)\end{aligned} \tag{10.19}$$

$$\eta_1^o\rho_1^o\,\partial_t\vec{v}_1 = \eta_1^o\nabla P_1 + \mu_{M11}\left(\nabla^2\vec{v}_1 + \frac{1}{3}\nabla = (\nabla\cdot\vec{v}_1)\right) + \eta_1^o\xi_1\nabla(\nabla\cdot\vec{v}_1) + \xi_1\nabla\partial_t\eta_1$$
$$-\mu_{M1s}\left(\nabla^2\vec{v}_s + \frac{1}{3}\nabla(\nabla\cdot\vec{v}_s)\right) + \mu_{M12}\left(\nabla^2\vec{v}_2 + \frac{1}{3}\nabla(\nabla\cdot\vec{v}_2)\right) \tag{10.20}$$
$$-Q_{11}(v_1 - v_s) + Q_{12}(v_1 - v_2) + \rho_{11}\,\partial_t(v_1 - v_s) + \rho_{12}\,\partial_t(v_1 - v_2)$$

$$\eta_2^o\rho_2^o\,\partial_t\vec{v}_2 = \eta_2^o\nabla P_2 + \mu_{M22}\left(\nabla^2\vec{v}_2 + \frac{1}{3}\nabla(\nabla\cdot\vec{v}_2)\right) + \eta_2^o\xi_2\nabla(\nabla\cdot\vec{v}_2) + \xi_2\nabla\partial_t\eta_2$$
$$-\mu_{M2s}\left(\nabla^2\vec{v}_s + \frac{1}{3}\nabla(\nabla\cdot\vec{v}_s)\right) + \mu_{M21}\left(\nabla^2\vec{v}_1 + \frac{1}{3}\nabla(\nabla\cdot\vec{v}_1)\right) \tag{10.21}$$
$$-Q_{21}(v_2 - v_1) + Q_{22}(v_2 - v_s) + \rho_{21}\,\partial_t(v_2 - v_1) + \rho_{22}\,\partial_t(v_2 - v_s)$$

In these equations, μ_{Mss} is the megascopic shear modulus and μ_{Mij} are the megascopic viscosities.

As well, to completely specify Newton's second law, the process-dependent equations for porosity and saturation are obtained. Note that these are actually statements about concentration, but in the present case these statements may be made in terms of porosity and saturation.

$$\frac{\partial\eta}{\partial t} = \delta_s\nabla\cdot\vec{v}_s - \delta_{f1}\nabla\cdot\vec{v}_{f1} - \delta_{f2}\nabla\cdot\vec{v}_{f2} \tag{10.22}$$

$$\frac{\partial S_1}{\partial t} = \vartheta_s\nabla\cdot\vec{v}_s + \vartheta_2\nabla\cdot\vec{v}_2 - \vartheta_1\nabla\cdot\vec{v}_1 \tag{10.23}$$

Dynamic processes that occur at the megascale are a composite of the motions of the component phases. These motions consist of almost in-phase motions and almost out-of-phase motions of the component phases. The equation of motion for the three components moving almost in phase is given by:

$$\rho_m\frac{\partial}{\partial t}\vec{v}_m = \eta_s\rho_s\frac{\partial}{\partial t}\vec{v}_s + \eta_1\rho_{f1}\frac{\partial}{\partial t}\vec{v}_{f1} + \eta_2\rho_{f2}\frac{\partial}{\partial t}\vec{v}_{f2} \tag{10.24}$$

Note that there are three momentum equations for the component phases, but the interactions of the components is non-linear and there are three ways in which the phases may interact out of phase. Thus there are four momentum equations.

$$\rho_m\frac{\partial}{\partial t}\vec{v}^{(i)}_{s(12)} = \eta_s\rho_s\frac{\partial}{\partial t}\vec{v}_s - \eta_1\rho_{f1}\frac{\partial}{\partial t}\vec{v}_{f1} - \eta_2\rho_{f2}\frac{\partial}{\partial t}\vec{v}_{f2} \tag{10.25}$$

$$\rho_m \frac{\partial}{\partial t}\vec{v}_{s2(1)}^{(i)} = \eta_s\rho_s \frac{\partial}{\partial t}\vec{v}_s - \eta_1\rho_{f1}\frac{\partial}{\partial t}\vec{v}_{f1} + \eta_2\rho_{f2}\frac{\partial}{\partial t}\vec{v}_{f2} \tag{10.26}$$

$$\rho_m \frac{\partial}{\partial t}\vec{v}_{s1(2)}^{(i)} = \eta_s\rho_s \frac{\partial}{\partial t}\vec{v}_s + \eta_1\rho_{f1}\frac{\partial}{\partial t}\vec{v}_{f1} - \eta_2\rho_{f2}\frac{\partial}{\partial t}\vec{v}_{f2} \tag{10.27}$$

$$v_m = \frac{\eta_s\rho_s v_s + \eta_{f1}\rho_{f1}v_{f1} + \eta_{f2}\rho_{f2}v_{f2}}{\eta_s\rho_s + \eta_{f1}\rho_{f1} + \eta_{f2}\rho_{f2}} \tag{10.28}$$

$$\rho^{(m)} = \eta_s\rho_s + \eta_{f1}\rho_{f1} + \eta_{f2}\rho_{f2} \tag{10.29}$$

$$v_{s(12)} = \frac{\eta_s\rho_s v_s - \eta_{f1}\rho_{f1}v_{f1} - \eta_{f2}\rho_{f2}v_{f2}}{\eta_s\rho_s + \eta_{f1}\rho_{f1} + \eta_{f2}\rho_{f2}} \tag{10.30}$$

$$v_{s1(2)} = \frac{\eta_s\rho_s v_s + \eta_{f1}\rho_{f1}v_{f1} - \eta_{f2}\rho_{f2}v_{f2}}{\eta_s\rho_s + \eta_{f1}\rho_{f1} + \eta_{f2}\rho_{f2}} \tag{10.31}$$

$$v_{s2(1)} = \frac{\eta_s\rho_s v_s - \eta_{f1}\rho_{f1}v_{f1} + \eta_{f2}\rho_{f2}v_{f2}}{\eta_s\rho_s + \eta_{f1}\rho_{f1} + \eta_{f2}\rho_{f2}} \tag{10.32}$$

$$\rho_{s(12)}^{(i)} = \eta_s\rho_s - \eta_{f1}\rho_{f1} - \eta_{f2}\rho_{f2} \tag{10.33}$$

$$\rho_{s1(2)}^{(i)} = \eta_s\rho_s + \eta_{f1}\rho_{f1} - \eta_{f2}\rho_{f2} \tag{10.34}$$

$$\rho_{s2(1)}^{(i)} = \eta_s\rho_s - \eta_{f1}\rho_{f1} + \eta_{f2}\rho_{f2} \tag{10.35}$$

The mass fractions and the velocities of momentum flux are related through the relations

$$\frac{\partial c_{fk}}{\partial t} = \tilde{\alpha}_1\nabla\cdot\vec{v}_m + \tilde{\alpha}_2\nabla\cdot\vec{v}_{s(12)}^{(i)} + \tilde{\alpha}_3\nabla\cdot\vec{v}_{s1(2)}^{(i)} + \tilde{\alpha}_4\nabla\cdot\vec{v}_{s2(1)}^{(i)} \tag{10.36}$$

$$\frac{\partial c^{(is(12))}}{\partial t} = \tilde{\beta}_1\nabla\cdot\vec{v}_m - \tilde{\beta}_2\nabla\cdot\vec{v}_{s(12)}^{(i)} + \tilde{\beta}_3\nabla\cdot\vec{v}_{s1(2)}^{(i)} + \tilde{\beta}_4\nabla\cdot\vec{v}_{s2(1)}^{(i)} \tag{10.37}$$

$$\frac{\partial c^{(is1(2))}}{\partial t}=\tilde{\gamma}_1\nabla\cdot\vec{v}_m+\tilde{\gamma}_2\nabla\cdot\vec{v}^{(i)}_{s(12)}-\tilde{\gamma}_3\nabla\cdot\vec{v}^{(i)}_{s1(2)}+\tilde{\gamma}_4\nabla\cdot\vec{v}^{(i)}_{s2(1)} \tag{10.38}$$

$$\frac{\partial c^{(is2(1))}}{\partial t}=\tilde{\delta}_1\nabla\cdot\vec{v}_m+\tilde{\delta}_2\nabla\cdot\vec{v}^{(i)}_{s(12)}+\tilde{\delta}_3\nabla\cdot\vec{v}^{(i)}_{s1(2)}-\tilde{\delta}_4\nabla\cdot\vec{v}^{(i)}_{s2(1)} \tag{10.39}$$

These are the generalized porosity and saturation equations discussed previously and, in the limiting case where no molecular diffusion is occurring, Equation 10.36 becomes the saturation equation, Equation 10.37 becomes the porosity equation and the final two equations contain no new information. This was observed previously in the case of miscible displacement with two phases; the two concentration–momentum flux equations reduced to one saturation equation under this condition.

The equations in terms of the four momenta

$$\begin{aligned}\rho^{(m)}\frac{\partial\vec{v}_m}{\partial t}&=\alpha_1\nabla(\nabla\cdot\vec{v}_m)+\alpha_2\nabla(\nabla\cdot\vec{v}^{(i)}_{s(12)})+\alpha_3\nabla(\nabla\cdot\vec{v}^{(i)}_{s1(2)})+\alpha_4\nabla(\nabla\cdot\vec{v}^{(i)}_{s2(1)})\\&-\alpha_5\nabla c_{fk}-\alpha_6\nabla c^{(is(12))}-\alpha_7\nabla c^{(is1(2))}-\alpha_8\nabla c^{(is2(1))}\\&+\alpha_9[\nabla^2\vec{v}_m+\frac{1}{3}\nabla(\nabla\cdot\vec{v}_m)]+\alpha_{10}[\nabla^2\vec{v}^{(i)}_{s(12)}+\frac{1}{3}\nabla(\nabla\cdot\vec{v}^{(i)}_{s(12)})]\\&+\alpha_{11}[\nabla^2\vec{v}^{(i)}_{s1(2)}+\frac{1}{3}\nabla(\nabla\cdot\vec{v}^{(i)}_{s1(2)})]+\alpha_{12}[\nabla^2\vec{v}^{(i)}_{s2(1)}+\frac{1}{3}\nabla(\nabla\cdot\vec{v}^{(i)}_{s2(1)})]\\&+\alpha_{13}(\vec{v}_m-\vec{v}^{(i)}_{s(12)})+\alpha_{14}(\vec{v}_m-\vec{v}^{(i)}_{s1(2)})+\alpha_{15}(\vec{v}_m-\vec{v}^{(i)}_{s2(1)})\\&+\alpha_{16}\left(\frac{\partial\vec{v}_m}{\partial t}-\frac{\partial\vec{v}^{(i)}_{s(12)}}{\partial t}\right)+\alpha_{17}\left(\frac{\partial\vec{v}_m}{\partial t}-\frac{\partial\vec{v}^{(i)}_{s1(2)}}{\partial t}\right)+\alpha_{18}\left(\frac{\partial\vec{v}_m}{\partial t}-\frac{\partial\vec{v}^{(i)}_{s2(1)}}{\partial t}\right)\end{aligned} \tag{10.40}$$

$$\begin{aligned}\rho_m\frac{\partial\vec{v}^{(i)}_{s(12)}}{\partial t}&=\beta_1\nabla(\nabla\cdot\vec{v}_m)+\beta_2\nabla(\nabla\cdot\vec{v}^{(i)}_{s(12)})+\beta_3\nabla(\nabla\cdot\vec{v}^{(i)}_{s1(2)})+\beta_4\nabla(\nabla\cdot\vec{v}^{(i)}_{s2(1)})\\&-\beta_5\nabla c_{fk}-\beta_6\nabla c^{(is(12))}-\beta_7\nabla c^{(is1(2))}-\beta_8\nabla c^{(is2(1))}\\&+\beta_9[\nabla^2\vec{v}_m+\frac{1}{3}\nabla(\nabla\cdot\vec{v}_m)]+\beta_{10}[\nabla^2\vec{v}^{(i)}_{s(12)}+\frac{1}{3}\nabla(\nabla\cdot\vec{v}^{(i)}_{s(12)})]\\&+\beta_{11}[\nabla^2\vec{v}^{(i)}_{2s}+\frac{1}{3}\nabla(\nabla\cdot\vec{v}^{(i)}_{s1(2)})]+\beta_{12}[\nabla^2\vec{v}^{(i)}_{s2(1)}+\frac{1}{3}\nabla(\nabla\cdot\vec{v}^{(i)}_{s2(1)})]\\&+\beta_{13}(\vec{v}^{(i)}_{s(12)}-\vec{v}_m)+\beta_{14}(\vec{v}^{(i)}_{s(12)}-\vec{v}^{(i)}_{s1(2)})+\beta_{15}(\vec{v}^{(i)}_{s(12)}-\vec{v}^{(i)}_{s2(1)})\\&+\beta_{16}\left(\frac{\partial\vec{v}^{(i)}_{s(12)}}{\partial t}-\frac{\partial\vec{v}_m}{\partial t}\right)+\beta_{17}\left(\frac{\partial\vec{v}^{(i)}_{s(12)}}{\partial t}-\frac{\partial\vec{v}^{(i)}_{s1(2)}}{\partial t}\right)+\beta_{18}\left(\frac{\partial\vec{v}^{(i)}_{s(12)}}{\partial t}-\frac{\partial\vec{v}^{(i)}_{s2(1)}}{\partial t}\right)\end{aligned} \tag{10.41}$$

$$\begin{aligned}
\rho_m \frac{\partial \vec{v}^{(i)}_{s1(2)}}{\partial t} &= \chi_1 \nabla(\nabla\cdot\vec{v}_m) + \chi_2 \nabla(\nabla\cdot\vec{v}^{(i)}_{s(12)}) + \chi_3 \nabla(\nabla\cdot\vec{v}^{(i)}_{s1(2)}) + \chi_4 \nabla(\nabla\cdot\vec{v}^{(i)}_{s2(1)}) \\
&- \chi_5 \nabla c_{fk} - \chi_6 \nabla c^{(is(12))} - \chi_7 \nabla c^{(is1(2))} - \chi_8 \nabla c^{(is2(1))} \\
&+ \chi_9 [\nabla^2 \vec{v}_m + \frac{1}{3}\nabla(\nabla\cdot\vec{v}_m)] + \chi_{10}[\nabla^2 \vec{v}^{(i)}_{s(12)} + \frac{1}{3}\nabla(\nabla\cdot\vec{v}^{(i)}_{s(12)})] \\
&+ \chi_{11}[\nabla^2 \vec{v}^{(i)}_{s1(2)} + \frac{1}{3}\nabla(\nabla\cdot\vec{v}^{(i)}_{s1(2)})] + \chi_{12}[\nabla^2 \vec{v}^{(i)}_{s2(1)} + \frac{1}{3}\nabla(\nabla\cdot\vec{v}^{(i)}_{s2(1)})] \\
&+ \chi_{13}(\vec{v}^{(i)}_{s1(2)} - \vec{v}^{(i)}_{m}) + \chi_{14}(\vec{v}^{(i)}_{s1(2)} - \vec{v}^{(i)}_{s(12)}) + \chi_{15}(\vec{v}^{(i)}_{s1(2)} - \vec{v}^{(i)}_{s2(1)}) \\
&+ \chi_{16}\left(\frac{\partial \vec{v}^{(i)}_{s1(2)}}{\partial t} - \frac{\partial \vec{v}_m}{\partial t}\right) + \chi_{17}\left(\frac{\partial \vec{v}^{(i)}_{s1(2)}}{\partial t} - \frac{\partial \vec{v}^{(i)}_{s(12)}}{\partial t}\right) + \chi_{18}\left(\frac{\partial \vec{v}^{(i)}_{s1(2)}}{\partial t} - \frac{\partial \vec{v}^{(i)}_{s2(1)}}{\partial t}\right)
\end{aligned} \tag{10.42}$$

$$\begin{aligned}
\rho_m \frac{\partial \vec{v}^{(i)}_{s2(1)}}{\partial t} &= \lambda_1 \nabla(\nabla\cdot\vec{v}_m) + \lambda_2 \nabla(\nabla\cdot\vec{v}^{(i)}_{s(12)}) + \lambda_3 \nabla(\nabla\cdot\vec{v}^{(i)}_{s1(2)}) + \lambda_4 \nabla(\nabla\cdot\vec{v}^{(i)}_{s2(1)}) \\
&- \lambda_5 \nabla c_{fk} - \lambda_6 \nabla c^{(is(12))} - \lambda_7 \nabla c^{(is1(2))} - \lambda_8 \nabla c^{(is2(1))} \\
&+ \lambda_9 [\nabla^2 \vec{v}_m + \frac{1}{3}\nabla(\nabla\cdot\vec{v}_m)] + \lambda_{10}[\nabla^2 \vec{v}^{(i)}_{s(12)} + \frac{1}{3}\nabla(\nabla\cdot\vec{v}^{(i)}_{s(12)})] \\
&+ \lambda_{11}[\nabla^2 \vec{v}^{(i)}_{s1(2)} + \frac{1}{3}\nabla(\nabla\cdot\vec{v}^{(i)}_{s1(2)})] + \lambda_{12}[\nabla^2 \vec{v}^{(i)}_{s2(1)} + \frac{1}{3}\nabla(\nabla\cdot\vec{v}^{(i)}_{s2(1)})] \\
&+ \lambda_{13}(\vec{v}^{(i)}_{s1(2)} - \vec{v}^{(i)}_{m}) + \lambda_{14}(\vec{v}^{(i)}_{s1(2)} - \vec{v}^{(i)}_{s(12)}) + \lambda_{15}(\vec{v}^{(i)}_{s1(2)} - \vec{v}^{(i)}_{s2(1)}) \\
&+ \lambda_{16}\left(\frac{\partial \vec{v}^{(i)}_{s1(2)}}{\partial t} - \frac{\partial \vec{v}_m}{\partial t}\right) + \lambda_{17}\left(\frac{\partial \vec{v}^{(i)}_{s1(2)}}{\partial t} - \frac{\partial \vec{v}^{(i)}_{s(12)}}{\partial t}\right) + \lambda_{18}\left(\frac{\partial \vec{v}^{(i)}_{s1(2)}}{\partial t} - \frac{\partial \vec{v}^{(i)}_{s2(1)}}{\partial t}\right)
\end{aligned} \tag{10.43}$$

Substituting into these equations for the motions of the component phases yields

$$\begin{aligned}
&\eta_s^o \rho_s^o \partial_t v_s + \eta_1^o \rho_1^o \partial_t \vec{v}_1 + \eta_2^o \rho_2^o \partial_t \vec{v}_2 = \eta_s^o K_s \nabla(\nabla\cdot u_s) \\
&+ \eta_s^o \mu_{Mss}[\nabla^2 u_s + \frac{1}{3}\nabla(\nabla\cdot u_s)] - (\mu_{M1s} + \mu_{M2s})[\nabla^2 \vec{v}_s + \frac{1}{3}\nabla(\nabla\cdot\vec{v}_s)] \\
&- K_s \nabla\eta + \xi_1 \nabla \partial_t \eta_1 + \xi_2 \nabla \partial_t \eta_2 + (\mu_{M11} + \mu_{M21})[\nabla^2 \vec{v}_1 + \frac{1}{3}\nabla(\nabla\cdot\vec{v}_1)] + \eta_1^o \xi_1 \nabla(\nabla\cdot\vec{v}_1) \\
&+ (\mu_{M12} + \mu_{M22})[\nabla^2 \vec{v}_2 + \frac{1}{3}\nabla(\nabla\cdot\vec{v}_2)] + \eta_2^o \xi_2 \nabla(\nabla\cdot\vec{v}_2) + \eta_1^o \nabla P_1 + \eta_2^o \nabla P_2 \\
&+ (Q_{s1} - Q_{11})(v_1 - v_s) + (Q_{s2} + Q_{22})(v_2 - v_s) + (Q_{12} + Q_{21})(v_1 - v_2) \\
&+ (\rho_{s1} + \rho_{11})\partial_t(v_1 - v_s) + (\rho_{s2} + \rho_{22})\partial_t(v_2 - v_s) + (\rho_{12} - \rho_{21})\partial_t(v_1 - v_2)
\end{aligned} \tag{10.44}$$

$$\begin{aligned}
&\eta_s\rho_s \frac{\partial}{\partial t}\vec{v}_s - \eta_1\rho_{f1}\frac{\partial}{\partial t}\vec{v}_{f1} - \eta_2\rho_{f2}\frac{\partial}{\partial t}\vec{v}_{f2} = \{\eta_s^o K_s \nabla(\nabla\cdot u_s) \\
&+(\eta_s^o\mu_{Mss}+\mu_{M1s}+\mu_{M2s})[\nabla^2 u_s + \frac{1}{3}\nabla(\nabla\cdot u_s] \\
&-K_s\nabla\eta - \xi_1\nabla\partial_t\eta_1 - \xi_2\nabla\partial_t\eta_2 - \eta_1^o\nabla P_1 - \eta_2^o\nabla P_2 \\
&-(\mu_{M11}+\mu_{M21})[\nabla^2\vec{v}_1 + \frac{1}{3}\nabla(\nabla\cdot\vec{v}_1)] - \eta_1^o\xi_1\nabla(\nabla\cdot\vec{v}_1) \\
&-(\mu_{M12}+\mu_{M22})[\nabla^2\vec{v}_2 + \frac{1}{3}\nabla(\nabla\cdot\vec{v}_2)] - \eta_2^o\xi_2\nabla(\nabla\cdot\vec{v}_2) \\
&+(Q_{s1}+Q_{11})(v_1-v_s)+(Q_{s2}-Q_{22})(v_2-v_s)-(Q_{12}+Q_{21})(v_1-v_2) \\
&+(\rho_{s1}-\rho_{11})\partial_t(v_1-v_s)+(\rho_{s2}-\rho_{22})\partial_t(v_2-v_s)-(\rho_{12}+\rho_{21})\partial_t(v_1-v_2)\}
\end{aligned} \tag{10.45}$$

$$\begin{aligned}
&\eta_s\rho_s \frac{\partial}{\partial t}\vec{v}_s + \eta_1\rho_{f1}\frac{\partial}{\partial t}\vec{v}_{f1} - \eta_2\rho_{f2}\frac{\partial}{\partial t}\vec{v}_{f2} = \{\eta_s^o K_s \nabla(\nabla\cdot u_s) \\
&+(\eta_s^o\mu_{Mss}-\mu_{M1s}+\mu_{M2s})[\nabla^2 u_s + \frac{1}{3}\nabla(\nabla\cdot u_s)] \\
&-K_s\nabla\eta + \xi_1\nabla\partial_t\eta_1 - \xi_2\nabla\partial_t\eta_2 + \eta_1^o\nabla P_1 - \eta_2^o\nabla P_2 \\
&+(\mu_{M11}-\mu_{M21})[\nabla^2\vec{v}_1 + \frac{1}{3}\nabla(\nabla\cdot\vec{v}_1)] + \eta_1^o\xi_1\nabla(\nabla\cdot\vec{v}_1) \\
&+(\mu_{M12}-\mu_{M22})[\nabla^2\vec{v}_2 + \frac{1}{3}\nabla(\nabla\cdot\vec{v}_2)] - \eta_2^o\xi_2\nabla(\nabla\cdot\vec{v}_2) \\
&+(Q_{s1}-Q_{11})(v_1-v_s)+(Q_{s2}-Q_{22})(v_2-v_s)+(Q_{12}-Q_{21})(v_1-v_2) \\
&+(\rho_{s1}+\rho_{11})\partial_t(v_1-v_s)+(\rho_{s2}-\rho_{22})\partial_t(v_2-v_s)+(\rho_{12}-\rho_{21})\partial_t(v_1-v_2)\}
\end{aligned} \tag{10.46}$$

$$\begin{aligned}
&\eta_s\rho_s \frac{\partial}{\partial t}\vec{v}_s - \eta_1\rho_{f1}\frac{\partial}{\partial t}\vec{v}_{f1} + \eta_2\rho_{f2}\frac{\partial}{\partial t}\vec{v}_{f2} = \{\eta_s^o K_s \nabla(\nabla\cdot u_s) \\
&+(\eta_s^o\mu_{Mss}+\mu_{M1s}-\mu_{M2s})[\nabla^2 u_s + \frac{1}{3}\nabla(\nabla\cdot u_s)] \\
&-K_s\nabla\eta - \xi_1\nabla\partial_t\eta_1 + \xi_2\nabla\partial_t\eta_2 - \eta_1^o\nabla P_1 + \eta_2^o\nabla P_2 \\
&-(\mu_{M11}-\mu_{M21})[\nabla^2\vec{v}_1 + \frac{1}{3}\nabla(\nabla\cdot\vec{v}_1)] + \eta_1^o\xi_1\nabla(\nabla\cdot\vec{v}_1) \\
&-(\mu_{M12}-\mu_{M22})[\nabla^2\vec{v}_2 + \frac{1}{3}\nabla(\nabla\cdot\vec{v}_2)] - \eta_2^o\xi_2\nabla(\nabla\cdot\vec{v}_2) \\
&+(Q_{s1}+Q_{11})(v_1-v_s)+(Q_{s2}+Q_{22})(v_2-v_s)-(Q_{12}-Q_{21})(v_1-v_2) \\
&+(\rho_{s1}+\rho_{11})\partial_t(v_1-v_s)+(\rho_{s2}-\rho_{22})\partial_t(v_2-v_s)+(\rho_{12}-\rho_{21})\partial_t(v_1-v_2)\}
\end{aligned} \tag{10.47}$$

Now, taking the divergence of these equations and eliminating $\nabla\cdot\vec{u}_s$, $\nabla\cdot\vec{v}_s$, $\nabla\cdot\vec{v}_1$ and $\nabla\cdot\vec{v}_2$, four equations are obtained for the variables η, S_1, P_1 and P_2. These equations describe the four coupled porosity–pressure–saturation waves. This reformulation is constructed in the next section.

DILATATIONAL WAVES

Now, taking the divergence of the equations of motion, one obtains

$$\begin{aligned}
&(\eta_s^o\rho_s^o+\rho_{s1}+\rho_{11}+\rho_{s2}+\rho_{22})\partial_t\nabla\cdot v_s+(\eta_1^o\rho_1^o-\rho_{s1}-\rho_{11}-\rho_{12}+\rho_{21})\partial_t\nabla\cdot\vec{v}_1\\
&+(\eta_2^o\rho_2^o-\rho_{s2}-\rho_{22}+\rho_{12}-\rho_{21})\partial_t\nabla\cdot\vec{v}_2=(\eta_s^oK_s+\frac{4}{3}\eta_s^o\mu_{Mss})\nabla^2(\nabla\cdot u_s)\\
&-\frac{4}{3}(\mu_{M1s}+\mu_{M2s})\nabla^2\nabla\cdot\vec{v}_s+\left(\frac{4}{3}(\mu_{M11}+\mu_{M21})+\eta_1^o\xi_1\right)\nabla^2(\nabla\cdot\vec{v}_1)\\
&+\left(\frac{4}{3}(\mu_{M12}+\mu_{M22})+\eta_2^o\xi_2\right)\nabla^2\nabla\cdot\vec{v}_2-K_s\nabla^2\eta+\xi_1\nabla^2\partial_t\eta_1+\xi_2\nabla^2\partial_t\eta_2\\
&+\eta_1^o\nabla^2P_1+\eta_2^o\nabla^2P_2+(Q_{11}-Q_{s1}-Q_{s2}-Q_{22})\nabla\cdot v_s\\
&-(Q_{11}-Q_{s1}-Q_{12}-Q_{21})\nabla\cdot v_1+(Q_{22}+Q_{s2}-Q_{12}-Q_{21})\nabla\cdot v_2
\end{aligned}\tag{10.48}$$

$$\begin{aligned}
&(\eta_s\rho_s+\rho_{s1}-\rho_{11}+\rho_{s2}-\rho_{22})\frac{\partial}{\partial t}\nabla\cdot\vec{v}_s-(\eta_1\rho_{f1}+\rho_{s1}-\rho_{11}-\rho_{12}-\rho_{21})\frac{\partial}{\partial t}\nabla\cdot\vec{v}_{f1}\\
&-(\eta_2\rho_{f2}+\rho_{s2}-\rho_{22}+\rho_{12}+\rho_{21})\frac{\partial}{\partial t}\nabla\cdot\vec{v}_{f2}=[\eta_s^oK_s+\frac{4}{3}(\eta_s^o\mu_{Mss}+\mu_{M1s}+\mu_{M2s})]\nabla^2\nabla\cdot u_s\\
&-\left(\frac{4}{3}(\mu_{M11}+\mu_{M21})+\eta_1^o\xi_1\right)\nabla^2(\nabla\cdot\vec{v}_1)-\left(\frac{4}{3}(\mu_{M12}+\mu_{M22})+\eta_2^o\xi_2\right)\nabla^2(\nabla\cdot\vec{v}_2)\\
&-K_s\nabla^2\eta-\xi_1\nabla^2\partial_t\eta_1-\xi_2\nabla^2\partial_t\eta_2-\eta_1^o\nabla^2P_1-\eta_2^o\nabla^2P_2\\
&+(Q_{22}-Q_{s2}-Q_{s1}-Q_{11})\nabla\cdot v_s+(Q_{11}+Q_{s1}-Q_{12}-Q_{21})\nabla\cdot v_1-(Q_{22}-Q_{s2}-Q_{12}-Q_{21})\nabla\cdot v_2
\end{aligned}\tag{10.49}$$

$$\begin{aligned}
&(\eta_s\rho_s+\rho_{s1}-\rho_{11}+\rho_{s2}+\rho_{22})\frac{\partial}{\partial t}\nabla\cdot\vec{v}_s-(\eta_1\rho_{f1}+\rho_{s1}-\rho_{11}-\rho_{12}+\rho_{21})\frac{\partial}{\partial t}\nabla\cdot\vec{v}_{f1}\\
&+(\eta_2\rho_{f2}-\rho_{s2}-\rho_{22}-\rho_{12}+\rho_{21})\frac{\partial}{\partial t}\nabla\cdot\vec{v}_{f2}\\
&=[\eta_s^oK_s+\frac{4}{3}(\eta_s^o\mu_{Mss}+\mu_{M1s}-\mu_{M2s})]\nabla^2(\nabla\cdot u_s)\\
&-\left(\frac{4}{3}(\mu_{M11}-\mu_{M21})-\eta_1^o\xi_1\right)\nabla^2(\nabla\cdot\vec{v}_1)-\left(\frac{4}{3}(\mu_{M12}-\mu_{M22})+\eta_2^o\xi_2\right)\nabla^2(\nabla\cdot\vec{v}_2)\\
&-K_s\nabla^2\eta-\xi_1\nabla^2\partial_t\eta_1+\xi_2\nabla^2\partial_t\eta_2-\eta_1^o\nabla^2P_1+\eta_2^o\nabla^2P_2\\
&-(Q_{11}+Q_{s1}+Q_{s2}+Q_{22})\nabla\cdot\vec{v}_s+(Q_{11}+Q_{s1}-Q_{12}+Q_{21})\nabla\cdot\vec{v}_1\\
&+(Q_{22}+Q_{s2}+Q_{12}-Q_{21})\nabla\cdot\vec{v}_2
\end{aligned}\tag{10.50}$$

$$(\eta_s\rho_s+\rho_{s1}+\rho_{11}+\rho_{s2}-\rho_{22})\frac{\partial}{\partial t}\nabla\cdot\vec{v}_s+(\eta_1\rho_{f1}-\rho_{s1}-\rho_{11}-\rho_{12}+\rho_{21})\frac{\partial}{\partial t}\nabla\cdot\vec{v}_{f1}$$

$$-(\eta_2\rho_{f2}+\rho_{s2}-\rho_{22}-\rho_{12}+\rho_{21})\frac{\partial}{\partial t}\nabla\cdot\vec{v}_{f2}=[\eta_s^oK_s+\frac{4}{3}(\eta_s^o\mu_{Mss}-\mu_{M1s}+\mu_{M2s})]\nabla^2(\nabla\cdot u_s)$$

$$\left(\frac{4}{3}(\mu_{M11}-\mu_{M21})\right)+\eta_1^o\xi_1\nabla^2(\nabla\cdot\vec{v}_1)+\left(\frac{4}{3}(\mu_{M12}-\mu_{M22})\right)-\eta_2^o\xi_2\nabla^2(\nabla\cdot\vec{v}_2) \tag{10.51}$$

$$-K_s\nabla^2\eta+\xi_1\nabla^2\partial_t\eta_1-\xi_2\nabla^2\partial_t\eta_2+\eta_1^o\nabla^2P_1-\eta_2^o\nabla^2P_2$$

$$+(Q_{22}-Q_{s2}+Q_{11}-Q_{s1})\nabla\cdot\vec{v}_s+(Q_{s1}-Q_{11}+Q_{12}-Q_{21})\nabla\cdot\vec{v}_1+(Q_{s2}-Q_{22}-Q_{12}+Q_{21})\nabla\cdot\vec{v}_2$$

Now consider a static compression of the medium, where

$$\frac{\eta-\eta_o}{\eta_o}=-\frac{\beta_s}{K_s}(P_s-P_o)+\frac{\beta_1}{K_1}(P_1-P_o)+\frac{\beta_2}{K_2}(P_2-P_o) \tag{10.52}$$

Substituting Equation 10.7 yields

$$\frac{\eta-\eta_o}{\eta_o}=\beta_s(\nabla\cdot\vec{u}_s-\frac{(\eta-\eta_o)}{(1-\eta_o)})+\frac{\beta_1}{K_1}(P_1-P_o)+\frac{\beta_2}{K_2}(P_2-P_o) \tag{10.53}$$

Solving for $\nabla\cdot\vec{u}_s$ yields the solid dilation equation

$$\nabla\cdot\vec{u}_s=\left(\frac{1}{\beta_s}+\frac{\eta_o}{(1-\eta_o)}\right)\frac{\eta-\eta_o}{\eta_o}-\frac{\beta_1}{\beta_sK_1}(P_1-P_o)-\frac{\beta_2}{\beta_sK_2}(P_2-P_o) \tag{10.54}$$

Eliminating $\nabla^2\nabla\cdot u_s$ using the equation

$$\nabla^2\nabla\cdot\vec{u}_s=\left(\frac{1}{\beta_s}+\frac{\eta_o}{(1-\eta_o)}\right)\nabla^2\eta-\frac{\beta_1}{\beta_sK_1}\nabla^2P_1-\frac{\beta_2}{\beta_sK_2}\nabla^2P_2 \tag{10.55}$$

Now define

$$\alpha_\eta=\frac{1}{\beta_s}+\frac{\eta_o}{(1-\eta_o)} \tag{10.56}$$

$$\alpha_{p_1}=\frac{\beta_1}{\beta_s} \tag{10.57}$$

$$\alpha_{p_2}=\frac{\beta_2}{\beta_s} \tag{10.58}$$

Therefore

$$\nabla^2 \nabla \cdot \vec{u}_s = \alpha_\eta \nabla^2 \eta - \frac{\alpha_{P1}}{K_1} \nabla^2 P_1 - \frac{\alpha_{P2}}{K_2} \nabla^2 P_2 \tag{10.59}$$

Upon substituting Equations 10.9 and 10.10 into the equation in the static limit

$$\delta_s \nabla \cdot \vec{v}_s = \frac{\partial \eta}{\partial t} + \delta_1 \nabla \cdot \vec{v}_1 + \delta_2 \nabla \cdot \vec{v}_2 \tag{10.60}$$

it is observed that

$$\alpha_\eta = \frac{(\eta_o - \delta_1 - \delta_2)}{\delta_s} \tag{10.61}$$

$$\alpha_{P1} = \frac{\delta_1}{\delta_s} \tag{10.62}$$

$$\alpha_{P2} = \frac{\delta_2}{\delta_s} \tag{10.63}$$

Taking the static limit of Equation 10.23 yields

$$\vartheta_s = \vartheta_2 = \vartheta_1 = 0 \tag{10.64}$$

Thus Equation 10.23 only describes dynamic processes.
Eliminating $\nabla \cdot v_s$ using the equation

$$\delta_s \nabla \cdot \vec{v}_s = \frac{\partial \eta}{\partial t} + \delta_1 \nabla \cdot \vec{v}_1 + \delta_2 \nabla \cdot \vec{v}_2 \tag{10.65}$$

then substitution of the divergences $\nabla \cdot \vec{v}_1$ and $\nabla \cdot \vec{v}_2$ given by Equations 10.9 and 10.10 into these intermediate equations yields four equations for the four unknowns P_1, P_2, η and S_1. This construction in terms of the components and volume fractions is possible because the fluids and solid are only mixed at one scale, the pore scale. These components and volume fractions are actually describing the momenta and mass fractions at the megascale.

$$\begin{aligned}
\tilde{a}_{\eta\eta} \frac{\partial^2 \eta}{\partial t^2} + \tilde{a}_{ss} \frac{\partial^2 S_1}{\partial t^2} + \tilde{a}_{pp1} \frac{1}{K_1} \frac{\partial^2 P_1}{\partial t^2} + \tilde{a}_{pp2} \frac{1}{K_2} \frac{\partial^2 P_2}{\partial t^2} &= \tilde{a}_\eta \nabla^2 \eta + \tilde{a}_{p1} \frac{1}{K_1} \nabla^2 P_1 + \tilde{a}_{p2} \frac{1}{K_2} \nabla^2 P_2 \\
&+ \tilde{a}_{\eta\nabla t} \nabla^2 \frac{\partial \eta}{\partial t} + \tilde{a}_{s\nabla t} \nabla^2 \frac{\partial S_1}{\partial t} + \tilde{a}_{pf1} \frac{1}{K_1} \nabla^2 \frac{\partial P_1}{\partial t} + \tilde{a}_{pf2} \frac{1}{K_2} \nabla^2 \frac{\partial P_2}{\partial t} \\
&+ \tilde{a}_{\eta t} \frac{\partial \eta}{\partial t} + \tilde{a}_{st} \frac{\partial S_1}{\partial t} + \tilde{a}_{p1t} \frac{1}{K_1} \frac{\partial P_1}{\partial t} + \tilde{a}_{p2t} \frac{1}{K_2} \frac{\partial P_2}{\partial t}
\end{aligned} \tag{10.66}$$

$$\tilde{b}_{\eta\eta}\frac{\partial^2\eta}{\partial t^2}+\tilde{b}_{ss}\frac{\partial^2 S_1}{\partial t^2}+\tilde{b}_{pp1}\frac{1}{K_1}\frac{\partial^2 P_1}{\partial t^2}+\tilde{b}_{pp2}\frac{1}{K_2}\frac{\partial^2 P_2}{\partial t^2}=\tilde{b}_{\eta}\nabla^2\eta+\tilde{b}_{p1}\frac{1}{K_1}\nabla^2 P_1+\tilde{b}_{p2}\frac{1}{K_2}\nabla^2 P_2$$
$$+\tilde{b}_{\eta\nabla t}\nabla^2\frac{\partial\eta}{\partial t}+\tilde{b}_{s\nabla t}\nabla^2\frac{\partial S_1}{\partial t}+\tilde{b}_{pf1}\frac{1}{K_1}\nabla^2\frac{\partial P_1}{\partial t}+\tilde{b}_{pf2}\frac{1}{K_2}\nabla^2\frac{\partial P_2}{\partial t} \quad (10.67)$$
$$+\tilde{b}_{\eta t}\frac{\partial\eta}{\partial t}+\tilde{b}_{st}\frac{\partial S_1}{\partial t}+\tilde{b}_{p1t}\frac{1}{K_1}\frac{\partial P_1}{\partial t}+\tilde{b}_{p2t}\frac{1}{K_2}\frac{\partial P_2}{\partial t}$$

$$\tilde{c}_{\eta\eta}\frac{\partial^2\eta}{\partial t^2}+\tilde{c}_{ss}\frac{\partial^2 S_1}{\partial t^2}+\tilde{c}_{pp1}\frac{1}{K_1}\frac{\partial^2 P_1}{\partial t^2}+\tilde{c}_{pp2}\frac{1}{K_2}\frac{\partial^2 P_2}{\partial t^2}=\tilde{c}_{\eta}\nabla^2\eta+\tilde{c}_{p1}\frac{1}{K_1}\nabla^2 P_1+\tilde{c}_{p2}\frac{1}{K_2}\nabla^2 P_2$$
$$+\tilde{c}_{\eta\nabla t}\nabla^2\frac{\partial\eta}{\partial t}+\tilde{c}_{s\nabla t}\nabla^2\frac{\partial S_1}{\partial t}+\tilde{c}_{pf1}\frac{1}{K_1}\nabla^2\frac{\partial P_1}{\partial t}+\tilde{c}_{pf2}\frac{1}{K_2}\nabla^2\frac{\partial P_2}{\partial t} \quad (10.68)$$
$$+\tilde{c}_{\eta t}\frac{\partial\eta}{\partial t}+\tilde{c}_{st}\frac{\partial S_1}{\partial t}+\tilde{c}_{p1t}\frac{1}{K_1}\frac{\partial P_1}{\partial t}+\tilde{c}_{p2t}\frac{1}{K_2}\frac{\partial P_2}{\partial t}$$

$$\tilde{d}_{\eta\eta}\frac{\partial^2\eta}{\partial t^2}+\tilde{d}_{ss}\frac{\partial^2 S_1}{\partial t^2}+\tilde{d}_{pp1}\frac{1}{K_1}\frac{\partial^2 P_1}{\partial t^2}+\tilde{d}_{pp2}\frac{1}{K_2}\frac{\partial^2 P_2}{\partial t^2}=\tilde{d}_{\eta}\nabla^2\eta+\tilde{d}_{p1}\frac{1}{K_1}\nabla^2 P_1+\tilde{d}_{p2}\nabla^2 P_2$$
$$+\tilde{d}_{\eta\nabla t}\nabla^2\frac{\partial\eta}{\partial t}+\tilde{d}_{s\nabla t}\nabla^2\frac{\partial S_1}{\partial t}+\tilde{d}_{pf1}\frac{1}{K_1}\nabla^2\frac{\partial P_1}{\partial t}+\tilde{d}_{pf2}\frac{1}{K_2}\nabla^2\frac{\partial P_2}{\partial t} \quad (10.69)$$
$$+\tilde{d}_{\eta t}\frac{\partial\eta}{\partial t}+\tilde{d}_{st}\frac{\partial S_1}{\partial t}+\tilde{d}_{p1t}\frac{1}{K_1}\frac{\partial P_1}{\partial t}+\tilde{d}_{p2t}\frac{1}{K_2}\frac{\partial P_2}{\partial t}$$

$$\tilde{a}_{\eta\eta}=(\eta_s^o\rho_s^o+\rho_{s1}+\rho_{11}+\rho_{s2}+\rho_{22})\frac{1}{\eta}\alpha_\eta-(\eta_1^o\rho_1^o+\eta_2^o\rho_2^o-\rho_{s1}-\rho_{s2}-\rho_{11}-\rho_{22})\frac{1}{\eta} \quad (10.70)$$

$$\tilde{a}_{ss}=-(\eta_s^o\rho_s^o+\rho_{s1}+\rho_{11}+\rho_{s2}+\rho_{22})\left(\frac{\alpha_{p1}}{S_1}-\frac{\alpha_{p2}}{S_2}\right)-(\eta_1^o\rho_1^o-\rho_{s1}-\rho_{11}-\rho_{12}+\rho_{21})\frac{1}{S_1}$$
$$+(\eta_2^o\rho_2^o-\rho_{s2}-\rho_{22}+\rho_{12}-\rho_{21})\frac{1}{S_2} \quad (10.71)$$

$$\tilde{a}_{pp1}=-(\eta_s^o\rho_s^o+\rho_{s1}+\rho_{11}+\rho_{s2}+\rho_{22})\alpha_{p1}-(\eta_1^o\rho_1^o-\rho_{s1}-\rho_{11}-\rho_{12}+\rho_{21}) \quad (10.72)$$

$$\tilde{a}_{pp2}=-(\eta_s^o\rho_s^o+\rho_{s1}+\rho_{11}+\rho_{s2}+\rho_{22})\alpha_{p2}-(\eta_2^o\rho_2^o-\rho_{s2}-\rho_{22}+\rho_{12}-\rho_{21}) \quad (10.73)$$

$$\tilde{a}_{\eta}=\alpha_\eta\left(\eta_s^o K_s+\frac{4}{3}\eta_s^o\mu_{Mss}\right)-K_s \quad (10.74)$$

$$\tilde{a}_{p1} = -\alpha_{p_1}\left(\eta_s^o K_s + \frac{4}{3}\eta_s^o \mu_{Mss}\right) + \eta_1^o K_1 \tag{10.75}$$

$$\tilde{a}_{p2} = -\alpha_{p_2}\left(\eta_s^o K_s + \frac{4}{3}\eta_s^o \mu_{Mss}\right) + \eta_2^o K_2 \tag{10.76}$$

$$\tilde{a}_{\eta\nabla t} = -\frac{4}{3}(\mu_{M1s} + \mu_{M2s})\frac{\alpha_\eta}{\eta} + \left(\frac{4}{3}(\mu_{M11} + \mu_{M21} + \mu_{M12} + \mu_{M22}) + \eta_1^o \xi_1 + \eta_2^o \xi_2\right) \tag{10.77}$$

$$\tilde{a}_{s\nabla t} = \frac{4}{3}(\mu_{M1s} + \mu_{M2s})\left(\frac{\alpha_{p1}}{S_1} - \frac{\alpha_{p2}}{S_2}\right) - \left(\frac{4}{3}(\mu_{M11} + \mu_{M21})_1\right)\frac{1}{S_1} + \left(\frac{4}{3}(\mu_{M12} + \mu_{M22})\right)\frac{1}{S_2} \tag{10.78}$$

$$\tilde{a}_{pf1} = \frac{4}{3}(\mu_{M1s} + \mu_{M2s})\alpha_{p1} - \left(\frac{4}{3}(\mu_{M11} + \mu_{M21}) + \eta_1^o \xi_1\right) \tag{10.79}$$

$$\tilde{a}_{pf2} = \frac{4}{3}(\mu_{M1s} + \mu_{M2s})\alpha_{p2} - \left(\frac{4}{3}(\mu_{M12} + \mu_{M22}) + \eta_2^o \xi_2\right) \tag{10.80}$$

$$\tilde{a}_{\eta t} = (Q_{11} - Q_{s1} - Q_{s2} - Q_{22})\frac{\alpha_\eta}{\eta} + (Q_{11} - Q_{22} - Q_{s1} - Q_{s2})\frac{1}{\eta} \tag{10.81}$$

$$\begin{aligned}\tilde{a}_{st} = &-(Q_{11} - Q_{s1} - Q_{s2} - Q_{22})\left(\frac{\alpha_{p1}}{S_1} - \frac{\alpha_{p2}}{S_2}\right) + (Q_{11} - Q_{s1} - Q_{12} - Q_{21})\frac{1}{S_1} \\ &+ (Q_{22} + Q_{s2} - Q_{12} - Q_{21})\frac{1}{S_2}\end{aligned} \tag{10.82}$$

$$\tilde{a}_{p1t} = -(Q_{11} - Q_{s1} - Q_{s2} - Q_{22})\alpha_{p1} + (Q_{11} - Q_{s1} - Q_{12} - Q_{21}) \tag{10.83}$$

$$\tilde{a}_{p2t} = -(Q_{11} - Q_{s1} - Q_{s2} - Q_{22})\alpha_{p2} - (Q_{22} + Q_{s2} - Q_{12} - Q_{21}) \tag{10.84}$$

$$\tilde{b}_{\eta\eta} = (\eta_s\rho_s + \rho_{s1} - \rho_{11} + \rho_{s2} - \rho_{22})\frac{1}{\eta}\alpha_\eta + (\eta_1\rho_{f1} + \eta_2\rho_{f2} + \rho_{s1} + \rho_{s2} - \rho_{11} - \rho_{22})\frac{1}{\eta} \tag{10.85}$$

$$\begin{aligned}\tilde{b}_{ss} = &-(\eta_s\rho_s + \rho_{s1} - \rho_{11} + \rho_{s2} - \rho_{22})\left(\frac{\alpha_{p1}}{S_1} - \frac{\alpha_{p2}}{S_2}\right) + (\eta_1\rho_{f1} + \rho_{s1} - \rho_{11} - \rho_{12} - \rho_{21})\frac{1}{S_1} \\ &-(\eta_2\rho_{f2} + \rho_{s2} - \rho_{22} + \rho_{12} + \rho_{21})\frac{1}{S_2}\end{aligned} \tag{10.86}$$

$$\tilde{b}_{pp1} = -(\eta_s\rho_s + \rho_{s1} - \rho_{11} + \rho_{s2} - \rho_{22})\alpha_{p1} + (\eta_1\rho_{f1} + \rho_{s1} - \rho_{11} - \rho_{12} - \rho_{21}) \tag{10.87}$$

$$\tilde{b}_{pp2} = -(\eta_s\rho_s + \rho_{s1} - \rho_{11} + \rho_{s2} - \rho_{22})\alpha_{p2} + (\eta_2\rho_{f2} + \rho_{s2} - \rho_{22} + \rho_{12} + \rho_{21}) \tag{10.88}$$

$$\tilde{b}_{\eta} = \alpha_{\eta}\left(\eta_s^o K_s + \frac{4}{3}(\eta_s^o \mu_{Mss} + \mu_{M1s} + \mu_{M2s})\right) - K_s \tag{10.89}$$

$$\tilde{b}_{p1} = -\alpha_{p_1}\left(\eta_s^o K_s + \frac{4}{3}(\eta_s^o \mu_{Mss} + \mu_{M1s} + \mu_{M2s})\right) - \eta_1^o K_1 \tag{10.90}$$

$$\tilde{b}_{p2} = -\alpha_{p_2}\left(\eta_s^o K_s + \frac{4}{3}(\eta_s^o \mu_{Mss} + \mu_{M1s} + \mu_{M2s})\right) - \eta_2^o K_2 \tag{10.91}$$

$$\tilde{b}_{\eta\nabla t} = \frac{1}{\eta}\left(\frac{4}{3}(\mu_{M11} + \mu_{M21} + \mu_{M12} + \mu_{M22}) + \eta_1^o\xi_1 + \eta_2^o\xi_2\right) \tag{10.92}$$

$$\tilde{b}_{s\nabla t} = \left(\frac{4}{3}(\mu_{M11} + \mu_{M21})\right)\frac{1}{S_1} - \left(\frac{4}{3}(\mu_{M12} + \mu_{M22})\right)\frac{1}{S_2} \tag{10.93}$$

$$\tilde{b}_{pf1} = \left(\frac{4}{3}(\mu_{M11} + \mu_{M21}) + \eta S_1 \xi_1\right) \tag{10.94}$$

$$\tilde{b}_{pf2} = \left(\frac{4}{3}(\mu_{M12} + \mu_{M22}) + \eta S_2 \xi_2\right) \tag{10.95}$$

$$\tilde{b}_{\eta t} = (Q_{22} - Q_{s2} - Q_{s1} - Q_{11})\frac{1}{\eta}\alpha_{\eta} - (Q_{11} - Q_{22} + Q_{s1} + Q_{s2})\frac{1}{\eta} \tag{10.96}$$

$$\begin{aligned}\tilde{b}_{st} = &-(Q_{22} - Q_{s2} - Q_{s1} - Q_{11})\left(\frac{\alpha_{p1}}{S_1} - \frac{\alpha_{p2}}{S_2}\right) - (Q_{11} + Q_{s1} - Q_{12} - Q_{21})\frac{1}{S_1} \\ &-(Q_{22} - Q_{s2} - Q_{12} - Q_{21})\frac{1}{S_2}\end{aligned} \tag{10.97}$$

$$\tilde{b}_{p1t} = -(Q_{22} - Q_{s2} - Q_{s1} - Q_{11})\alpha_{p1} - (Q_{11} + Q_{s1} - Q_{12} - Q_{21}) \tag{10.98}$$

$$\tilde{b}_{p2t} = -(Q_{22} - Q_{s2} - Q_{s1} - Q_{11})\alpha_{p2} + (Q_{22} - Q_{s2} - Q_{12} - Q_{21}) \tag{10.99}$$

$$\tilde{c}_{\eta\eta} = (\eta_s\rho_s + \rho_{s1} - \rho_{11} + \rho_{s2} + \rho_{22})\frac{\alpha_\eta}{\eta} + (\eta_1\rho_{f1} - \eta_2\rho_{f2} + \rho_{s1} + \rho_{s2} - \rho_{11} + \rho_{22})\frac{1}{\eta} \quad (10.100)$$

$$\tilde{c}_{ss} = -(\eta_s\rho_s + \rho_{s1} - \rho_{11} + \rho_{s2} + \rho_{22})\left(\frac{\alpha_{p1}}{S_1} - \frac{\alpha_{p2}}{S_2}\right) + (\eta_1\rho_{f1} + \rho_{s1} - \rho_{11} - \rho_{12} + \rho_{21})\frac{1}{S_1}$$
$$+ (\eta_2\rho_{f2} - \rho_{s2} - \rho_{22} - \rho_{12} + \rho_{21})\frac{1}{S_2} \quad (10.101)$$

$$\tilde{c}_{pp1} = -(\eta_s\rho_s + \rho_{s1} - \rho_{11} + \rho_{s2} + \rho_{22})\alpha_{p1} + (\eta_1\rho_{f1} + \rho_{s1} - \rho_{11} - \rho_{12} + \rho_{21}) \quad (10.102)$$

$$\tilde{c}_{pp2} = -(\eta_s\rho_s + \rho_{s1} - \rho_{11} + \rho_{s2} + \rho_{22})\alpha_{p2} - (\eta_2\rho_{f2} - \rho_{s2} - \rho_{22} - \rho_{12} + \rho_{21}) \quad (10.103)$$

$$\tilde{c}_{\eta} = \alpha_\eta\left(\eta_s^o K_s + \frac{4}{3}(\eta_s^o\mu_{Mss} + \mu_{M1s} - \mu_{M2s})\right) - K_s \quad (10.104)$$

$$\tilde{c}_{p1} = -\alpha_{p_1}\left(\eta_s^o K_s + \frac{4}{3}(\eta_s^o\mu_{Mss} + \mu_{M1s} - \mu_{M2s})\right) - \eta_1^o K_1 \quad (10.105)$$

$$\tilde{c}_{p2} = -\alpha_{p_2}\left(\eta_s^o K_s + \frac{4}{3}(\eta_s^o\mu_{Mss} + \mu_{M1s} - \mu_{M2s})\right) + \eta_2^o K_2 \quad (10.106)$$

$$\tilde{c}_{\eta\nabla t} = \frac{1}{\eta}\left(\frac{4}{3}(\mu_{M11} - \mu_{M21} + \mu_{M12} - \mu_{M22}) - \eta_1^o\xi_1 + \eta_2^o\xi_2\right) \quad (10.107)$$

$$\tilde{c}_{s\nabla t} = \left(\frac{4}{3}(\mu_{M11} - \mu_{M21})\right)\frac{1}{S_1} - \left(\frac{4}{3}(\mu_{M12} - \mu_{M22})\right)\frac{1}{S_2} - 2\eta\xi_1 - 2\eta\xi_2 \quad (10.108)$$

$$\tilde{c}_{pf1} = \left(\frac{4}{3}(\mu_{M11} - \mu_{M21}) - \eta_1^o\xi_1\right) \quad (10.109)$$

$$\tilde{c}_{pf2} = \left(\frac{4}{3}(\mu_{M12} - \mu_{M22}) + \eta_2^o\xi_2\right) \quad (10.110)$$

$$\tilde{c}_{\eta t} = -(Q_{11} + Q_{s1} + Q_{s2} + Q_{22})\frac{\alpha_\eta}{\eta} - (Q_{11} + Q_{22} + Q_{s1} + Q_{s2})\frac{1}{\eta} \quad (10.111)$$

$$\tilde{c}_{st} = (Q_{11} + Q_{s1} + Q_{s2} + Q_{22})\left(\frac{\alpha_{p1}}{S_1} - \frac{\alpha_{p2}}{S_2}\right) - (Q_{11} + Q_{s1} - Q_{12} + Q_{21})\frac{1}{S_1} + (Q_{22} + Q_{s2} + Q_{12} - Q_{21})\frac{1}{S_2} \tag{10.112}$$

$$\tilde{c}_{p1t} = (Q_{11} + Q_{s1} + Q_{s2} + Q_{22})\alpha_{p1} - (Q_{11} + Q_{s1} - Q_{12} + Q_{21}) \tag{10.113}$$

$$\tilde{c}_{p2t} = (Q_{11} + Q_{s1} + Q_{s2} + Q_{22})\alpha_{p2} - (Q_{22} + Q_{s2} + Q_{12} - Q_{21}) \tag{10.114}$$

$$\tilde{d}_{\eta\eta} = (\eta_s\rho_s + \rho_{s1} + \rho_{11} + \rho_{s2} - \rho_{22})\frac{\alpha_\eta}{\eta} - (\eta_1\rho_{f1} - \eta_2\rho_{f2} - \rho_{s1} - \rho_{s2} - \rho_{11} + \rho_{22})\frac{1}{\eta} \tag{10.115}$$

$$\tilde{d}_{ss} = -(\eta_s\rho_s + \rho_{s1} + \rho_{11} + \rho_{s2} - \rho_{22})\left(\frac{\alpha_{p1}}{S_1} - \frac{\alpha_{p2}}{S_2}\right) - (\eta_1\rho_{f1} - \rho_{s1} - \rho_{11} - \rho_{12} + \rho_{21})\frac{1}{S_1} - (\eta_2\rho_{f2} + \rho_{s2} - \rho_{22} - \rho_{12} + \rho_{21})\frac{1}{S_2} \tag{10.116}$$

$$\tilde{d}_{pp1} = -(\eta_s\rho_s + \rho_{s1} + \rho_{11} + \rho_{s2} - \rho_{22})\alpha_{p1} - (\eta_1\rho_{f1} - \rho_{s1} - \rho_{11} - \rho_{12} + \rho_{21}) \tag{10.117}$$

$$\tilde{d}_{pp2} = -(\eta_s\rho_s + \rho_{s1} + \rho_{11} + \rho_{s2} - \rho_{22})\alpha_{p2} + (\eta_2\rho_{f2} + \rho_{s2} - \rho_{22} - \rho_{12} + \rho_{21}) \tag{10.118}$$

$$\tilde{d}_\eta = \alpha_\eta\left(\eta_s^o K_s + \frac{4}{3}(\eta_s^o \mu_{Mss} - \mu_{M1s} + \mu_{M2s})\right) - K_s \tag{10.119}$$

$$\tilde{d}_{p1} = -\alpha_{p_1}\left(\eta_s^o K_s + \frac{4}{3}(\eta_s^o \mu_{Mss} - \mu_{M1s} + \mu_{M2s})\right) + \eta_1^o K_1 \tag{10.120}$$

$$\tilde{d}_{p2} = -\alpha_{p_2}\left(\eta_s^o K_s + \frac{4}{3}(\eta_s^o \mu_{Mss} - \mu_{M1s} + \mu_{M2s})\right) - \eta_2^o K_2 \tag{10.121}$$

$$\tilde{d}_{\eta\nabla t} = -\left(\frac{4}{3}(\mu_{M11} - \mu_{M21} + \mu_{M12} - \mu_{M22}) + \eta_1^o\xi_1 - \eta_2^o\xi_2\right)\frac{1}{\eta} \tag{10.122}$$

$$\tilde{d}_{s\nabla t} = -\left(\frac{4}{3}(\mu_{M11} - \mu_{M21})\right)\frac{1}{S_1} + \left(\frac{4}{3}(\mu_{M12} - \mu_{M22})\right)\frac{1}{S_2} \tag{10.123}$$

$$\tilde{d}_{pf1} = -\left(\frac{4}{3}(\mu_{M11} - \mu_{M21}) + \eta_1^o \xi_1\right) \tag{10.124}$$

$$\tilde{d}_{pf2} = -\left(\frac{4}{3}(\mu_{M12} - \mu_{M22}) - \eta_2^o \xi_2\right) \tag{10.125}$$

$$\tilde{d}_{\eta t} = (Q_{22} - Q_{s2} + Q_{11} - Q_{s1})\frac{\alpha_\eta}{\eta} - (Q_{s1} + Q_{s2} - Q_{11} - Q_{22})\frac{1}{\eta} \tag{10.126}$$

$$\begin{aligned}\tilde{d}_{st} = &-(Q_{22} - Q_{s2} + Q_{11} - Q_{s1})\left(\frac{\alpha_{p1}}{S_1} - \frac{\alpha_{p2}}{S_2}\right) - (Q_{s1} - Q_{11} + Q_{12} - Q_{21})\frac{1}{S_1} \\ &+(Q_{s2} - Q_{22} - Q_{12} + Q_{21})\frac{1}{S_2}\end{aligned} \tag{10.127}$$

$$\tilde{d}_{p1t} = -(Q_{22} - Q_{s2} + Q_{11} - Q_{s1})\alpha_{p1} - (Q_{s1} - Q_{11} + Q_{12} - Q_{21}) \tag{10.128}$$

$$\tilde{d}_{p2t} = -(Q_{22} - Q_{s2} + Q_{11} - Q_{s1})\alpha_{p2} - (Q_{s2} - Q_{22} - Q_{12} + Q_{21}) \tag{10.129}$$

These equations are the dilatational wave equations. They describe the seismic P waves, the first due to the total momentum (or in-phase motions of the components) and the next three due to out-of-phase motion and thus highly attenuating. They also describe four highly non-linear waves moving at less than a tenth of the speed of the first P wave, very much like the porosity–pressure waves considered in Chapter 3. These waves are also all strongly coupled and able to suppress viscous fingering, thus causing dispersion of the fluids to be the dominate flow process. At first sight, this system of equations appears quite complex, so as a very naive first approximation assume that all four waves' porosity, saturation, P_1 and P_2 are so strongly coupled they move at approximately the same speed. This assumption does not seem unreasonable because it was observed that the porosity–pressure waves were so strongly coupled they could be approximated as a single wave and experimental observations of the strong coupling between the porosity and saturation waves have been made. This allows for a Fourier decomposition of these equations to obtain the speed versus wavelength. Here, it is assumed the solid is silica sand and the fluids are water and a light oil.

TRIAL SOLUTION

Upon combining these four equations with the porosity and saturation equations, six coupled wave equations are now obtained, which may be solved by the same methods as considered previously. The difference is that now a sufficient number of equations have been obtained to solve for coupled porosity–saturation waves. Note that, as shown on the web page associated with the book, solving the equations of motion for the components coupled to the saturation and porosity equations solutions are only obtained when the waves are decoupled. However, simple experimental observations show the waves are strongly coupled. As well, it is shown on the web page associated with the book that if one attempts to obtain Onsager's relations from this system of equations it is observed that the system of equations is not complete. The current formulation yields four coupled wave equations in terms of mass and momentum.

Once again, define a differential wave operator basis W_b by letting

$$W_b = \left[\frac{\partial^2}{\partial t^2}, \frac{\partial}{\partial t}, \nabla^2 \frac{\partial}{\partial t}, \nabla^2\right] \tag{10.130}$$

Again, let $\psi = \psi(\vec{x}, t)$ be a wave variable associated with a wave operator W_ψ, defined formally by

$$\begin{aligned} W_\psi &= [w_o, w_1, w_2, w_3] \cdot W_b \\ &= w_o \frac{\partial^2}{\partial t^2} + w_1 \frac{\partial}{\partial t} + w_2 \nabla^2 \frac{\partial}{\partial t} + w_3 \nabla^2 \end{aligned} \tag{10.131}$$

Again, the application of the wave operator W_ψ to the variable ψ is

$$\begin{aligned} W_\psi \cdot \psi &= [w_o, w_1, w_2, w_3] \cdot W_b \cdot \psi \\ &= w_o \frac{\partial^2}{\partial t^2}\psi + w_1 \frac{\partial}{\partial t}\psi + w_2 \nabla^2 \frac{\partial}{\partial t}\psi + w_3 \nabla^2 \psi \end{aligned} \tag{10.132}$$

The four dilatational wave equations may now be assembled into a wave operator matrix equation.

$$\begin{bmatrix} W_{m\eta} & W_{mS_1} & W_{mP_1} & W_{mP_2} \\ W_{i12\eta} & W_{i12S_1} & W_{i12P_1} & W_{i12P_2} \\ W_{i1\eta} & W_{i1S_1} & W_{i1P_1} & W_{i1P_2} \\ W_{i2\eta} & W_{i2S_1} & W_{i2P_1} & W_{i2P_2} \end{bmatrix} \begin{bmatrix} \eta \\ S_1 \\ P_1 \\ P_2 \end{bmatrix} = \begin{bmatrix} 0 \\ 0 \\ 0 \\ 0 \end{bmatrix} \tag{10.133}$$

Let the trial solution be

$$\begin{bmatrix} \eta & S_1 & P_1 & P_2 \end{bmatrix} = \begin{bmatrix} A_\eta^o & A_{S_1}^o & A_{P_1}^o & A_{P_2}^o \end{bmatrix} e^{nx+wt} \tag{10.134}$$

Substituting this trial solution into equation (10.134) yields the dispersion matrix equation

$$\begin{bmatrix} \mathcal{P}_{m\eta} & \mathcal{P}_{mS_1} & \mathcal{P}_{mP_1} & \mathcal{P}_{mP_2} \\ \mathcal{P}_{i12\eta} & \mathcal{P}_{i12S_1} & \mathcal{P}_{i12P_1} & \mathcal{P}_{i12P_2} \\ \mathcal{P}_{i1\eta} & \mathcal{P}_{i1S_1} & \mathcal{P}_{i1P_1} & \mathcal{P}_{i1P_2} \\ \mathcal{P}_{i2\eta} & \mathcal{P}_{i2S_1} & \mathcal{P}_{i2P_1} & \mathcal{P}_{i2P_2} \end{bmatrix} \begin{bmatrix} \mathcal{P}_{\eta}^{o} \\ \mathcal{P}_{S_1}^{o} \\ \mathcal{P}_{P_1}^{o} \\ \mathcal{P}_{P_2}^{o} \end{bmatrix} = \begin{bmatrix} 0 \\ 0 \\ 0 \\ 0 \end{bmatrix} \tag{10.135}$$

where

$$\mathcal{P}_{m\eta} = \tilde{a}_{\eta\eta} w^2 - \tilde{a}_{\eta} n^2 - \tilde{a}_{\eta\nabla t} n^2 w - \tilde{a}_{\eta t} w \tag{10.136}$$

$$\mathcal{P}_{mS_1} = \tilde{a}_{ss} w^2 - \tilde{a}_{s\nabla t} n^2 w - \tilde{a}_{st} w \tag{10.137}$$

$$\mathcal{P}_{mP_1} = \tilde{a}_{pp1} \frac{1}{K_1} w^2 - \tilde{a}_{p1} \frac{1}{K_1} n^2 - \tilde{a}_{pf1} \frac{1}{K_1} n^2 w - \tilde{a}_{p1t} w \tag{10.138}$$

$$\mathcal{P}_{mP_2} = \tilde{a}_{pp2} \frac{1}{K_2} w^2 - \tilde{a}_{p2} \frac{1}{K_2} n^2 - \tilde{a}_{pf2} \frac{1}{K_2} n^2 w - \tilde{a}_{p2t} w \tag{10.139}$$

$$\mathcal{P}_{i12\eta} = \tilde{b}_{\eta\eta} w^2 - \tilde{b}_{\eta} n^2 - \tilde{b}_{\eta\nabla t} n^2 w - \tilde{b}_{\eta t} w \tag{10.140}$$

$$\mathcal{P}_{i12S_1} = \tilde{b}_{ss} w^2 - \tilde{b}_{s\nabla t} n^2 w - \tilde{b}_{st} w \tag{10.141}$$

$$\mathcal{P}_{i12P_1} = \tilde{b}_{pp1} \frac{1}{K_1} w^2 - \tilde{b}_{p1} \frac{1}{K_1} n^2 - \tilde{b}_{pf1} \frac{1}{K_1} n^2 w - \tilde{b}_{p1t} w \tag{10.142}$$

$$\mathcal{P}_{i12P_2} = \tilde{b}_{pp2} \frac{1}{K_2} w^2 - \tilde{b}_{p2} \frac{1}{K_2} n^2 - \tilde{b}_{pf2} \frac{1}{K_2} n^2 w - \tilde{b}_{p2t} w \tag{10.143}$$

$$\mathcal{P}_{i2\eta} = \tilde{c}_{\eta\eta} w^2 - \tilde{c}_{\eta} n^2 - \tilde{c}_{\eta\nabla t} n^2 w - \tilde{c}_{\eta t} w \tag{10.144}$$

$$\mathcal{P}_{i2S_1} = \tilde{c}_{ss} w^2 - \tilde{c}_{s\nabla t} n^2 w - \tilde{c}_{st} w \tag{10.145}$$

$$\mathcal{P}_{i2P_1} = \tilde{c}_{pp1} \frac{1}{K_1} w^2 - \tilde{c}_{p1} \frac{1}{K_1} n^2 - \tilde{c}_{pf1} \frac{1}{K_1} n^2 w - \tilde{c}_{p1t} w \tag{10.146}$$

$$\mathcal{P}_{i2P_2} = \tilde{c}_{pp2}\frac{1}{K_2}w^2 - \tilde{c}_{p2}\frac{1}{K_2}n^2 - \tilde{c}_{pf2}\frac{1}{K_2}n^2w - \tilde{c}_{p2t}w \tag{10.147}$$

$$\mathcal{P}_{i1\eta} = \tilde{d}_{\eta\eta}w^2 - \tilde{d}_{\eta}n^2 - \tilde{d}_{\eta\nabla t}n^2w - \tilde{d}_{\eta t}w \tag{10.148}$$

$$\mathcal{P}_{i1S_1} = \tilde{d}_{ss}w^2 - \tilde{d}_{s\nabla t}n^2w - \tilde{d}_{st}w \tag{10.149}$$

$$\mathcal{P}_{i1P_1} = \tilde{d}_{pp1}\frac{1}{K_1}w^2 - \tilde{d}_{p1}\frac{1}{K_1}n^2 - \tilde{d}_{pf1}\frac{1}{K_1}n^2w - \tilde{d}_{p1t}w \tag{10.150}$$

$$\mathcal{P}_{i1P_2} = \tilde{d}_{pp2}\frac{1}{K_2}w^2 - \tilde{d}_{p2}\frac{1}{K_2}n^2 - \tilde{d}_{pf2}\frac{1}{K_2}n^2w - \tilde{d}_{p2t}w \tag{10.151}$$

For a non-trivial solution to exist, the determinant of the matrix in Equation 10.135 must be 0. This solution allows for the strong coupling of the four waves' porosity, saturation, pressure of Fluid 1 and pressure of Fluid 2. Because of the strong coupling, the porosity waves cause the saturation waves to disperse as they pass. This is much different than the case of seismic waves, where the waves travel with little interaction.

SUMMARY

The equations of motion for dilatational waves in a porous medium consisting of a solid and two almost incompressible and immiscible fluids were constructed. The solutions of these wave equations appear to be coupled porosity and saturation waves. The equations of motion are actually expressing statements about momenta and mass fractions. However, since the components are macroscopically segregated at the macroscale (pore scale), these statements in terms of mass fractions and momentum may be reformulated in terms of the component phases and the volume fractions.

The porosity wave described previously is an almost incompressible fluid flow process where a dynamic change in the porosity of a porous medium is coupled to a dynamic pressure change in the fluid, which in turn induces the fluid to flow. The saturation wave is a fluid flow process where a dynamic change occurs in the proportion of two incompressible fluids in a porous medium. This description of the coupling between these waves is only possible because the components are distinct separate phases at the pore scale; thus this description is a special case of the more general description in terms of momenta and mass fractions.

The four momentum equations were rewritten as four coupled equations expressed in terms of the four variables porosity, saturation and the two fluid pressures. Here, it is important to note that if an attempt is made to solve the problem using the three-component equations the system of equations obtained yields three equations for these four unknowns, and the three-component problem cannot be solved. Here, the components cannot be varied independently and can only be described dynamically in terms of the four momenta. The four dilatational wave equations may now be assembled into a wave operator matrix equation. Because of the strong coupling, the porosity waves cause the saturation waves to disperse as they pass. This is much different than the case of seismic waves, where the waves travel with little interaction.

REFERENCES

Bear, J., 1972, *Dynamics of Fluids in Porous Media*, Dover, New York.

de la Cruz, V., and Spanos, T.J.T., 1985, Seismic wave propagation in a porous medium, *Geophysics*, 50(10), 1556–1565.

Dusseault, M.B., Shand, D., Meling, T., Spanos, T., and Davidson, B., 2002, Field applications of pressure pulsing in porous media, Poromechanics II, Proceedings of the Second Biot Conference on Poromechanics, Grenoble, France, 26–28 August 2002, A.A Balkema, Lisse.

Geilikman, M.B., Spanos, T.J.T., and Nyland, E., 1993, Porosity diffusion in fluid-saturated media, *Tectonophysics*, 217, 111–115.

Groenenboom, J., Wong, S., Meling, T., Zschuppe, R., and Davidson, B., 2003, Pulsed water injection during water flooding, in Proceedings SPE International Improved Oil Recovery Conference in Asia Pacific, 20–21 October 2003, Kuala Lumpur, Malaysia.

Hickey, C., 1994, Mechanics of porous media, PhD dissertation, University of Alberta, Edmonton.

Slattery, J.C., 1967, Flow of viscoelastic fluids through porous media, *AIChE J.*, 13, 1066–1071.

Spanos, T.J.T., de la Cruz, V., and Hube, J., 1988, An analysis of the theoretical foundations of relative permeability curves, *AOSTRA J. Res.*, 4(3), 181–192.

Spanos, T.J.T., 2002, *The Thermophysics of Porous Media*, Monographs and Surveys in Pure and Applied Mathematics 126, Chapman & Hall/CRC, Boca Raton, FL.

Spanos, T., Shand, D., Davidson, B., Dusseault, M., and Samaroo, M., 2003, Pressure pulsing at the reservoir scale: A new IOR approach, *J. Can. Petrol. Technol.*, 42(2), 16–28.

Whitaker, S., 1967, Diffusion and dispersion in porous media, *AIChE J.*, 13, 420–427.

Udey, N., 2009, Dispersion waves of two fluids in a porous medium, *Trans. Porous Media,* 79, 107–115.

Wang, J., Dusseault, M.B., Spanos, T.J.T., and Davidson, B., 1998, Fluid enhancement under liquid pressure pulsing at low frequency, Proceedings 7th UNITAR International Conference on Heavy Crude and Tar Sands– October 1998, Beijing, China.

Zschuppe, R.P., 2001, Pulse flow enhancement in two-phase media, MSc dissertation, University of Waterloo, Waterloo.

CHAPTER 11

Thermodynamic Automata

OBJECTIVES OF THIS CHAPTER

Cellular automata provide an alternative to mathematical equations when describing physical systems. The motion of particles is described on a lattice according to prescribed rules. In this chapter, it is shown that automata may be used to describe basic physics such as thermodynamics and thermomechanical processes. It is also shown that automata can be used to model dynamical processes in porous media. In the case of describing physical processes, the rules must be able to be used to rigorously derive the associated energy momentum tensor and basic thermodynamic results such as the Boltzmann equation. In the case of geophysical models, the rules are less basic and simply imitate observed processes.

Early models constrained the particles to move on a lattice, for example, Frisch et al. (1986) showed that particles having discrete mass and velocity populating a two-dimensional triangular lattice evolve according to the incompressible Navier–Stokes equation in the limit of large lattice size and low velocity. This class of automata models that requires particles to move on a lattice according to simple collision and propagation rules is referred to as *lattice gas models*. These models evolve such that in one time step the particles propagate by one lattice unit to neighbouring lattice sites. For these models no more than one particle may occupy a particular propagation direction at any given site, resulting in an exclusion principle. In order to achieve isotropy using these rules, one is restricted to a triangular lattice in two dimensions. In three dimensions, isotropy requires that one use a four-dimensional face-centred hypercubic lattice and then project the simulation results into three dimensions. Another problem that arises in this model is that it doesn't satisfy Galilean invariance. Another model that has been used to model fluid flow is the lattice Boltzmann method (Lawniczak and Kapral 1996; Chen and Doolen, 1998). This method allows for the use of square as well as triangular lattices, but the particles are still constrained by the lattice.

In the approach presented in this chapter, the particles are no longer constrained by the lattice. This work was first published in the proceedings of the Royal Society of London (Udey et al. 1999). Here, the lattice serves only as a bookkeeper. Particles are allowed to populate a discretized space that is equivalent to all particles and their properties in a volume of space being ascribed to a point in that volume, the lattice point. There is no exclusion principle (i.e. any number of particles may occupy the volume ascribed to a given lattice site). Since the particle motions are independent of the lattice, any lattice shape may be chosen. For simplicity, a square or cubic lattice is generally chosen, but the generalization to any other lattice shapes is trivial. Since the maximum distance a particle may move in one time step is one lattice step, this distance is taken as the speed of light. The particles are allowed to undergo Lorentz invariant elastic collisions. From the collision and propagation rules, it is shown that the energy momentum tensor and relativistic Boltzmann equation can be derived. Thus it is observed that the collision and propagation rules are more fundamental than the Boltzmann equation. An interesting numerical observation is that, for any initial distribution of particle velocities, the Boltzmann distribution evolves in very few time steps. Another interesting numerical observation is that, for average particle speeds less than 2/10th of the speed of light, the results of Lorentz invariant elastic and Newtonian elastic collisions are indistinguishable.

A LORENTZ INVARIANT THERMAL LATTICE GAS

In two dimensions, each site is labelled by (i,j) in a rectangular array of lattice sites, with $i_{min} \le i \le i_{max}$ and $j_{min} \le j \le j_{max}$. The position of site (i,j) is $\vec{x} = (x_i, y_j)$, where $x_i = L \times i$, $y_i = L \times j$ and $\vec{x}$ specifies the position at the centre of lattice cell (i,j), which has an area L^2. At each time step, a particle with velocity $\vec{v} = (v_x, v_y)$ and residing in cell (i,j) has a probability $M_{ab}(\vec{v})$ of moving to cell (i + a, j + b), where a, b can have the values −1, 0, 1.

The x and y components of motion are independent and therefore may be considered separately. So here the two-dimensional propagation rules are decomposed into two one-dimensional sets of rules. Consider a single particle with x components of velocity given by v_x. Now let $M_a\left(\frac{v_x}{v_{max}}\right)$ represent the probability of a particle moving from its current position along the *x*-axis (i) to the position (i + a). Thus $M_1\left(\frac{v_x}{v_{max}}\right)$ is the probability of moving right by one site (crossing the boundary between the volume element described by the original lattice point to the volume element described by the lattice point to the right), $M_0\left(\frac{v_x}{v_{max}}\right)$ is the probability of staying at the same site (not moving far enough to leave the original lattice site) and $M_{-1}\left(\frac{v_x}{v_{max}}\right)$ is the probability of moving left by one site. Here, the one-dimensional movement matrix is defined by

$$M_a\left(\frac{v_x}{v_{max}}\right) = \begin{cases} \theta(v)\times v & a=1 \\ \theta(v)\times(1-v)+\theta(-v)\times(1+v) & a=0 \\ \theta(-v)\times(-v) & a=-1 \end{cases} \tag{11.1}$$

where the Heaviside step function θ(x) is defined by

$$\theta(x)=\begin{cases}1 & x>0\\ \tfrac{1}{2} & x=0\\ 0 & x<0\end{cases} \tag{11.2}$$

and $M_a(v)$ has the properties

$$\sum_{a=-1}^{1} M_a(v)=1 \tag{11.3}$$

$$\sum_{a=-1}^{1} a\times M_a(v)=v \tag{11.4}$$

For a single particle, which starts at x = 0, its position after N time steps is determined by applying the movement algorithm N times. In the absence of collisions with other particles, the particle will have moved N_{-1} times to the left, N_1 times to the right and will not have moved N_0 times. The particle's position will then be

$$x=(N_1-N_{-1})\times L \tag{11.5}$$

The expected values for N_i would be

$$\langle N_a\rangle = N\times M_a\left(\frac{v_x}{v_{max}}\right) \tag{11.6}$$

and the average motion along the *x*-axis is given by

$$\langle x\rangle = v_x\times t \tag{11.7}$$

where $t = N\times\Delta t$. In two dimensions, the average motion is a straight line, given by $\vec{x}=\vec{v}\times t$.

The two-dimensional movement matrix may now be expressed in terms of the one-dimensional movement matrixes by

$$M_{ab}(\vec{v}) = M_a(v_x)M_b(v_y) \tag{11.8}$$

where

$$\sum_{a=-1}^{1}\sum_{b=-1}^{1} M_{ab}(\vec{v})=1 \tag{11.9}$$

The general case of an arbitrary velocity in two dimensions is reduced by rotations to the case $v_x \geq 0$ and $v_y \geq 0$. Thus, the movement matrix becomes

$$M_{ab}(\vec{v}) = \begin{pmatrix} 0 & (1-v_x)\times v_y & v_x \times v_y \\ 0 & (1-v_x)\times(1-v_y) & v_x(1-v_y) \\ 0 & 0 & 0 \end{pmatrix} \tag{11.10}$$

The ability of the particle to move in any direction removes the restriction that the particles move along the lattice and relegates the lattice to the role of bookkeeper. This ability also removes the restriction of standard lattice gas models that the equilibrium distribution be constrained to a Fermi–Dirac distribution. The space–time metric is $g_{\alpha\beta} = \text{diag}(-1, 1, 1)$, where Greek indices take on the values 0, 1, 2. The velocity is $\vec{v} = (v_x, v_y)$ where $v_x = \frac{p_x}{E}$ and $v_y = \frac{p_y}{E}$. The quantities p_x and p_y are the spatial components of the relativistic momentum and E is the energy. Here, the energy is related to the spatial components of the momenta by the formula $E^2 = p_x^2 + p_y^2 + m^2$ for an ideal gas.

For the particle collisions, special relativistic elastic binary collisions are adopted. A formal description of these rules is available in any introductory text on special relativity. Now allow two particles of different rest mass to collide. A Lorentz transform is taken of their momenta from the lattice (laboratory) frame to the centre of mass frame of the collision. These momenta are denoted by $\vec{p}_1^{\alpha} = (\vec{p}_1, E_1)$ and $\vec{p}_2^{\alpha} = (\vec{p}_2, E_2)$, where $\vec{p}_1 = -\vec{p}_2$. The outcome of the collision is found by rotating the initial spatial momenta $\vec{p}_1$ and $\vec{p}_2$ into the final spatial momenta $\vec{p}_1'$ and $\vec{p}_2'$ by an angle θ. A random value for θ is generated in the interval $[-\pi, \pi]$, yielding the outcome of the collision. The energies E_1' and E_2' of the particles are now calculated and the particles' new momenta are transformed back to the lattice (laboratory) frame.

Here, the notation $W(p_1^{\alpha}, p_2^{\alpha} | p_3^{\alpha}, p_4^{\alpha})$ is introduced to describe the collisions where the collisions are symmetric under an exchange of particles $[W(p_1^{\alpha}, p_2^{\alpha} | p_3^{\alpha}, p_4^{\alpha}) = W(p_2^{\alpha}, p_1^{\alpha} | p_4^{\alpha}, p_3^{\alpha})]$ and the reverse collision has the same probability as the forward collision.

These particle collisions in the centre-of-mass frame result in Lorentz invariance in the relativistic model, and if non-relativistic collisions are considered Galilean invariance is obtained. This distinguishes this thermodynamic automata from other automata models.

The evolution of particles on the lattice is obtained by allowing propagation and collisions to occur sequentially.

The Lorentz invariant distribution function for a gas is defined by stating that $N_{ij}^t(\vec{p})$ is the number of particles in cell (i,j) at time *t*. The total number of particles in cell (i,j) at time *t* is N_{0ij}^t and the number density is given by n_{0ij}^t, where

$$N_{0ij}^t = n_{0ij}^t V = \int_{\vec{p}=0}^{\infty} N_{ij}^t(\vec{p}) d\omega \tag{11.11}$$

where

$$d\omega = \frac{dp_x dp_y}{|p^4|} \tag{11.12}$$

is the Lorentz invariant volume element of momentum.

Let $\Delta_M N_{ij}^t$ represent the change of $N_{ij}^t(\vec{p})$ due to particles leaving and entering cell (i,j). The updated value may then be denoted by

$$M(N_{ij}^t) = N_{ij}^t + \Delta_M N_{ij}^t \tag{11.13}$$

This equation represents the application of the movement operator M on the distribution function N_{ij}^t. The change due to collisions is denoted by $\Delta_C N_{ij}^t$, so that the updated value of the distribution function is

$$C(N_{ij}^t) = N_{ij}^t + \Delta_C N_{ij}^t \tag{11.14}$$

This equation represents an application of the collision operator C on N_{ij}^t. The total change in the distribution function in one time step is obtained by applying the movement operation to the distribution function followed by the collision operation. These sequential operations may be represented mathematically by

$$N_{ij}^{t+1} = C[M(N_{ij}^t)] \tag{11.15}$$

Employing these two operations, one observes that the total change of the distribution function in a cell from one time step to the next is

$$N_{ij}^{t+1} = N_{ij}^t + \Delta N_{ij}^t \tag{11.16}$$

where the total change in the number of particles is

$$\Delta N_{ij}^t = \Delta_M N_{ij}^t + \Delta_C N_{ij}^t \tag{11.17}$$

Therefore the total change in distribution function in a cell may be written as

$$N_{ij}^{t+1} = N_{ij}^t + \Delta_M N_{ij}^t + \Delta_C N_{ij}^t \tag{11.18}$$

The change in number of particles in a cell consists of the number of particles entering from neighbouring cells $\Delta_M^+ N_{ij}^t$ minus the number of particles that leave the cell $\Delta_M^- N_{ij}^t$

$$\Delta_M N_{ij}^t = \Delta_M^+ N_V^t - \Delta_M^- N_{ij}^t \tag{11.19}$$

The number of particles that leave an arbitrary neighbour cell (i + a, j + b) to enter (i,j) is the number of particles in that cell multiplied by the probability of moving to cell (i,j). Summing this number of particles, $N^t_{i+a,j+b}(\vec{p}) \times M_{-a,-b}(\vec{v})$, over all neighbouring cells gives the net influx of particles into cell (i,j), which yields

$$\Delta^+_M N^t_{ij}(\vec{p}) = \sum_{a=-1}^{1}\sum_{b=-1}^{1} N^t_{i+a,j+b}(\vec{p}) \times M_{-a,-b}(\vec{v}) - N^t_{ij}(\vec{p}) \times M_{0,0}(\vec{v}) \tag{11.20}$$

The probability of leaving the cell is 1 minus the probability of staying in the cell.

$$\Delta^-_M N^t_{ij}(\vec{p}) = N^t_{ij}(\vec{p}) \times [1 - M_{0,0}(\vec{v})] \tag{11.21}$$

The total change due to particle propagation is therefore

$$\Delta_M N^t_{ij}(\vec{p}) = \sum_{a=-1}^{1}\sum_{b=-1}^{1} N^t_{i+a,j+b}(\vec{p}) \times M_{-a,-b}(\vec{v}) - N^t_{ij}(\vec{p}) \tag{11.22}$$

If the gas is in equilibrium, then the particle distribution should be the same in each cell, that is,

$$\Delta_M N^t_{ij}(\vec{p}) \approx 0 \tag{11.23}$$

This condition expresses a lack of net particle and heat flow in equilibrium.

Now let $L(a, b) = x_{i+a,j+b} - x_{ij}$, where L is the lattice spacing. If the particle distribution is a slowly varying function of position then

$$N^t_{i+a,j+b} = N^t_{ij} + L(a,b) \cdot \nabla N^t_{ij} \tag{11.24}$$

Now, substituting Equation 11.24 into Equation 11.22 and using the relations in Equations 11.3, 11.4, 11.8 and 11.9

$$\Delta_M N^t_{ij} = -L\vec{v} \cdot \nabla N_{ij} \tag{11.25}$$

is obtained. This result may be substituted into Equation 11.17, which in turn may be substituted into Equation 11.16. The result is the lattice Boltzmann equation:

$$N^{t+1}_{ij} - N^t_{ij} + L\vec{v} \cdot \nabla N_{ij} = \Delta_C N^t_{ij} \tag{11.26}$$

Thus the Boltzmann equation has been derived from the collision and propagation rules. Assuming that the distribution function is a slowly varying function of time, then the Boltzmann equation may be converted into an equation of the evolution of the particle distribution.

$$N_{ij}^{t+1} = N_{ij}^{t} + \frac{\partial N_{ij}^{t}}{\partial t} \Delta t \tag{11.27}$$

Substituting this equation into Equation 11.26 yields

$$\Delta t \frac{\partial N_{ij}^{t}}{\partial t} + L\vec{v} \cdot \nabla N_{ij} = \Delta_C N_{ij}^{t} \tag{11.28}$$

Equation 11.28 is the equation of evolution of the lattice gas in the frame of reference of the lattice. This equation can be converted into a relativistically invariant form. Here $\vec{v} = \vec{\omega}/\omega^4$ in special relativity, where $p^\alpha = m(\vec{\omega}, \omega^4)$ and $x^4 = ct$. Then since the time and distance scales are related by $L = c\Delta t$, Equation 11.28 may be multiplied by the factor $\omega^4/(c\Delta t)$, yielding

$$\frac{\partial N_{ij}^{t} \omega^\alpha}{\partial x^\alpha} = \omega^4 \frac{\Delta_C N_{ij}^{t}}{c\Delta t} \tag{11.29}$$

The expression on the right-hand side of Equation 11.29 is the relativistic collision term, which may be written as

$$D_c N_{ij}^{t} = \omega^4 \frac{\Delta_C N_{ij}^{t}}{c\Delta t} \tag{11.30}$$

This collision term represents the net change in the number of particle in the momentum state p^α due to binary collisions. In terms of the collision transition probability, the collision term is as follows (Israel 1972):

$$\begin{aligned} D_c N_{ij}^{t}(p^\alpha) = &\int N_{ij}^{t}(p_1^\alpha) N_{ij}^{t}(p_2^\alpha) W(p_1^\alpha, p_2^\alpha | p_3^\alpha, p^\alpha) d\omega_1 d\omega_2 d\omega_3 \\ &- \int N_{ij}^{t}(p^\alpha) N_{ij}^{t}(p_1^\alpha) W(p^\alpha, p_1^\alpha | p_2^\alpha, p_3^\alpha) d\omega_1 d\omega_2 d\omega_3 \end{aligned} \tag{11.31}$$

This expression consists of the total number of particles gained by state p^α due to collisions minus the total number of particles lost from state p^α.

When Equation 11.29 is rewritten using Equation 11.30, the relativistic Boltzmann equation for a lattice gas in the absence of external forces is obtained:

$$(N_{ij}^{t} \omega^\alpha)_{|\alpha} = D_c N_{ij}^{t} \tag{11.32}$$

where the vertical stroke denotes the covariant derivative, which is equivalent to the partial derivative with respect to flat space–time Cartesian coordinates x^{α}. This is the equation of evolution for the lattice gas.

Now that the relativistic Boltzmann equation has been firmly established, the well-understood theory and techniques of relativistic thermodynamics may be applied. A detailed construction of this theory may be found in Israel (1972).

To simplify notation, the time and space indices will be dropped unless otherwise specified, that is, $N_{ij}^{t} \rightarrow N$, indicating the variables refer to cell (i,j) at time *t*.

A master balance equation may be obtained from the relativistic Boltzmann equation. Multiplying the Boltzmann equation by $\xi = \Psi d\Phi / N$, where Ψ is any tensor function of position or momentum and Φ is a function of N, then integrate over all momenta and divide by the volume. Then the master balance equation is

$$\frac{1}{V}\frac{\partial}{\partial x^{\alpha}}\int \Phi\Psi\omega^{\alpha} d\omega = \frac{1}{V}\int \Phi \frac{\partial(\Psi\omega^{\alpha})}{\partial x^{\alpha}} d\omega + \frac{1}{V}\int \xi D_c N d\omega \tag{11.33}$$

The collision integral in Equation 11.33 may be expressed as

$$\frac{1}{V}\int \xi D_c N(p^{\alpha}) d\omega = \frac{1}{2}\int [\xi] N(p^{\alpha}) N(p_1^{\alpha}) W(p_1^{\alpha}, p_2^{\alpha} | p_3^{\alpha}, p_4^{\alpha}) d\omega_1 d\omega_2 d\omega_3 d\omega \tag{11.34}$$

where

$$[\xi] = \xi(p^{\alpha}) + \xi(p_1^{\alpha}) - \xi(p_2^{\alpha}) - \xi(p_3^{\alpha}) \tag{11.35}$$

The macroscopic conservation laws correspond to various choices of Φ and Ψ. For example, the mass flux

$$M^{\alpha} = \frac{1}{V}\int N p^{\alpha} d\omega \tag{11.36}$$

corresponds to the choice $\Phi = N$ and $\Psi = m$ so Equation 11.33 becomes

$$M^{\alpha}_{|\alpha} = 0 \tag{11.37}$$

which expresses conservation of mass flux. The energy momentum tensor is defined by

$$T^{\alpha\beta} = \frac{1}{V}\int N p^{\alpha} p^{\beta} d\omega \tag{11.38}$$

which corresponds to the choice $\Phi = N$ and $\Psi = mp^{\alpha}$. With these choices, Equation 11.33 expresses conservation of the energy momentum tensor,

$$T^{\alpha\beta}_{|\beta} = 0 \tag{11.39}$$

The entropy flux of the lattice gas is defined by

$$S^{\alpha} = -\frac{k}{V}\int [N\ln(N) - N]\omega^{\alpha} d\omega \tag{11.40}$$

is specified by the choice $\Phi = N\ln(N) - N$ and $\Psi = -k$, where k is Boltzmann's constant. These choices make $\xi = -k\ln(N)$. Then Equation 11.33 becomes the equation of entropy production

$$S^{\alpha}_{|\alpha} = -\frac{k}{V}\int \ln(N) D_c N\, d\omega. \tag{11.41}$$

Symmetry of the collisions along with Equation 11.34 can be used to transform Equation 11.41 into

$$S^{\alpha}_{|\alpha} = -\frac{k}{2V}\int [\Xi - \ln(\Xi) - 1] N(p^{\alpha}) N(p_1^{\alpha}) W(p^{\alpha}, p_1^{\alpha} | p_2^{\alpha}, p_3^{\alpha}) d\omega_1\, d\omega_2\, d\omega_3\, d\omega \tag{11.42}$$

where

$$\Xi = \frac{N(p_2^{\alpha}) N(p_3^{\alpha})}{N(p^{\alpha}) N(p_1^{\alpha})} \tag{11.43}$$

Since

$$\Xi - \ln(\Xi) - 1 \geq 0 \quad \forall \Xi \tag{11.44}$$

the entropy production satisfies the constraint

$$S^{\alpha}_{|\alpha} \geq 0 \tag{11.45}$$

which is Boltzmann's H-theorem: in an isolated system the entropy remains the same or increases.

Equilibrium is achieved when the entropy production is 0. This is achieved by the condition $\Xi = 1$, which means that $\ln(N)$ is a collisional invariant and therefore has the form

$$\ln(N) = \alpha + \beta^{\lambda} p_{\lambda} \tag{11.46}$$

The integrals for the mass flux and energy momentum tensor will converge only if β^{λ} is time like so we set

$$\beta^{\lambda} = \beta u^{\lambda} \tag{11.47}$$

$$\beta = \frac{1}{kT} \tag{11.48}$$

$$u^{\lambda} u_{\lambda} = -1 \tag{11.49}$$

which define a local fluid velocity $u^{\lambda} = \gamma(u)(\vec{u},1) = \gamma(u)(u_x, u_y, 1)$, temperature T and inverse temperature β. Equation 11.46 makes $D_c N = 0$ in equilibrium and the Boltzmann equation imposes the constraints

$$\alpha_{|\mu} = 0 \tag{11.50}$$

$$\beta_{\mu|\nu} + \beta_{\nu|\mu} = 0 \tag{11.51}$$

Equation 10.49 states that the potential is constant in time and space and is therefore the same for all cells at all times. The general decomposition for the derivative of the fluid velocity is as follows (Ryan and Shepley 1975):

$$u_{\mu|\nu} = -\dot{u}_{\mu} u_{\nu} + \omega_{\mu\nu} + \sigma_{\mu\nu} + \frac{1}{2} \theta_V \Delta_{\mu\nu} \tag{11.52}$$

where the spatial projection operator $\Delta_{\mu\nu}$ is

$$\Delta_{\mu\nu} = g_{\mu\nu} + u_{\mu} u_{\nu} \tag{11.53}$$

and the fluid acceleration $\dot{u}_{\mu}$, vorticity $\omega_{\mu\nu}$, shear stress $\sigma_{\mu\nu}$ and volume expansion θ_V are defined by

$$\dot{u}_{\mu} = u_{\mu|\nu} u^{\nu} \tag{11.54}$$

$$\omega_{\mu\nu} = \Delta^{\lambda}_{\mu} \Delta^{\tau}_{\nu} \frac{1}{2} (u_{\lambda|\tau} - u_{\tau|\lambda}) \tag{11.55}$$

$$\sigma_{\mu\nu} = (\Delta^{\lambda}_{\mu} \Delta^{\tau}_{\nu} - \frac{1}{2} \Delta_{\mu\nu} \Delta^{\lambda\tau}) \frac{1}{2} (u_{\lambda|\tau} - u_{\tau|\lambda}) \tag{11.56}$$

$$\theta_V = u^{\lambda}_{|\lambda} \tag{11.57}$$

Note that $\Delta^{\lambda}_{|\lambda} = 2$. Equations 11.47, 11.51 and 11.52 now imply that the fluid is shear free, $\sigma_{\mu\nu} = 0$, expansionless, $\theta_V = 0$ and that

$$\beta_{|\mu} = \beta \dot{u}_{\mu} \tag{11.58}$$

Since rotating systems are not being considered and there are no external forces, vorticity and acceleration are 0, that is, $\omega_{\mu\nu} = 0$ and $\dot{u}_\mu = 0$. Consequently

$$\beta_{|\mu} = 0 \tag{11.59}$$

$$u_{\mu|\nu} = 0 \tag{11.60}$$

Thus the temperature is the same in all cells, as is the fluid flow, and both quantities are constant in time. Consequently, the distribution function is the same everywhere, which is consistent with Equation 11.23.

The expressions for the number density (Equation 11.11), the mass flux (Equation 11.36) and the energy momentum tensor (Equation 11.38) may now be evaluated. The momentum may be expressed in hyperspherical coordinates (φ, χ) in a Lorentz frame as

$$p^\lambda = m(\sinh\chi\cos\varphi, \sinh\chi\sin\varphi, \cosh\chi) \tag{11.61}$$

Then $d\omega = m^2 \sinh \chi^d \chi^d \varphi$ and $E = m \cosh \chi$, which yields $d\omega = mdEd\,\varphi$. Now substituting the equilibrium distribution function (Equation 11.46) into the expression for the number density (Equation 11.11) yields

$$n_o = \frac{m}{V} e^\alpha \int e^{-\beta E(1-\vec{u}\cdot\vec{v})}\, dE\, d\varphi \tag{11.62}$$

where the particle velocity is $\vec{v} = \vec{p}/E = (\tanh\chi\cos\varphi, \tanh\chi\sin\varphi)$. The Lorentz frame is now chosen to coincide with the fluid flow frame, that is, $\vec{u} = \vec{0}$. Then Equation 11.62 may be easily integrated to yield

$$e^\alpha = \frac{n_o \beta V}{2\pi m} e^{\beta m} \tag{11.63}$$

and thus the distribution function is

$$N = e^{(\alpha+\beta^\lambda p_\lambda)} = \frac{n_o \beta V}{2\pi m} e^{-\beta E(1-\vec{u}\cdot\vec{v})} \tag{11.64}$$

The mass flux is

$$M^\alpha = \rho u^\alpha \tag{11.65}$$

where

$$\rho = n_o m\left(1 + \frac{1}{m\beta}\right) \tag{11.66}$$

is the mass density. The energy momentum tensor is

$$T^{\mu\nu} = \varepsilon u^\mu u^\nu + P\Delta^{\mu\nu} \tag{11.67}$$

where the energy density, ε, and the pressure, P, are given by

$$\varepsilon = \frac{n_0\beta}{m}\left(\frac{2}{\beta^3} + \frac{2m}{\beta^2} + \frac{m^2}{\beta}\right) \tag{11.68}$$

$$P = \frac{\rho}{m\beta} \tag{11.69}$$

Here, the theoretical expressions for the equilibrium form of the distribution function, the mass flux and the energy momentum tensor provide quantities that can be measured in a simulation.

SIMULATIONS OF THE LORENTZ INVARIANT LATTICE GAS

To verify that the lattice gas model produces the equilibrium configuration discussed, a number of simulations have been performed. Some implementation details are discussed here prior to presenting the simulations. The programming language C is used to implement the lattice gas model (this differs from the simulator for the porosity–pressure and saturation waves, which was done in Python) to accommodate multiple species of particles and consequently the size of a vector of masses as a computational tool. The relevant code fragment looks like

```
#define NUM_SPECIES 1
Float mass[NUM_SPECIES] ;
```

where the mass of each species of particle is initialized separately, for example mass[0] = 1.0. In this model, particles are implemented as an abstract data type, which are called 'a_ptcl'. Since the particles can move in any direction, a bit field representation for particle directions cannot be used, so the C definition for 'a_ptcl' is

```
typedef struct a_ptcl_struct {
    short int m ;
    short int i, j, k ;
    float px, py ;
    struct a_ptcl_struct* Next ;
} a_ptcl ;
```

The program was implemented on a UNIX system, where the size of a 'short int' is 2 bytes and the size of a 'float' is 4 bytes. In this structure, the short integer m is an index into the vector 'mass[ ]', that is, the value of the particle's mass is mass[m]. The short integers i and j are used to specify what cell the particle is in. The short integer k is reserved for use in three dimensions when a cell needs to be specified by three variables. Typically, this variable does not consume any extra memory space because of the way C compilers perform memory alignment. The float numbers px and py are the spatial components of the relativistic momentum.

$$p^\alpha = (p^1, p^2, p^4) = (p_x, p_y, E) = (\vec{p}, E) \tag{11.70}$$

For three dimensions, an extra float variable is added, p_z.

An unlimited number of particles is allowed to occupy a lattice site or reside in the same cell by using a linked list to keep track of the particles in a cell; this is implemented using the Next variable in the definition of a_ptcl. As noted previously, this feature of the model is very important because it removes a restriction present in other lattice gas models:

$$\frac{\rho_{pf}}{K_f}\left(\frac{\partial^2 p_f}{\partial t^2}+2a_{pf}\frac{\partial p_f}{\partial t}-2b_{pf}\nabla^2\frac{\partial p_f}{\partial t}-v_{pf}^2\nabla^2 p_f\right)$$

and the equilibrium distribution is no longer constrained to have the form of a Fermi–Dirac distribution.

The velocity of a particle is

$$\vec{v}=(v_x,v_y)=\left(\frac{p_x}{E},\frac{p_y}{E}\right) \tag{11.71}$$

The momentum is used in this definition of a_ptcl rather than velocity because it has proven to be more numerically accurate and stable. The calculation of the velocity always guarantees that the particle's speed is less than unity (as required) and the momentum is more readily available for performing collisions.

At each time step in the simulation, all particles must be subject to a propagation rule. First, the particle's velocity is found, then the case of an arbitrary velocity can be reduced by rotations to the case $v_x \geq 0$ and $v_y \geq 0$. Equations 11.1, 11.2 and 11.8 may be used to calculate the movement matrix and find

$$M_{ab}(\vec{v})=\begin{pmatrix} 0 & (1-v_x)(v_y) & (v_x)(v_y) \\ 0 & (1-v_x)(1-v_y) & (v_x)(1-v_y) \\ 0 & 0 & 0 \end{pmatrix} \tag{11.72}$$

With the movement matrix specified, the algorithm for moving the particle is based on Equation 11.9; this algorithm is expressed in pseudo-code as

```
x=random number in [0,1)
if (x<M00)
    do not move
else if (x<M00+M10)
    move to (i+1,j)
else if (x< M00+M10+ M01)
    move to (i,j+1)
else
    move to (i+1,j+1)
```

The collision and propagation rules require the generation of random numbers from a uniform distribution, and to perform this an abstract data type called 'a_rng' has been implemented. This data type is based on the 'rngl' random number generator of Press et al. (1988). In their implementation, the parameter sets that define the random number generator are constants and can only be redefined by editing the source code and doing a recompilation. That implementation has been extended here by incorporating those parameter sets as variables in a C structure, namely

```
typedef struct {
    unsigned long m, a, c ;
} a_rng_param ;
```

and all of the available parameter sets are stored in a table. Furthermore, instead of using static variables as part of the random number generator function, those variables are incorporated as part of a_rng, which is defined by

```
typedef struct {
    a_rng_param p[3] ;
    float rm[3] ;
    unsigned long ix[3] ;
    float r[97] ;
    } a_rng ;
```

Here, an attempt has been made to maintain the same variable names used by Press et al. (1988). The reason for defining this abstract type is that this allows a random number generator to be defined by specifying three different indices in the parameter-set table. Instead of being restricted to a single or perhaps a few statistically declared random number generators, a large number of random number generators may now be used. The function used to generate a random number now takes a pointer to a_rng as a parameter. Furthermore, since an integer index can be randomly generated, a function has been implemented; this allows the use of a random number generator to pick three different parameter-set indices, which is used to create another random number generator. Thus a random number generator is used to randomly generate (i.e. initialize) other random number generators. This implementation is essential because when a very large number of collisions are considered the lack of randomness in a single random number generator quickly becomes apparent in observations of unphysical behaviour. However, when the procedure for generating random number generators described here is used, such effects are not observed.

When the simulator starts, it first enters a set-up phase. The first action is to use a statically defined random number generator to randomly initialize 100 random number generators; these are stored in a one-dimensional array. The lattice is dynamically allocated as a two-dimensional array of cells, where each cell is a data structure containing an index of a random number generator, a list of particles, plus some auxiliary variables for analysis. Cyclic boundary conditions are imposed by requiring that $i_{max}+1$ be equivalent to i_{min} and $j_{max}+1$ be equivalent to j_{min}. Each cell is randomly assigned a random number generator from the set of random number generators. The randomness of the assignment means it is unlikely that a cell's neighbours will have the same random number generator; furthermore this minimizes any correlation between cells. The last action of the setup phase is to create and initialize a set of particles. If N_o is the average number of particles per cell, then $N_T = N_o N_c$ particles are created, where N_c is the number of cells in the lattice. Each particle is initialized by randomly assigning it to a cell and giving it momentum. How the momentum is specified depends on the type of numerical experiment being conducted.

The simulator now enters the time evolution phase. This consists of a series of time steps implemented as a single program loop. Each time step starts by scanning the set of particles; for each particle it is determined what cell it is in and that particle is appended to the cell's particle list. Then for each cell collision and propagation actions are performed. The collisions are performed by selecting and deleting two particles at random from the cell's particle list. The two particles are made to collide using the binary elastic that has been discussed, and then they are added to the 'collided' list, which is a local program variable that keeps track of which particles have undergone a collision. Once the cell's particle list contains one or fewer particles, all remaining particles are removed from the cell's particle list and added to the collided list. At this point, some additional analysis may be applied to the particles in the cell, such as recording the number of particles and finding the fluid velocity. The propagation action for the cell is performed by visiting each particle on the collided list and moving it according to the propagation rules that have been discussed. The simulation then proceeds to the next time step. At regular intervals, say every 1000 time steps, the accumulated results are saved to files for later analysis. The simulation ends with a final save of results to files.

To verify that the lattice gas model produces the equilibrium configuration that has been discussed, a number of numerical simulation have been performed. The quantities measured in these simulations as a function of temperature are the distribution function via its isotropy and energy density, the mass density, the energy density and the pressure. Finally, it is verified that the equilibrium distribution is invariant under a Lorentz transformation.

The simulations presented here were performed on a rectangular lattice whose dimensions were 100 cells by 100 cells, where $i_{min} = j_{min} = 0$ and $i_{max} = j_{max} = 99$. Here, m = 1 and each cell contains six particles so the total number of particles was $N_T = 60{,}000$. Here, an even number of particles per cell was deliberately chosen so the particles could be initialized in pairs to have equal and opposite momenta. The Lorenz frame of reference is taken as the rest frame of the lattice. All particles are initialized with the same speed v_o. For each pair of particles, a random angle φ is generated from a uniform distribution in $[0, 2\pi)$. The momentum of the first particle is $m\gamma(v_o)(v_o\cos\varphi,\ v_o \sin\varphi, 1)$ and the momentum of the second particle is $m\gamma(v_o)(v_o\cos(\varphi + \pi),\ v_o \sin(\varphi + \pi), 1)$. Then each particle is randomly assigned to a cell in the lattice.

This initialization method is designed to guarantee that the fluid velocity will be 0 and thus the temperature can be predicted. The average mass flux for a discrete set of particles is

$$M^{\alpha} = \frac{1}{N_T}\sum_i p_i^{\alpha} \tag{11.73}$$

which evaluates to

$$M^{\alpha} = m\gamma(v_o)(0,0,1) \tag{11.74}$$

using this initialization scheme. This expression corresponds to the mass flux density, Equations 11.65 and 11.66 with number density n_0 set to 1. Since the mass flux is conserved, the temperature of the gas may be calculated and yields

$$\beta = \frac{1}{m[\gamma(v_o)-1]} \tag{11.75}$$

Nine simulations were performed for $v_o = 0.1, 0.2, \ldots, 0.9$ to cover the range from a non-relativistic gas to a highly relativistic gas. The simulations were run for 1100 time steps. The first 100 time steps were used to allow the particle distribution to relax to an equilibrium configuration. The next 1000 time steps were the analysis phase of the simulation and were used to accumulate data about the particle distribution, the average mass flux and the average energy momentum tensor.

The differential form of Equation 11.62 with $\vec{u} = \vec{0}$ states that the number of particles $dN(E, \varphi)$ in the range $dEd\varphi$ in equilibrium is given by

$$dN(E,\varphi) = N(E,\varphi)dEd\varphi = \frac{n_o\beta}{2\pi m}e^{(\beta m-\beta E)}dEd\varphi \tag{11.76}$$

To measure the distribution function, the energy and angle set domain is partitioned into sets of bins. The energy domain is partitioned into n_E bins of size $\Delta E = (E_{max} - m)/n_E$; similarly, the angle domain is partitioned into n_φ bins of size $\Delta_\varphi = 2\pi/n_\varphi$. Then the energy and angle domain $[m, E_{max}]\otimes[0,2\pi)$ consists of $n_E n_\varphi$bins of size $\Delta_E\,\Delta_\varphi$. A good choice for E_{max} has been found to be $m+3/\beta$, where β is given by Equation 11.75. When a simulation first starts, all the particles are located in one bin, but as the simulation proceeds the distribution rapidly evolves to a Boltzmann distribution. At each time step, the number of particles in each bin is counted and added to a running total $\Delta N(E, \varphi)$. This summation is performed over 1000 iterations. At the end of the simulation, $\Delta N(E, \varphi)$ is divided by the number of particles in the accumulation, that is, $1000N_T$ and the size of the bin. This yields a time-averaged estimate of the distribution function $N(E, \varphi)$ corresponding to the choice $n_o = 1$.

Equation 11.76 shows that the distribution function is an independent angle, that is, it should be isotropic. So, for fixed values of energy, a plot of $\ln(N(E,\varphi))$ versus φ should be a flat line. The simulations show this is true. This is illustrated in Figure 11.1, which shows a plot of some specific values of energy for the simulation with $v_o = 0.3$. The lines

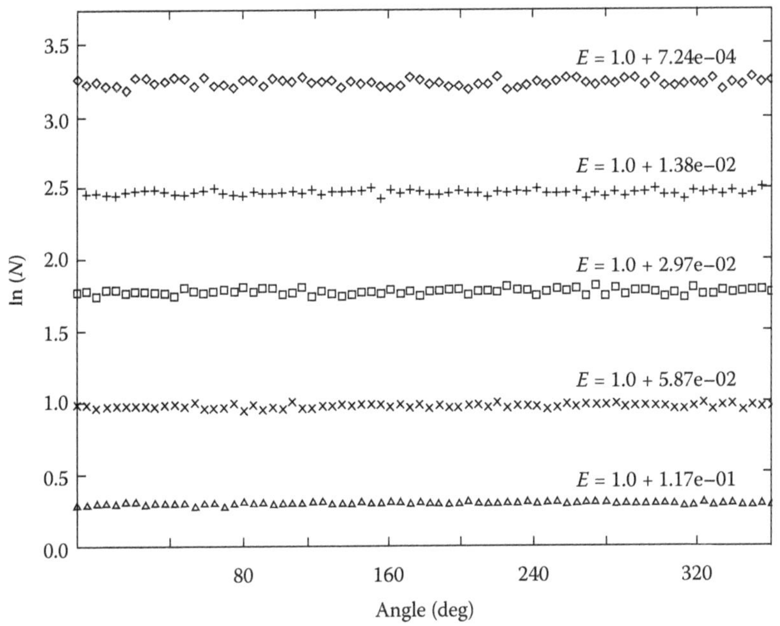

FIGURE 11.1 Angular dependence of the distribution function. For five selected values of energy, ln(N(E, φ)) is plotted versus angle φ. The initial speed of all particles is $v_o = 0.3$, temperature is $\beta = 20.710$ and fluid flow is $\vec{u} = \vec{0}$. The angular bin size is 5° and the energy range $[1, 1 + 3/\beta]$ is divided into 100 bins. (From Udey, N., Shim, D., and Spanos, T.J.T., 1999, A Lorentz invariant thermal lattice gas model, *Proc. Roy. Soc. A*, A455, 3565–3587, 1999.)

are not perfectly flat because of statistical fluctuations. When the number of time steps is increased, these fluctuations are smoothed out.

Now that it has been shown that the distribution function is isotropic, an integration is performed over angle, or equivalently, sum all bin results over angle. Then the distribution in energy with $n_o = 1$ should be

$$\ln(N(E)) = \ln(\beta / m) - \beta(E - m) \tag{11.77}$$

A plot of ln(N(E)) versus (E-m) should be a straight line with intercept ln(β/m) and slope −β. Figure 11.2 shows a plot of ln(N(E)) versus (E−m)for some of the simulations. All plots are straight lines. The data can be analysed by using the method of least squares for a straight line (Taylor 1982). However, the intercept and slope of these lines are not independent, since they are both functions of β. Thus the least-squares analysis leads to a single equation to be solved for β; and this equation can be easily solved using Newton's method. The results presented in Table 11.1 show an excellent agreement between theory and experiment.

An estimate of the energy–momentum tensor at each time step in a simulation is given by

$$T^{\mu\nu} = \frac{1}{N_T} \sum_i p_i^\mu p_i^\nu \tag{11.78}$$

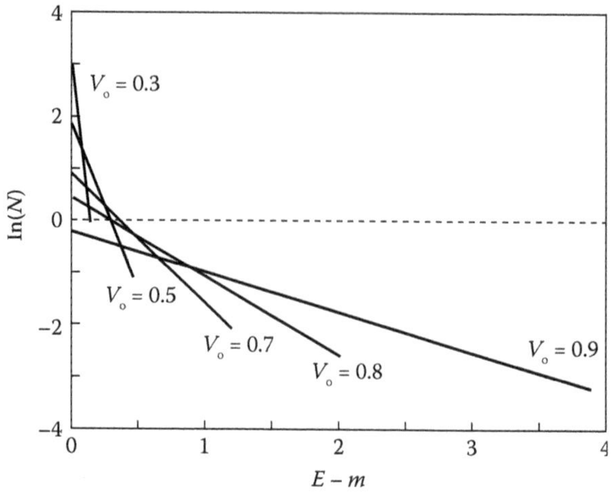

FIGURE 11.2 Energy dependence of the distribution function. This is a scatter plot of the data for some selected simulation results. In all cases, the fluid flow is $\vec{u} = \vec{0}$ and the energy range [1, 1+3/β] is divided into 300 bins. (From Udey, N., Shim, D., and Spanos, T.J.T., 1999, A Lorentz invariant thermal lattice gas model, *Proc. Roy.* Soc. *A*, A455, 3565–3587, 1999.)

TABLE 11.1 Experimental β versus Initial Speed v_o

Initial Speed $_{vo}$	Theoretical ß	Experimental ß	Discrepancy (%)
0.1	198.50	198.38	0.060
0.2	48.495	48.497	0.004
0.3	20.710	20.708	0.010
0.4	10.978	10.977	0.009
0.5	6.4641	6.4627	0.022
0.6	4.0000	4.0006	0.015
0.7	2.4983	2.4986	0.012
0.8	1.5000	1.5004	0.027
0.9	0.77270	0.77255	0.019

Source: From Udey, N., Shim, D., and Spanos, T.J.T., 1999, A Lorentz invariant thermal lattice gas model, *Proc. Roy. Soc. A*, A455, 3565–3587, 1999.

The time average of this quantity over 1000 time steps of the analysis phase of each simulation should correspond precisely to the energy–momentum tensor, Equations 11.67, 11.68 and 11.69 with $n_o = 1$. This yields a method to measure the energy density per particle and pressure per particle for the lattice gas. For example, the time average and standard deviation of $T^{\mu\nu}$ for the simulation with $v_o = 0.5$ is

$$T^{\mu\nu} = \begin{pmatrix} 0.178655 \pm 0.000776 & -0.000037 \pm 0.000798 & 0.000060 \pm 0.000426 \\ -0.000037 \pm 0.000798 & 0.178631 \pm 0.000786 & -0.000029 \pm 0.000414 \\ 0.000060 \pm 0.000426 & -0.000029 \pm 0.000414 & 1.357286 \pm 0.000183 \end{pmatrix} \quad (11.79)$$

which compares favourably with the theoretical value

$$T^{\mu\nu} = \begin{pmatrix} 0.178633 & 0.0 & 0.0 \\ 0.0 & 0.178633 & 0.0 \\ 0.0 & 0.0 & 1.357266 \end{pmatrix} \quad (11.80)$$

A single pressure for the lattice gas can be assigned as the average value of the two pressure values appearing in the energy–momentum tensor estimate. Thus for each simulation an experimental value is obtained for the pressure and energy density. This also provides an alternative method for finding the temperature.

The Lorentz invariance of the model is examined by investigating the form of the equilibrium distribution function when the fluid's velocity is non-zero. The same particle initialization procedure defined previously is used. Once all particles have been initialized a velocity $\vec{u} = u(\cos\theta, \sin\theta)$ is applied, that is, a Lorentz boost, to each particle. Thus the fluid velocity is $\vec{u}$. In these simulations, the initial speed of the particles before the Lorentz boost is applied is $v_o = 0.1$. Two types of simulations are performed. In the first, the fluid direction θ is kept constant and the fluid speed u is varied. In the second type the fluid speed u is kept constant and the fluid direction θ is varied.

The differential form of Equation 11.62 with $\vec{u} \neq \vec{0}$ and $n_o = 1$ is

$$dN(E,\varphi) = N(E,\varphi)dEd\varphi = \frac{\beta}{2\pi m} e^{(\beta m - \beta\gamma(u)E(1-\vec{u}\cdot\vec{v})}dEd\varphi. \quad (11.81)$$

A three-dimensional plot of $N(E, \varphi)$ versus E and φ would show a strong angular dependence. However, a plot of $\ln(N(E, \varphi))$ versus $E(1-\vec{u}\cdot\vec{v})$ should yield a straight line with intercept $\beta m + \ln(\beta/2\pi m)$ and slope $-\beta\gamma$. Figure 11.3 is a plot of the results of a simulation for fluid velocity $u = 0.1$, $\theta = 30°$. The scatter in the figure results because only 1000 time steps are used in the analysis phase. As the number is increased, the scatter is reduced. All simulations produced results similar to Figure 11.3.

The method of least squares may be used to determine β. If the lattice is Lorentz invariant, then the value of β should be the same in all simulations. In the simulations, the lower energy states are more easily accessible by particles than the higher energy states. Thus the lower energy states are more statistically representative of the distribution function and should display less scatter than the higher energy states. This effect is evident in Figure 11.3 as the scatter increases from upper left to lower right. Therefore, to obtain more accurate results, the application of the least-squares analysis is limited to the range $\ln(N) \geq 0$.

In the first type of experiment, the fluid speed is varied from 0.1 to 0.9 in increments of 0.1 and kept in the direction constant $\theta = 30°$. The results of these simulations are shown in Table 11.2. Here, the experimental values of β differ from the theoretical value of 198.50 by at most 0.156%. For an analysis phase consisting of only 1000 time steps, these results are quite good. When the number of time steps in the analysis phase is increased, the results get better.

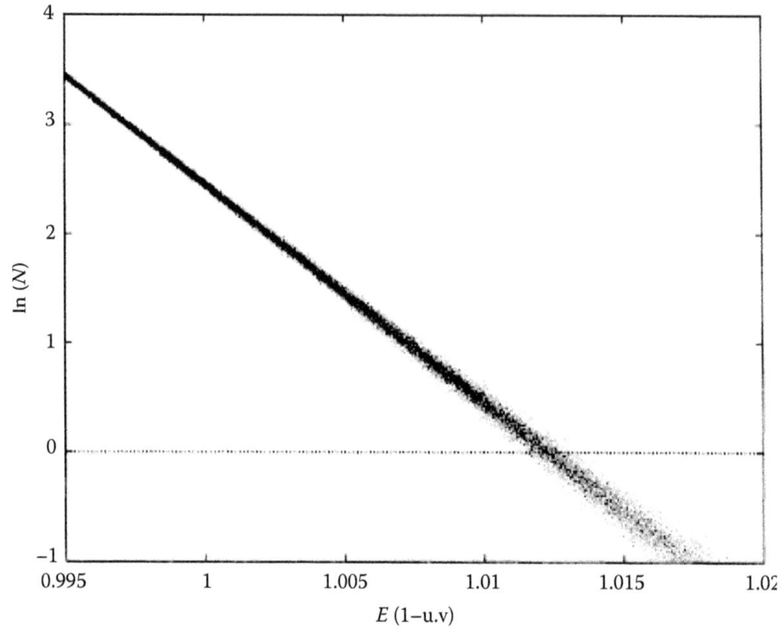

FIGURE 11.3 Dependence of the distribution function on $\vec{u} = \vec{0}$. This is a scatter plot of the data for the simulation with $v_o = 0.1$, theoretical $\beta = 198.50$ and fluid velocity $(u, \theta) = (0, 1, 30°)$. The angular bin size was 1.2° and the energy range $[1, 1+3/\beta]$ is divided into 300 bins. (From Udey, N., Shim, D., and Spanos, T.J.T., 1999, A Lorentz invariant thermal lattice gas model, *Proc. Roy. Soc. A*, A455, 3565–3587, 1999.)

TABLE 11.2 Experimental β versus Fluid Flow Speed u[a]

Fluid Flow-Speed u	**Experimental** β	**Discrepancy** (%)
0.1	198.31	0.096
0.2	198.34	0.081
0.3	198.25	0.126
0.4	198.19	0.156
0.5	198.23	0.136
0.6	198.25	0.126
0.7	198.31	0.096
0.8	198.38	0.060
0.9	198.35	0.076

Source: From Udey, N., Shim, D., and Spanos, T.J.T., 1999, A Lorentz invariant thermal lattice gas model, *Proc. Roy. Soc. A*, A455, 3565–3587, 1999.

[a] The theoretical value of β is 198.50.

In the second type of experiment, the fluid speed is kept constant, $u = 0.1$, but the fluid flow direction is varied from $\theta = 0°$ to 45° in increments of 5°. Because of the symmetry of the lattice, only these angles are required. These results are shown in Figure 11.3. Again, there is good agreement with the theoretical value of β (Table 11.3).

These results demonstrate that the equilibrium distribution is isotropic and independent of the underlying lattice. Certainly the number relativistic problems that can now be studied is huge, but most such problems are outside of the scope of this book.

TABLE 11.3 Experimental β versus Fluid Flow Direction θ[a]

Fluid Flow Direction θ (deg)	Experimental β	Discrepancy (%)
0.0	198.29	0.106
5.0	198.20	0.151
10.0	198.18	0.161
15.0	198.53	0.015
20.0	198.11	0.196
25.0	198.24	0.131
30.0	198.31	0.096
35.0	198.33	0.086
40.0	198.40	0.050
45.0	198.34	0.081

Source: From Udey, N., Shim, D., and Spanos, T.J.T., 1999, A Lorentz invariant thermal lattice gas model, *Proc. Roy. Soc. A*, A455, 3565–3587, 1999.

[a] The theoretical value of β is 198.50.

APPLICATIONS IN THE NON-RELATIVISTIC LIMIT

An identical construction may be made with Newtonian elastic scattering. The first problem that arises in this case is that there is no longer a constraint that particles cannot exceed the speed of light. This problem is addressed by simply discarding such events. Numerical comparisons of the two simulators yields the observation that if the average particle speeds are taken to be 0.2 or less, then the simulations are indistinguishable. It is also observed that the probability of a particle with a speed greater than 1 in the Newtonian simulation with average particle speeds less than 0.2 is so small that discarding such events is not statistically significant.

One possible application of these simulators would be to study thermodynamic fluid flow in the atmosphere. Other straightforward models are turbulence and relativistic shock waves. However, comparing such simulations with experiments is quite difficult.

Note that the model presented here differs from lattice gas automata models that are discrete in both space and momentum and from molecular dynamics models that are continuous in both space and momentum. The current thermodynamic automata is discrete in space but continuous in momentum.

Plane Poiseuille was modelled by the conventional lattice gas FHP model (Frisch et al. 1986), with particle reversing applied at the walls (e.g. Rothman 1988) to simulate a no-slip boundary condition. In general, particle reversing can be used to model the effect of an impulse on the particle and thus can also be used to model the effect of external forces. Since the conventional model did not incorporate thermal effects, the influence of heat generated by viscous dissipation could not be addressed. Chen et al. (1989) utilized a multispeed lattice model, an extension of the HLF model (D'Humières et al. 1986), to conduct an isothermal channel flow simulation with a no-slip boundary. Although temperature was included in their model, the effect of viscous heating was not observed. Yang et al. (1999a) simulated a thermodynamic process of plane flow with a similar insulating no-slip boundary condition and demonstrated the effect of viscous heating on the temperature of the fluid and thus the fluid viscosity. The objective of this chapter is to discuss the use of the thermodynamic automata in describing flow through porous media. The effect of boundaries, of course, becomes a dominant phenomenon when dealing with porous media, and thus obtaining proper boundary conditions is essential.

Thermal boundary conditions that can be considered as a heat bath have been used in both molecular dynamics (Tenenbaum et al. 1982; Trozzi and Ciccotti 1984) and in a lattice gas simulation for heat conduction processes (Chopard and Droz 1988; Chen et al. 1989).

In the present model, to mimic a pressure gradient when modelling fluid flow, momentum is added to each particle at each iteration. Here, the first simulations to be considered are

1. Poiseuille flow with all conditions identical except the boundary conditions, that is, no-slip boundary and thermodynamic boundary conditions
2. Fluid flow with thermal boundary conditions and the same pressure drop but different temperatures

Here, the no-slip boundary condition is implemented by simply reversing the particle direction when a collision with a boundary occurs. This boundary condition is then essentially identical to the boundary condition implemented in standard lattice gas automata models. The thermodynamic boundary condition is implemented by taking all particles that reach the outer cells and placing them in the neighbouring cell in the interior with a random orientation ranging from 0° to 180° and a range of speeds randomly selected from a characteristic Boltzmann distribution. This specifies the temperature of the boundary.

In these simulation experiments, 100 particles were initialized in each cell of a 101 by 11 lattice with a thermal equilibrium state at a normalized temperature of 0.005 (i.e. the average velocity of the particles is 0.1). Periodic boundary conditions were applied at the left and right boundaries of the lattice and no-slip or thermal boundary conditions were employed at the top and bottom boundaries. For each row of the lattice, the flow velocity, mean flow velocity, total energy, thermal energy and density of particles were recorded at every 200 iterations. Figure 11.4 shows the results of these simulations. Here, the mean velocity is given by

$$\bar{u}_x = -\frac{h^2}{3\mu}\frac{\partial p}{\partial x} \tag{11.82}$$

Figure 11.4 indicates that under the no-slip condition the thermal energy (i.e. temperature) increases with time (this is because the boost simulating the pressure drop adds energy and the boundary condition thermally insulates the system) and the mean flow velocity decreases with increasing viscosity (the viscosity of a gas increases with increasing temperature). For the thermodynamic boundary conditions, the thermal energy remains constant, as does the mean flow velocity. Both flows show parabolic velocity distribution consistent with Poiseuille flow at their respective viscosities. Here, the fluid viscosities are proportional to $\sqrt{T}$ (Yang et al. 1999a).

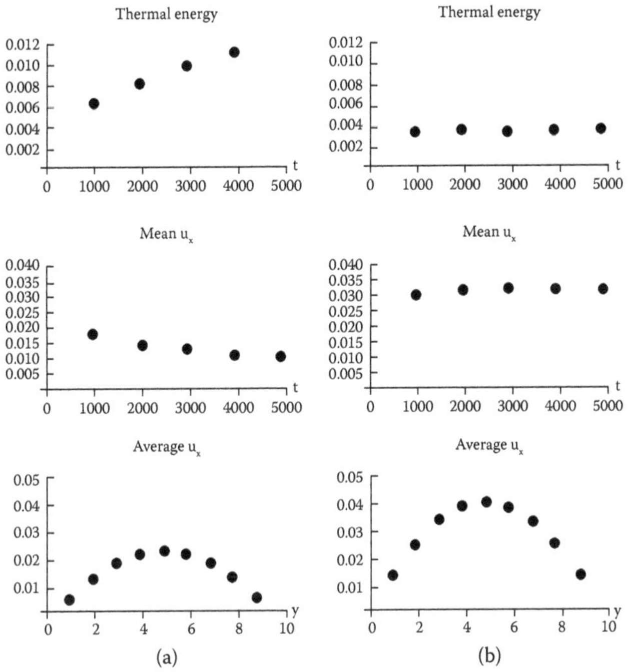

FIGURE 11.4 Simulation and comparison of the no-slip boundary condition, (a) (thermal insulator), and the thermodynamic boundary condition, (b) (heat bath). (From Yang, D., Udey, N., and Spanos, T.J.T., *Can. J. Phys.*, 77, 473–479, 1999.)

The next simulations to be performed with this model are fluid flow and diffusion in porous media. Here, fluid flow is modelled at a scale orders of magnitude lager than the pore scale, so the fluid–solid interactions are introduced by allowing for the probability of a solid collision at each cell in the lattice. The permeability is adjusted by changing that probability, and heterogeneity can be incorporated by allowing that probability to change in space. Although this approach does not allow observation of the details of the pore structure, it does permit simulations of flow at the megascale without computational difficulty. In this study a random number is generated for each particle in a cell. If the random number is larger than the solid probability (SP, the probability of collision with a solid object) in that cell, the fluid particle is allowed to undergo a solid collision; moreover, that particle is not permitted another collision until it moves out of that cell. The rest of the particles that do not collide with the solid randomly collide with each other.

As a first step, Darcy flow is simulated and the simulation results are compared with theoretical predictions. In the initial setup, 100 particles were initialized in each cell of a 100 by 11 lattice with a thermal equilibrium state at a temperature of 0.005 (average particle velocity of 0.1). For Darcy flow, thermodynamic boundary conditions were utilized at the top and bottom boundaries to remove extra heat caused by the pump action. The average flow velocity of each row of the lattice over 2000–5000 iterations was computed and the distribution of particle numbers for each column was recorded at 100-iteration intervals. Figure 11.5 shows the simulation results for Darcy flow. It can be seen that for the same pressure drop (0.0005)

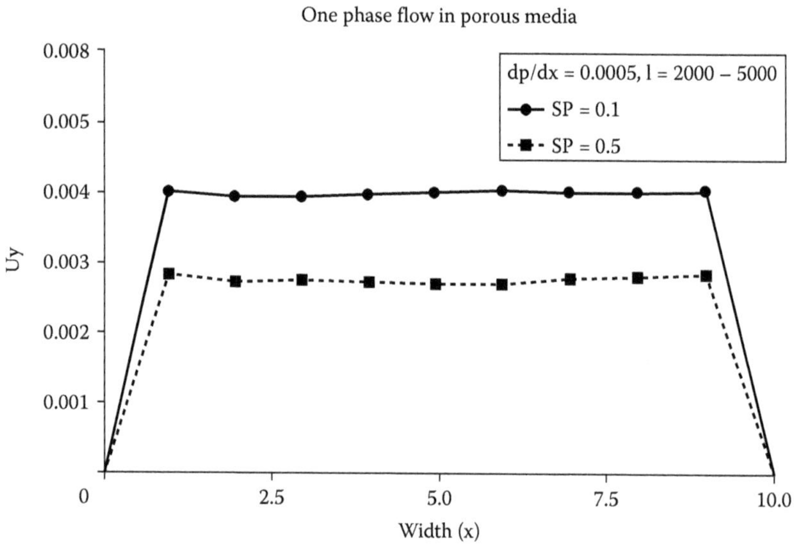

FIGURE 11.5 Flow velocity profiles for two different probabilities of solid collision at the lattice sites. (From Yang, D., Udey, N., and Spanos, T.J.T., *Transp. Porous Media*, 35, 37–47, 1999b.)

TABLE 11.4 Average Flow Rate versus Pressure Drop

Pressure Drop	Average Flow Rate
0.0001	0.0010
0.0002	0.0018
0.0003	0.0026
0.0004	0.0033
0.0005	0.0040

a decrease in the SP results in an increase in flow rate and thus an increase in permeability. Here, for Darcy flow, the profile is flat due to solid collisions as opposed to Poiseuille flow, which is parabolic. Balasubramanian et al. (1987) also presented a study of Darcy's law using lattice-gas hydrodynamics. They obtained an effective Darcy's law by allowing a damping term (a function of velocity) in the Navier–Stokes equation.

Table 11.4 shows simulation results for Darcy flow with the same permeability (SP=0.1) but different pressure drops. A straight line is obtained in Figure 11.6 when the simulation data points (in Table 11.4) are plotted.

Figure 11.7 shows that an increase in permeability (decrease in SP) results in a flow profile changing from flat to parabolic. In the limit as SP goes to zero Poiseuille flow is obtained.

When the flow velocity is reduced to extremely small values, the Darcy profile breaks down. This is consistent with experimental observations, where Darcy flow beaks down in the slow flow limit (Figure 11.8).

In order to model a diffusion process, the initial configuration is that the tracer (100 particles per cell) is initialized in the centre of the lattice (columns 49, 50 and 51). The tracer then spreads with time. Figure 11.9 shows that as SP is increased the rate of diffusion is decreased. Pore scale models have been constructed by considering a number of two dimensional structures of

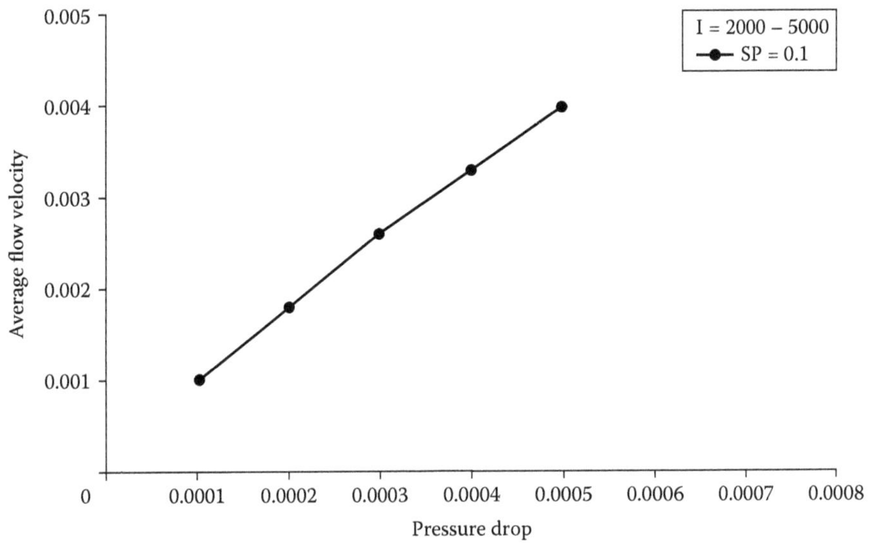

FIGURE 11.6 Flow velocity versus pressure drop along the direction of flow for one phase in a porous medium. (From Yang, D., Udey, N., and Spanos, T.J.T., *Transp. Porous Media*, 35, 37–47, 1999b.)

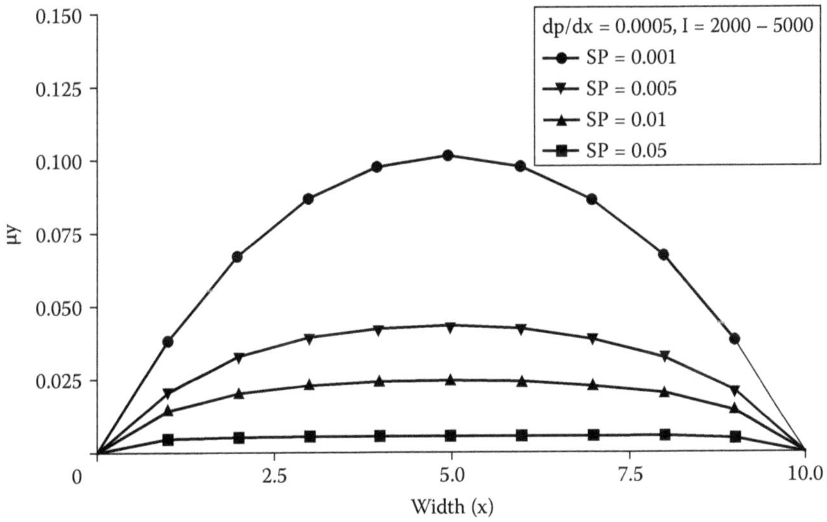

FIGURE 11.7 The change in flow velocity profiles with increasing permeability. (From Yang, D., Udey, N., and Spanos, T.J.T., *Transp. Porous Media*, 35, 37–47, 1999b.)

various shapes (Olson and Rothman, 1997). The simplest model of a porous medium is supplied by parallel capillaries (Yang et al. 1998). The effect that this well-controlled pore structure has on miscible flow in a porous media may then be compared with a megascale model that incorporates the pore scale information through the collision rules. The observations of this model clearly differ from the predictions of the convection diffusion model.

Laboratory results of Sternberg et al. (1996) provided evidence indicating that the conventional convection diffusion equation fails to adequately predict dispersion in porous media.

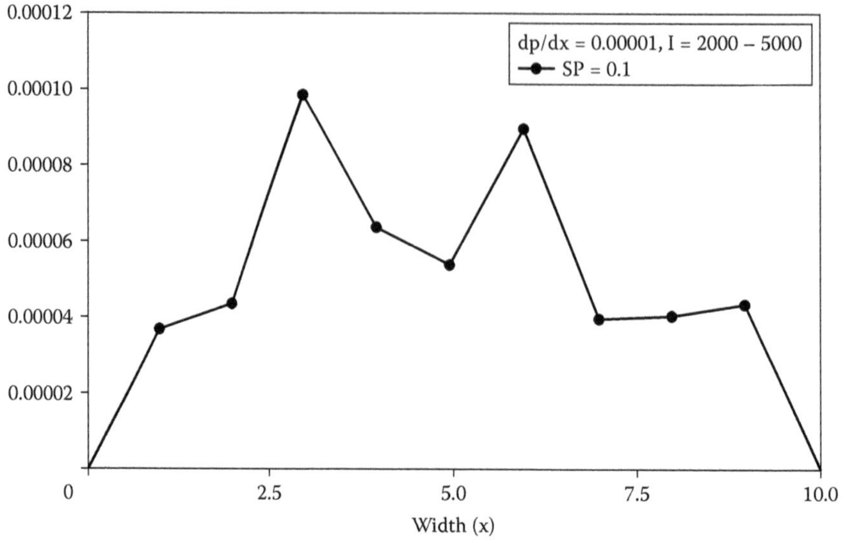

FIGURE 11.8 Flow velocity profile when the distance moved by a particle in a time step is many orders of magnitude less than a lattice unit (velocity = 0.00006). (From Yang, D., Udey, N., and Spanos, T.J.T., *Transp. Porous Media*, 35, 37–47, 1999b.)

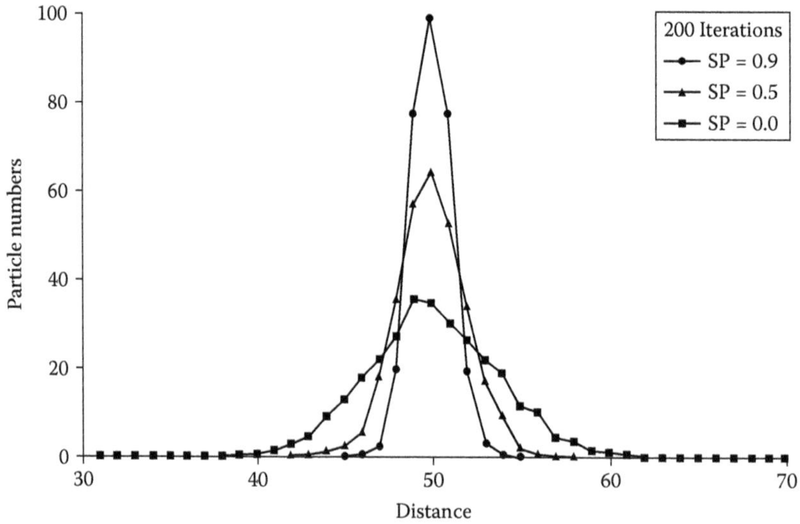

FIGURE 11.9 Diffusion in a homogeneous porous medium. The concentration profiles are plotted after 200 iterations, with the probabilities of a solid collision being 0.9, 0.5 and 0.0. (From Yang, D., Udey, N., and Spanos, T.J.T., *Transp. Porous Media*, 35, 37–47, 1999b.)

Theoretically, two approaches (equations) have been used to replace the convection diffusion equation when describing dispersion. One approach involves the use of nonlocal equations (Edelen, 1976) that allow information from a region of space to be included in order to determine the effect at any particular point in the system. This approach incorporates memory of the past history of the flow. The other approach is described by Udey and Spanos (1993) and in chapters 7 and 8 in which additional degree of freedom (i.e., an additional equation and variable; a dynamic megascopic pressure difference between the phases) is obtained.

Of course, the number of simulations that can be presented is limitless, but what is required to make this tool much more powerful is remove the restriction of an ideal gas and introduce a phase transition.

Initial attempts have shown that the introduction of a binding energy gives promising results but that is all. A Lorentz invariant lattice liquid model describing the phase transitions could be a powerful computational tool.

SUMMARY

Cellular automata provide an alternative to mathematical equations when describing physical systems. Particles are allowed to populate a discretized space that is equivalent to all particles and their properties in a volume of space being ascribed to a point in that volume, the lattice point. Since the particle motions are independent of the lattice, any lattice shape may be chosen. Since the maximum distance a particle may move in one time step is one lattice step, this distance in one time step is taken as the speed of light. The particles are allowed to undergo Lorentz invariant elastic collisions. From the collision and propagation rules it is shown that the energy momentum tensor and relativistic Boltzmann equation can be derived.

An identical construction may be made with Newtonian elastic scattering. The first problem that arises in this case is that there is no longer a constraint that particles cannot exceed the speed of light. This problem is addressed by simply discarding such events. Numerical comparisons of the two simulators yields the observation that if the average particle speeds are taken to be 0.2 or less that the simulations are indistinguishable.

A thermodynamic no slip boundary condition is implemented by taking all particles that reach the outer cells and placing them in the neighbouring cell in the interior with a random orientation ranging from to and a range of speeds randomly selected from a characteristic Boltzmann distribution. This specifies the temperature of the boundary. Descriptions of Poiseuille flow, Darcy flow and diffusion in porous media are presented.

REFERENCES

Balasubramanian, K., Hayot, F., and Saam, W.F., 1987, Darcy's law from lattice gas hydrodynamics, *Phys. Rev. A.*, **36**, 2248–2253.

Chen, S., and Doolen, G.D., 1998, Lattice Boltzmann Method for Fluid Flows. *Ann. Rev. Fluid Mech.*, 30, 329–364.

Chen, S., Lee, M., Zhao, K.H., and Doolen, G.D., 1989, A lattice gas model with temperature, *Physica D*, **37**, 42–59.

Chopard, B., and Droz, M., 1988, Cellular automata model for heat conduction in a fluid, *Phys. Lett. A*, **126**(8,9), 476–480.

D'Humières, D., Lallemand, P., and Frisch, U., 1986, Lattice gas model for 3D hydrodynamics, *Europhys. Lett.*, **2**(4), 291–297.

Edelen, D.G.B., 1976, Nonlocal field theories, in *Continuum Physics*, edited by A.C. Eringen, Vol. 4, Academic Press, New York.

Frisch, U., Hasslacher, B., and Pomeau, Y., 1986, Lattice-gas automata for the Navier-Stokes equation, *Phys. Rev. Lett.*, **56**(14), 1505–1508.

Israel, W., 1972, *The Relativistic Boltzmann Equation, General Relativity: papers in honour of J.L. Synge*, edited by L. O'Raifeartaigh, pp. 201–241, Clarendon, Oxford.

Lawniczak, A.T., and Kapral, R., 1996, *Pattern Formation and Lattice Gas Automata*, Fields Institute Communications, American Mathematical Society, Washington D.C., 346 pp.

Olson, J., and Rothman, D.H., 1997, Two phase flow in sedimentary rock: Simulation, transport, and complexity, *J. Fluid Mech.*, **341**, 343–370.

Press, W.H., Flannery, B.P., Teukolsky, S.A., and Vetterling, W.T., 1988, Numerical recipes in C, ch 7, Cambridge University Press, pp. 210–211.

Rothman, D.H., 1988, Cellular-automata fluids: A model for flow in porous media, *Geophysics*, **53**(4), 509–518.

Ryan, M.P., and Shepley, L.C., 1975, Homogeneous Relativistic Cosmologies, Princeton University Press, Princeton, New Jersey.

Spanos, T.J.T., de la Cruz, V., and Hube, J., 1988, An analysis of the theoretical foundations of relative permeability curves, *AOSTRA J. Res.*, **4**, 181–192.

Sternberg, S.P.K., Cushman, J.H., and Greenkorn, R.A., 1996, Laboratory observation of nonlocal dispersion, *Trans. Porous Media*, **23**, 135–151.

Taylor, J.R., 1982, An introduction to error analysis, Mill Valley: University Science Books, pp. 153–157.

Tenenbaum, A., Ciccotti, G., and Gallico, R., 1982, Stationary nonequilibrium states by molecular dynamics. Fourier's Law, *Phys. Rev. A*, **25**(5), 2778–2787.

Trozzi, C., and Ciccotti, G., 1984, Stationary nonequilibrium states by molecular dynamics. II. Newton's law. *Phys. Rev. A*, **29**(2), 916–925.

Udey, N., Shim, D., and Spanos T.J.T., 1999, A Lorentz invariant thermal lattice gas model, *Proc. Roy. Soc. A*, **455**, 3565–3587.

Udey, N., and Spanos, T.J.T., 1993, The equations of miscible flow with negligible molecular diffusion, *Trans. Porous Media*, **10**, 1–41.

Yang, D., Udey, N., and Spanos, T.J.T., 1998, Automaton simulations of dispersion in porous media, *Trans. Porous Media*, **32**, 187–198.

Yang, D., Udey, N., and Spanos, T.J.T., 1999a, A thermodynamic lattice gas model of Poiseuille flow, *Can. J. Phys.*, **77**, 473–479.

Yang, D., Udey, N., and Spanos, T.J.T., 1999b, Thermodynamic automaton simulations of fluid flow and diffusion in porous media, *Trans. Porous Media*, **35**, 37–47.

Index

Milton Keynes UK
Ingram Content Group UK Ltd.
UKHW050308111024
449327UK00044B/2330

9 780367 875305